VORONOI DIAGRAMS AND DELAUNAY TRIANGULATIONS

VORONOI DIAGRAMS AND DELAUNAY TRIANGULATIONS

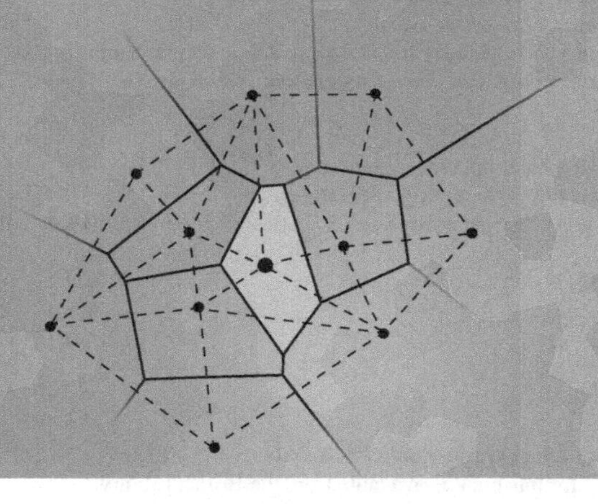

Franz Aurenhammer
Graz University of Technology, Austria

Rolf Klein
University of Bonn, Germany

Der-Tsai Lee
Academia Sinica, Taiwan

NEW JERSEY • LONDON • SINGAPORE • BEIJING • SHANGHAI • HONG KONG • TAIPEI • CHENNAI

Published by

World Scientific Publishing Co. Pte. Ltd.
5 Toh Tuck Link, Singapore 596224
USA office: 27 Warren Street, Suite 401-402, Hackensack, NJ 07601
UK office: 57 Shelton Street, Covent Garden, London WC2H 9HE

Library of Congress Cataloging-in-Publication Data
Aurenhammer, Franz, 1957–
Voronoi diagrams and Delaunay triangulations / Franz Aurenhammer, Graz University of Technology, Austria, Rolf Klein, University of Bonn, Germany, Der-Tsai Lee, Academia Sinica, Taiwan.
pages cm
Includes bibliographical references and index.
ISBN 978-9814447638 (hardcover : alk. paper)
1. Voronoi polygons. 2. Spatial analysis (Statistics) I. Klein, Rolf, 1953– II. Lee, Der-Tsai. III. Title.
QA278.2.A97 2013
516.22--dc23

2013018154

British Library Cataloguing-in-Publication Data
A catalogue record for this book is available from the British Library.

Typeset by Stallion Press
Email: enquiries@stallionpress.com

Printed in Singapore

CONTENTS

1. Introduction **1**

2. Elementary Properties **7**

2.1. Voronoi diagram 7
2.2. Delaunay triangulation 11

3. Basic Algorithms **15**

3.1. A lower time bound 16
3.2. Incremental construction 18
3.3. Divide & conquer 24
3.4. Plane sweep 28
3.5. Lifting to 3-space 31

4. Advanced Properties **35**

4.1. Characterization of Voronoi diagrams 35
4.2. Delaunay optimization properties 41

5. Generalized Sites **47**

5.1. Line segment Voronoi diagram 47
5.2. Convex polygons 53
5.3. Straight skeletons 54
5.4. Constrained Delaunay and relatives 62
5.5. Voronoi diagrams for curved objects 66
5.5.1. Splitting the Voronoi edge graph 67
5.5.2. Medial axis algorithm 70

6. Higher Dimensions **75**

6.1. Voronoi and Delaunay tessellations in 3-space 75
6.1.1. Structure and size 75

6.1.2. Insertion algorithm 77
6.1.3. Starting tetrahedron 79
6.2. Power diagrams 81
6.2.1. Basic properties 81
6.2.2. Polyhedra and convex hulls 83
6.2.3. Related diagrams 85
6.3. Regular simplicial complexes 87
6.3.1. Characterization 88
6.3.2. Polytope representation in weight space 89
6.3.3. Flipping and lifting cell complexes 90
6.4. Partitioning theorems 93
6.4.1. Least-squares clustering 94
6.4.2. Two algorithms 97
6.4.3. More applications 100
6.5. Higher-order Voronoi diagrams 103
6.5.1. Farthest-site diagram 103
6.5.2. Hyperplane arrangements and k-sets 106
6.5.3. Computing a single diagram 109
6.5.4. Cluster Voronoi diagrams 112
6.6. Medial axis in three dimensions 114
6.6.1. Approximate construction 114
6.6.2. Union of balls and weighted α-shapes 117
6.6.3. Voronoi diagram for spheres 120

7. General Spaces & Distances **123**

7.1. Generalized spaces 123
7.1.1. Voronoi diagrams on surfaces 123
7.1.2. Specially placed sites 128
7.2. Convex distance functions 129
7.2.1. Convex distance Voronoi diagrams 130
7.2.2. Shape Delaunay tessellations 136
7.2.3. Situation in 3-space 141
7.3. Nice metrics 144
7.3.1. The concept 144
7.3.2. Very nice metrics 148
7.4. Weighted distance functions 152
7.4.1. Additive weights 152
7.4.2. Multiplicative weights 156
7.4.3. Modifications 160
7.4.4. Anisotropic Voronoi diagrams 163
7.4.5. Quadratic-form distances 165
7.5. Abstract Voronoi diagrams 167
7.5.1. Voronoi surfaces 167
7.5.2. Admissible bisector systems 168
7.5.3. Algorithms and extensions 172
7.6. Time distances 175
7.6.1. Weighted region problems 175

7.6.2. City Voronoi diagram . . . 176
7.6.3. Algorithm and variants . . . 180

8. Applications and Relatives **183**

8.1. Distance problems . . . 183
8.1.1. Post office problem . . . 183
8.1.2. Nearest neighbors and the closest pair . . . 186
8.1.3. Largest empty and smallest enclosing circle . . . 189
8.2. Subgraphs of Delaunay triangulations . . . 194
8.2.1. Minimum spanning trees and cycles . . . 195
8.2.2. α-shapes and shape recovery . . . 200
8.2.3. β-skeletons and relatives . . . 202
8.2.4. Paths and spanners . . . 205
8.3. Supergraphs of Delaunay triangulations . . . 207
8.3.1. Higher-order Delaunay graphs . . . 207
8.3.2. Witness Delaunay graphs . . . 210
8.4. Geometric clustering . . . 211
8.4.1. Partitional clustering . . . 212
8.4.2. Hierarchical clustering . . . 214
8.5. Motion planning . . . 216
8.5.1. Retraction . . . 217
8.5.2. Translating polygonal robots . . . 219
8.5.3. Clearance and path length . . . 220
8.5.4. Roadmaps and corridors . . . 222

9. Miscellanea **225**

9.1. Voronoi diagram of changing sites . . . 225
9.1.1. Dynamization . . . 225
9.1.2. Kinetic Voronoi diagrams . . . 226
9.2. Voronoi region placement . . . 228
9.2.1. Maximizing a region . . . 229
9.2.2. Voronoi game . . . 232
9.2.3. Hotelling game . . . 235
9.2.4. Separating regions . . . 236
9.3. Zone diagrams and relatives . . . 238
9.3.1. Zone diagram . . . 238
9.3.2. Territory diagram . . . 241
9.3.3. Root finding diagram . . . 242
9.3.4. Centroidal Voronoi diagram . . . 244
9.4. Proximity structures on graphs . . . 246
9.4.1. Voronoi diagrams on graphs . . . 246
9.4.2. Delaunay structures for graphs . . . 248

10. Alternative Solutions in $\mathbf{R}^d$ **251**

10.1. Exponential lower size bound . . . 251
10.2. Embedding into low-dimensional space . . . 252
10.3. Well-separated pair decomposition . . . 254

10.4. Post office revisited 258
10.4.1. Exact solutions 258
10.4.2. Approximate solutions 259
10.5. Abstract simplicial complexes 261

11. Conclusions **267**
11.1. Sparsely covered topics 267
11.2. Implementation issues 269
11.3. Some open questions 272

Bibliography **275**

Index **329**

Chapter 1

INTRODUCTION

This book is devoted to an important and influential geometrical structure called the *Voronoi diagram*, whose origins in the Western literature date back to at least the 17th century. In his work on the principles of philosophy [257], René Descartes (Renatus Cartesius) claims that the solar system consists of vortices. His illustrations show a decomposition of space into convex regions, each consisting of matter revolving around one of the fixed stars; see Figure 1.1.

Even though Descartes has not explicitly defined the extension of these regions, the underlying idea seems to be the following. Let some *space*, and a set of *sites* in this space be given, together with a notion of the *influence* a site p exerts on every location x of the space. Then the *region* of the site p consists of all points x for which the influence of p is the strongest. Figure 1.2 illustrates the simplest case: Sites are points in the plane, and influence is modeled by Euclidean distance. The regions of the sites are convex polygons covering the entire plane.

'*The world is full of Voronoi diagrams*'...this phrase seems true wherever we look, and it is the intention of this book to lead the reader into the fascinating realm of Voronoi diagrams. Whether we look to the sky, where gravitational fields structure the macrocosmos, or consider the spread of animals in their habitats, simply watch the interference pattern of waves on a quiet pond, or even peer deep within the structure of matter, we will find a Voronoi diagram-like pattern which structures the world.

Indeed, this fundamental concept has emerged independently, and proven useful, in various fields of science. Most applications have their own notions of 'space', 'sites', and 'influence', resulting in Voronoi diagrams whose structures differ greatly. Understanding their properties is the key to their effective application and to the development of fast construction algorithms. Various names have been given to Voronoi diagrams, depending on the particular domain, such as *Thiessen polygons* in geography and

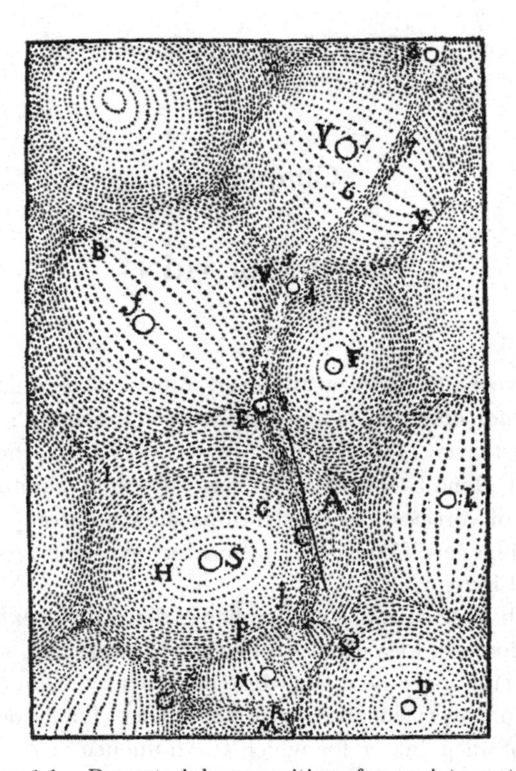

Figure 1.1. Descartes' decomposition of space into vortices.

meteorology (Thiessen [683], 1911), *domains of action* or *Wirkungsbereiche* in crystallography (Niggli [562], 1927), *Wigner–Seitz zones* in chemistry and physics (Wigner and Seitz [705], 1933), *Johnson–Mehl model* in mineralogy (Johnson and Mehl [433], 1939), and *medial axis transform* in biology and physiology (Blum [134], 1973).

Voronoi diagrams are important to both theory and applications, and play a unique interdisciplinary role. Several thousands of research articles have been published about them in different communities. Results thus dispersed might not always be widely known, and might fade into oblivion. While no single book can present all known results on Voronoi diagrams and their relatives, our aim is to thoroughly cover the structural and algorithmic viewpoints. In addition to being a versatile space partitioning structure, Voronoi diagrams are also aesthetically pleasing, and many people feel attracted to them, even regarding their artistic aspects. We have tried to communicate this quality in this book.

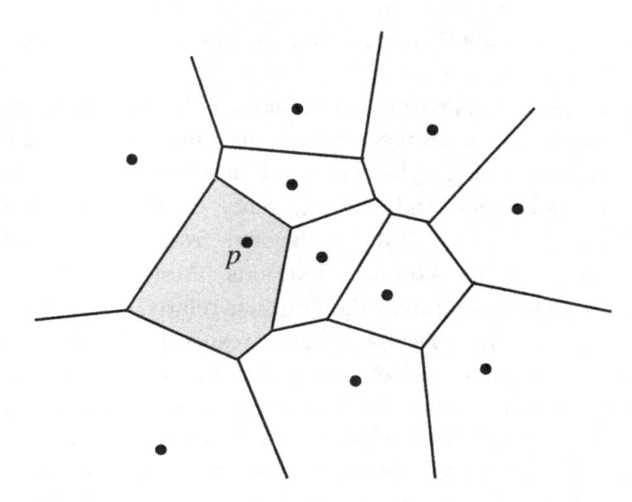

Figure 1.2. A Voronoi diagram of point sites in the Euclidean plane.

One and a half centuries ago, the mathematicians Carl Friedrich Gauß [353] (1840) and Gustav Lejeune Dirichlet [280] (1850), and later Georgi Feodosjewitsch Voronoi [693, 694] (1908) were the first to formally introduce this concept. They used it to study quadratic forms: Here the sites are integer lattice points, and influence is measured by Euclidean distance. The resulting structure was called a *Dirichlet tessellation* or *Voronoi diagram*, which became its standard name today.

Voronoi [693] also considered the *geometrical dual* of this structure, where any two (point) sites are connected whose regions have a boundary in common. Later, Boris Delaunay (Delone) [253] obtained the same structure by defining that two sites are connected if they lie on a circle whose interior contains no other sites. After him, the dual of the Voronoi diagram was denoted *Delaunay tessellation* or *Delaunay triangulation*.

With the advent of modern computers, the important role of Voronoi diagrams in structuring, representing, and displaying multidimensional data was rediscovered by computational geometers. Voronoi diagrams are a well-established geometric data structure nowadays. About one out of sixteen articles in computational geometry is dedicated to them, ever since Shamos and Hoey [637] introduced them to the field. In the two-dimensional case, for instance in the Euclidean plane, the Voronoi diagram does not require significantly more storage than does its underlying set of sites, and thus captures the inherent proximity information in a comprehensive and computationally useful manner. Its applications in more practically oriented areas of computer science are numerous, for example, in geographic information systems, robotics, computer graphics, and data classification and clustering, to name a few.

Besides its direct applications in diverse fields of science, the Voronoi diagram and its dual can be used for solving numerous, and surprisingly different, geometric and graph-theoretical problems. Due to their close relationship to polytopes and arrangements of hyperplanes in higher dimensions, many questions (and solutions) from convex and discrete geometry carry over to Voronoi diagrams. Moreover, the Delaunay triangulation, seen as a combinatorial graph, is related to several prominent connectivity graphs. We discuss various respective applications, often including alternative solutions for their special merits. Along the way, we give a state-of-the-art account of the literature on Voronoi diagrams in computational geometry. This fills the need for a technically sound and well-structured book, which is up-to-date on the theory and mathematical applications of Voronoi diagrams.

The reader is invited on a guided tour of gently increasing difficulty through a fascinating area. Insight will be given into the ideas and principles of Voronoi diagrams, without the baggage of too much technical detail. When later faced with a geometric partitioning problem, readers should find our book helpful in deciding whether their problem shows Voronoi characteristics and which type of Voronoi diagram applies. They might find a ready-to-use solution in our book, or follow up the links to the literature provided, or they might work out their own solution based on the algorithms they have seen. The book targets researchers in mathematics, computer science, natural and economical sciences, instructors and graduate students in those fields, as well as the ambitious engineers looking for alternative solutions. A brief discussion of algorithmic implementation questions, and of currently available geometric computation libraries, is included in the final chapter. Since human intuition is aided by visual perception, especially where geometric topics are concerned, the diversity and appeal of Voronoi diagrams is liberally illustrated with appropriate figures.

The presentation and structure of this book strives to highlight the intrinsic potential of Voronoi diagrams and Delaunay triangulations, which lies in their structural properties, in the existence of efficient algorithms for their construction, and in their relationship to seemingly unrelated concepts. We therefore organized topics by concept, rather than by application.

Another book which nicely complements ours, but from a much more applied perspective, is by Okabe *et al.* [571] (2000). It contains a wealth of applications of Voronoi diagrams, several of them not (or only marginally) covered here, like *Poisson Voronoi diagrams* and *locational optimization problems*. The more than one thousand and six hundred references listed in [571] constitute a bibliography still quite distinct from ours — demonstrating once more the broad scope of Voronoi diagrams. In addition, the book by Gavrilova [354] (2008) contains a collection of articles on diverse applications in the natural sciences, with numerous citations. A careful study of Delaunay triangulations, mainly oriented towards their algorithmic applications in *surface meshing* and *finite element methods*, is contained in George and Borouchaki [356] (1998). The recent book on Delaunay mesh generation by Cheng, Dey, and Shewchuk [202] (2013) takes into account the various new developments to date.

Shorter and early treatments of Voronoi diagrams and Delaunay triangulations, closer to the spirit of the present book, are the surveys by Aurenhammer [93], Fortune [343], and Aurenhammer and Klein [100]. Also, Chapters 5 and 6 of Preparata and Shamos [596], Chapter 13 of Edelsbrunner [300], Chapters 7 and 9 of de Berg *et al.* [244], and Part V of Boissonnat and Yvinec [151] could be consulted.

Although interesting and insightful in its own right, we decided to refrain from a detailed historical treatment of Voronoi diagrams in this book. The interested reader is encouraged to enjoy the historical perspectives presented in [93], and later in more detail, in [571].

We start in Chapter 2 with a simple case — the Voronoi diagram and the Delaunay triangulation of n points in the plane, under Euclidean distance. We state elementary structural properties that follow directly from the definitions. Further properties will be revealed in Chapter 3, where different algorithmic schemes for computing these structures are presented. In Chapter 4 we complete our presentation of the classical two-dimensional case, with advanced properties of planar Voronoi diagrams and Delaunay triangulations. We next turn to generalizations, to sites more general than points in Chapter 5, and to higher dimensions in Chapter 6. Generalized spaces and distances are treated elaborately in Chapter 7. In Chapter 8, important geometric applications of the Voronoi diagram and the Delaunay triangulation are discussed, along with respective related structures and concepts. The reader interested mainly in these applications can proceed directly to Chapter 8, after Chapter 2 or 3. Chapter 9 presents relevant topics which, for clarity of exposition, are best described separately. Chapter 10 offers alternative solutions in high dimensions, where the attractiveness of Voronoi diagrams is partially lost due to their high combinatorial and computational worst-case complexity. Finally, Chapter 11 concludes the book, gives a short discussion of algorithmic implementation issues, and mentions some important open problems.

Chapter 2

ELEMENTARY PROPERTIES

In this chapter, we present definitions and basic properties of the Voronoi diagram and its dual, the Delaunay triangulation.

Only the simplest case is considered — point sites in the plane under the Euclidean distance. Yet, the properties discussed are of general importance, as many (but not all) of them will carry over to other types of Voronoi diagrams and their relatives, presented later in this book.

2.1. Voronoi diagram

Let us start with giving some standard notation and explanations. Throughout this chapter, we will denote by S a set of $n \geq 3$ point sites $p, q, r, \ldots$ in the Euclidean plane, $\mathbf{R}^2$. For points $p = (p_1, p_2)$ and $x = (x_1, x_2)$, their *Euclidean distance* is given as

$$d(p, x) = \sqrt{(p_1 - x_1)^2 + (p_2 - x_2)^2}.$$

The straight-line segment that connects two points p and q will be written as $\overline{pq}$, or sometimes just as pq.

For $p, q \in S$, let $B(p, q)$ be the *bisector* of p and q (also called their *separator*), which is the locus of all points in $\mathbf{R}^2$ at equal distance from both p and q. $B(p, q)$ is the perpendicular line through the midpoint of the line segment $\overline{pq}$. It separates the halfplane

$$D(p, q) = \{x \mid d(p, x) \leq d(q, x)\}$$

closer to p from the halfplane $D(q, p)$ closer to q.

We next specify maybe the most important notion in this book. The *Voronoi region* of p among the given set S of sites, for short $\mathrm{VR}(p, S)$, is the intersection of the $n - 1$ halfplanes $D(p, q)$, where q ranges over all the other sites in S. More formally,

$$\mathrm{VR}(p, S) = \bigcap_{q \in S,\, q \neq p} D(p, q).$$

$\mathrm{VR}(p, S)$ consists of all points $x \in \mathbf{R}^2$ for which p is a nearest neighbor site. As being the finite intersection of halfplanes, which are convex sets in $\mathbf{R}^2$, the region $\mathrm{VR}(p, S)$ is a convex, possibly unbounded *polygon* in the plane. Different Voronoi regions are interior-disjoint, as they are separated by the bisector of the two sites that own them.

(A set M in $\mathbf{R}^2$, or in general d-space $\mathbf{R}^d$, is *convex* if it contains, with each pair of points $x, y \in M$, the line segment $\overline{xy}$. Set M is called *bounded* if there exists a circle, or sphere, respectively, of finite radius which encloses M. Otherwise, M is *unbounded*. We say that M is a *closed* set if M contains its (topological) boundary. M minus its boundary is an *open* set, called the *interior* of M.)

Definition 2.1. The common boundary part of two Voronoi regions is called a *Voronoi edge*, if it contains more than one point.

The *Voronoi diagram* of S, for short $V(S)$, is defined as the union of all Voronoi edges.

Endpoints of Voronoi edges are called *Voronoi vertices*; they belong to the common boundary of three or more Voronoi regions.

If a Voronoi edge e borders the regions of p and q then $e \subset B(p, q)$ holds. That is, $V(S)$ is a *planar straight-line graph* whose edges emanate from Voronoi vertices. $V(S)$ is sometimes also referred to as the *Voronoi edge graph* in the literature.

(Recall that a *graph* consists of a set of *vertices*, which in principle could be any objects, and a set of *edges* that pair up certain vertices, in order to display a relation between the two objects. A graph is called *planar* if it can be geometrically embedded in the plane without edge crossings. For sources on (combinatorial) graphs and graph-related algorithms, the books by Diestel [275], Gibbons [363], or West [703] may be consulted.)

There is an intuitive way of looking at the Voronoi diagram $V(S)$. Let x be an arbitrary point in the plane. We center a circle, C, at x and let its radius grow, from 0 on. At some stage the expanding circle will, for the first time, hit one or more sites of S. Now there are three different cases.

Lemma 2.1. *If the circle C expanding from point x hits exactly one site, p, then x belongs to the interior of region $\mathrm{VR}(p, S)$. If C hits exactly two sites, p and q, then x is an interior point of a Voronoi edge separating the* *regions of p and q. If C hits three or more sites simultaneously, then x is a Voronoi vertex adjacent to those regions whose sites have been hit.*

Proof. If only site p is hit then p is the unique element of S closest to x. Consequently, $x \in D(p,r)$ holds for each site $r \in S$ with $r \neq p$. If C hits exactly p and q, then x is contained in either halfplane $D(p,r), D(q,r)$, where $r \notin \{p,q\}$, and also in $B(p,q)$, the common boundary of $D(p,q)$ and $D(q,p)$. By Definition 2.1, x belongs to the closure of the regions of both p and q, but of no other site in S. In the third case, the argument is analogous. □

This lemma shows that the Voronoi diagram forms a *decomposition* (or *partition*) of the plane; see Figure 2.1: Regions do not overlap, and for each point x there is at least one closest site in S, meaning that x is covered by this region.

Conversely, if we imagine n circles expanding from the sites at the same speed, the fate of each point x of the plane is determined by those sites whose circles reach x first. This '*expanding waves*' view, or *wavefront model*, has been systematically used by Chew and Drysdale [215] and Thurston [685].

The Voronoi vertices are of degree at least three, by Lemma 2.1. (The *degree* of a vertex is the number of its incident edges.) Vertices of degree higher than three do not occur if no four point sites are cocircular. The

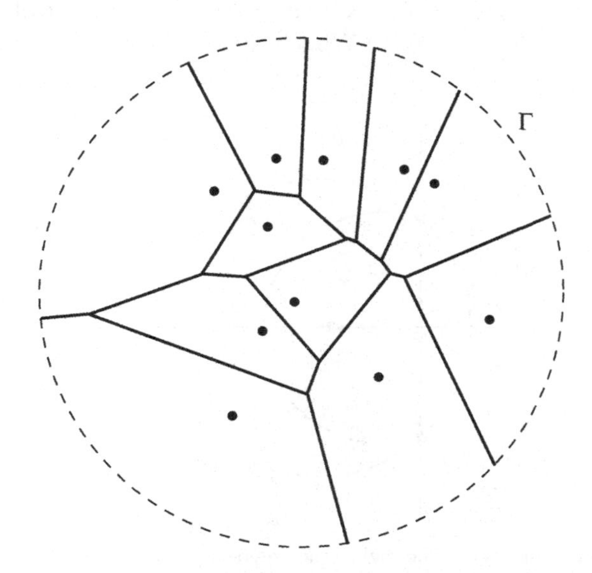

Figure 2.1. Voronoi diagram of 11 sites and bounding curve Γ.

Voronoi diagram $V(S)$ is disconnected if all point sites are collinear; in this case it consists of parallel lines.

From the Voronoi diagram of S one can easily derive the *convex hull* of S, which is defined as the smallest convex set containing S. As S is finite, its convex hull is a *convex polygon*, spanned by $h \leq n$ *extreme* points of S.

Lemma 2.2. *A point p of S lies on the boundary of the convex hull of S iff (i.e., if and only if) its Voronoi region* $\mathrm{VR}(p, S)$ *is unbounded.*

Proof. The Voronoi region of p is unbounded iff there exists some point $q \in S$ such that $V(S)$ contains an unbounded piece of $B(p, q)$ as a Voronoi edge. Let $x \in B(p, q)$, and let $C(x)$ denote the circle through p and q centered at x, as shown in Figure 2.2. Point x belongs to $V(S)$ iff $C(x)$ contains no other site. As we move x to the right along $B(p, q)$, the part of $C(x)$ contained in halfplane R keeps growing. If there is another site r in R, it will eventually be reached by $C(x)$, causing the Voronoi edge to end at x. Otherwise, all other sites of S must be contained in the closure of the left halfplane L. Then p and q both lie on the convex hull of S. □

If the point set S is in *convex position* (i.e., all its points are vertices of the convex hull of S) then all regions of $V(S)$ are unbounded, by Lemma 2.2, and their edges form a *cycle-free* and connected graph, that is to say, a *tree*.

Sometimes it is convenient to imagine a simple closed curve Γ around the 'interesting' part of the Voronoi diagram, so large that it intersects only the unbounded Voronoi edges; see Figure 2.1. (A curve is called *simple* if it contains no self-intersections.) While walking along Γ, the vertices of the

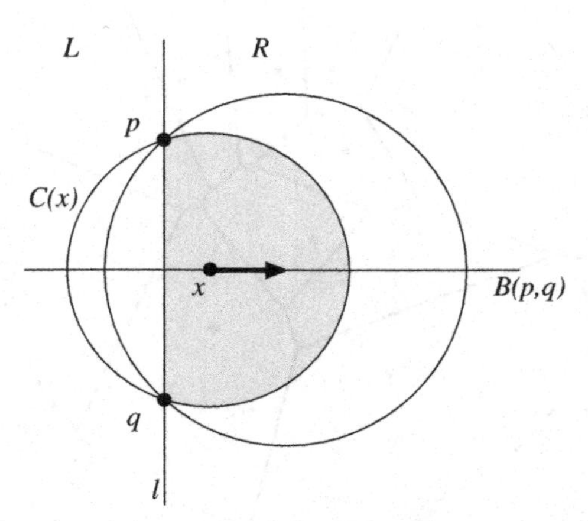

Figure 2.2. As x moves to the right, the intersection of circle $C(x)$ with the left halfplane L shrinks, while $C(x) \cap R$ grows.

convex hull of S can be reported in cyclic order. After removing the halflines outside Γ, a connected embedded planar graph with $n+1$ faces results. Its faces are the n Voronoi regions and the unbounded face outside Γ. We call this graph the *finite* Voronoi diagram.

One virtue of the Voronoi diagram is its small size.

Lemma 2.3. *The Voronoi diagram* $V(S)$ *has* $O(n)$ *many edges and vertices. The average number of edges in the boundary of a Voronoi region is less than* 6.

Proof. By *Euler's polyhedron formula* (see e.g. [363]) for planar graphs, the following relation holds for the numbers v, e, f, and c of vertices, edges, faces, and connected components, respectively

$$v - e + f = 1 + c.$$

We apply this formula to the finite Voronoi diagram. Each vertex has at least three incident edges; by adding up we obtain $e \geq 3v/2$, because each edge is counted twice. Substituting this inequality together with $c = 1$ and $f = n + 1$ yields

$$v \leq 2n - 2 \quad \text{and} \quad e \leq 3n - 3.$$

Adding up the numbers of edges contained in the boundaries of all $n+1$ faces results in $2e \leq 6n - 6$, because each edge is again counted twice. Thus, the average number of edges that border a region is at most $(6n-6)/(n+1) < 6$. The same bounds apply to $V(S)$. □

2.2. Delaunay triangulation

Now we turn to the Delaunay tessellation. In general, a *triangulation* of S is a planar graph with vertex set S and straight-line edges, which is maximal in the sense that no further straight-line edge can be added without crossing others. Triangulations are also called *triangular networks* in the literature.

Each triangulation of S contains the edges of the convex hull of S. Its bounded faces are triangles, due to maximality. Their number equals exactly $2n - h - 2$, where h counts the vertices of the convex hull. The number of edges is $3n - h - 3$. Note that both numbers, which easily follow from Euler's polyhedron formula, are independent from the way of triangulating the point set S, by the characteristics of this formula.

We call a connected subset of edges of a triangulation a *tessellation* of S if it contains the edges of the convex hull, and if each point of S has at least two incident edges.

Definition 2.2. The *Delaunay tessellation* $\mathrm{DT}(S)$ is obtained by connecting with a line segment any two points p, q of S, for which a circle exists that passes through p and q but does not contain any other site of S in its interior or boundary. The edges of $\mathrm{DT}(S)$ are called *Delaunay edges.*

Equivalently, and also common in the literature, $\mathrm{DT}(S)$ can be defined to contain all polygonal faces spanned by S whose *circumcircles* are empty of points of S. This is the so-called *empty-circle property* of the Delaunay tessellation. Another equivalent characterization of $\mathrm{DT}(S)$ is a direct consequence of Lemma 2.1.

Lemma 2.4. *Two points of S are joined by a Delaunay edge iff their Voronoi regions are edge-adjacent.*

Since each Voronoi region has at least two neighbors, at least two Delaunay edges must emanate from each point of S. By the proof of Lemma 2.2, each edge of the convex hull of S is Delaunay. Finally, two Delaunay edges can only intersect at their endpoints, because they allow for circumcircles whose respective closures do not contain other sites. This shows that $\mathrm{DT}(S)$ is in fact a *tessellation* of S.

Two Voronoi regions can share at most one Voronoi edge, by convexity. Therefore, Lemma 2.4 implies that $\mathrm{DT}(S)$ is the graph-theoretical *dual* of $V(S)$, realized by straight-line edges. Delaunay edges correspond to Voronoi edges, Delaunay faces to Voronoi vertices (the centers of their circumcircles), and Delaunay vertices (the n sites) to Voronoi regions — in a bijective way.

An example is depicted in Figure 2.3; the Voronoi diagram $V(S)$ is drawn by dashed lines, and $\mathrm{DT}(S)$ by solid lines. Note that a Voronoi vertex (like w) need not be contained in its associated face of $\mathrm{DT}(S)$. The sites p, q, r, s are cocircular, giving rise to a Voronoi vertex v of degree 4. Consequently, its corresponding Delaunay face is bordered by four edges. This cannot happen if the points of S are in *general position*, i.e., no three of them are collinear, and no four of them are cocircular.

Theorem 2.1. *If the point set S is in general position then* $\mathrm{DT}(S)$, *the dual of the Voronoi diagram $V(S)$, is a triangulation of S, called the* Delaunay triangulation. *Three points of S give rise to a Delaunay triangle exactly if the circle they define does not enclose any other point of S.*

Observe that $\mathrm{DT}(S)$ can be a triangulation even when S is not in general position. Its number of triangles varies, as mentioned before, with the number, h, of point sites which are *extreme* in S, i.e., which lie on the boundary of the convex hull of S. A maximum of $2n - 5$ is attained if the

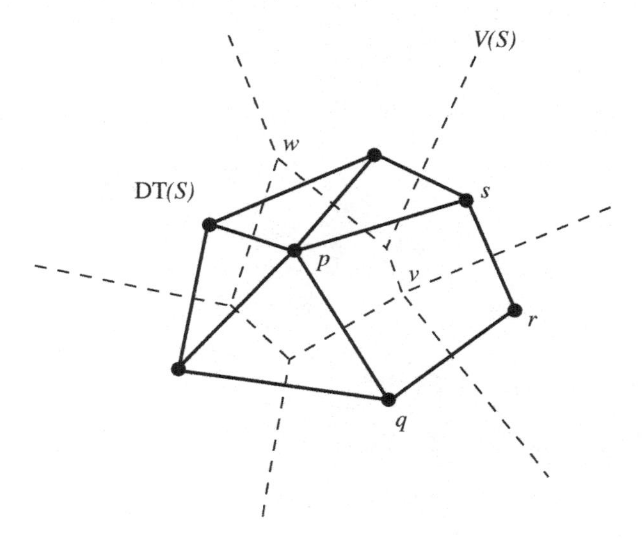

Figure 2.3. Voronoi diagram and Delaunay tessellation.

convex hull is triangular ($h = 3$). On the other end of the spectrum, if S is in *convex position* ($h = n$), then the convex hull is an n-gon, C, and DT(S) partitions C into only $n - 2$ triangles, using diagonals of C.

As in every triangulation of a planar point set S, vertices of high degree may occur in DT(S) for special positions of S. A site $p \in S$ can have $n - 1$ incident Delaunay edges (and triangles), because its Voronoi region VR(p, S) can be adjacent to all other regions in $V(S)$. Still, 'almost all' sites will be of small constant degree in DT(S), because their *average* degree is bounded by 6; see Lemma 2.3.

The Delaunay triangulation possesses various interesting features, concerning its optimization properties and subgraph structures, but also its flexibility in adapting to more general settings and higher dimensions. We will discuss this later at appropriate places, mainly in Sections 4.2 and 8.2 and, regarding algorithmic issues, in Section 3.2 and Subsection 6.1.2.

We have seen, in this short chapter, two structures for a finite planar point set S which are most fundamental and elementary, apart maybe from the convex hull of S. The Voronoi diagram $V(S)$ and the Delaunay triangulation DT(S) display the proximity influence that S exerts in a simple and intuitive way, with a description of only linear size. Their hidden intrinsic potential will be gradually revealed in the chapters to come. We start in Chapter 3, with showing that they lend themselves to several efficient methods of construction.

Chapter 3

BASIC ALGORITHMS

In this chapter we present several ways of computing a geometric representation of the Voronoi diagram and its dual, the Delaunay tessellation. For simplicity, we assume of the n point sites of S that no four of them are cocircular, and that no three of them are collinear. According to Theorem 2.1 we can then refer to DT(S) as the Delaunay triangulation. All algorithms presented herein can be made to run without this *general position* assumption. Also, they can be generalized to metrics other than the Euclidean, and to sites other than points. This will be discussed at appropriate places in later chapters.

Implementation issues, including degenerate configurations that might occur in the input or during the execution of algorithms, along with general numerical questions that arise in the computation of geometric objects, will be discussed in some detail in Section 11.2. They are only marginally addressed in the present chapter, not least for the sake of clarity in the presentation of the algorithmic techniques and their analyses. For a rich source on basic *data structures* (several of which will be used later on), we refer to the books by Cormen *et al.* [234] and Mehlhorn and Sanders [538].

We mainly seek for a representation of Voronoi diagrams in exact vector geometry, rather than as a binary image. Voronoi diagrams on *pixel maps* can be obtained by exact geometric algorithms followed by pixel extraction, or can be computed and visualized by graphics methods, beyond the scope of this book; see Section 11.1.

Data structures well suited for working with *planar graphs* like the Voronoi diagram are the *doubly connected edge list* (DCEL), by Muller and Preparata [552], and the *quad-edge* structure by Guibas and Stolfi [394]. In either structure, a record is associated with each edge e that stores the following information: the names of the two endpoints of e; references to the edges clockwise or counterclockwise next to e about its endpoints;

finally, the names of the faces to the left and to the right of e. The space requirement of both structures is $O(n)$.

Either structure allows to efficiently traverse the edges incident to a given vertex, or bordering a given face. The quad-edge structure offers the additional advantage of describing, at the same time, a planar graph and its *dual*, so that it can be used for constructing both the Voronoi diagram and the Delaunay triangulation. From the DCEL of $V(S)$ we can derive the set of triangles constituting the Delaunay triangulation in linear time. Conversely, from the set of all Delaunay triangles the DCEL of the Voronoi diagram can be constructed in time $O(n)$. Therefore, each algorithm for computing one of the two structures can be used for computing the other one, within $O(n)$ extra time effort.

3.1. A lower time bound

Before constructing the Voronoi diagram we want to establish a lower bound for its computational complexity.

Suppose that n real numbers $x_1, \ldots, x_n$ are given, which are to be sorted. We construct the point set $S = \{p_i = (x_i, {x_i}^2) \mid 1 \leq i \leq n\}$, which lies on the unit parabola and thus is in *convex position*; see Figure 3.1(i). By Lemma 2.2, the Voronoi diagram $V(S)$ forms a tree, from which one can derive, in $O(n)$ time, the vertices of the *convex hull* of S, in counterclockwise order. From the leftmost point in S on, this vertex sequence contains all points p_i, sorted by increasing values of x_i.

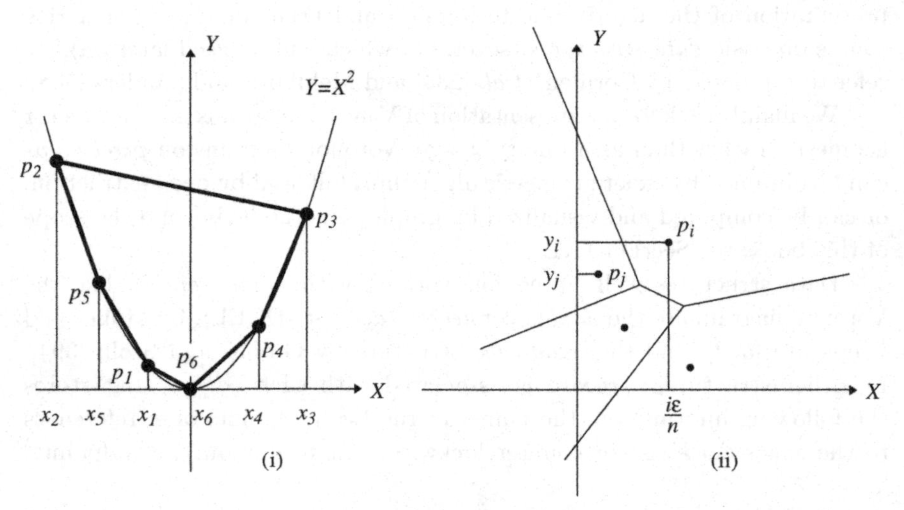

Figure 3.1. Proving the $\Omega(n \log n)$ lower bound for constructing the Voronoi diagram: by transformation (i) from sorting, and (ii) from ε-closeness.

This argument due to Shamos [636] shows that constructing the convex hull and, *a fortiori*, computing the Voronoi diagram, is at least as difficult as sorting n real numbers, which requires $\Omega(n \log n)$ time in the algebraic computation tree model [596].

However, a fine point is lost in this reduction. After sorting n points by their x-values, their convex hull can be computed in linear time [286], whereas sorting does not help in constructing the Voronoi diagram. The following result has been independently found by Djidjev and Lingas [281] and by Zhu and Mirzaian [716].

Theorem 3.1. *It takes time $\Omega(n \log n)$ to construct the Voronoi diagram of n points $p_1, \ldots, p_n$ whose x-coordinates are strictly increasing.*

Proof. The proof is by reduction from the ε-*closeness problem* [596], which is known to be in $\Omega(n \log n)$. Let $y_1, \ldots, y_n$ be positive real numbers, and let $\varepsilon > 0$. The question is if there exists $i \neq j$ such that $|y_i - y_j| < \varepsilon$ holds. We form the sequence of points

$$p_i = \left(\frac{i\varepsilon}{n}, y_i\right), \quad \text{for } 1 \leq i \leq n,$$

and compute their Voronoi diagram; see Figure 3.1(ii). In time $O(n)$, we can determine the Voronoi regions that are intersected by the y-axis, in bottom-up order (such techniques will be detailed in Section 3.3).

If, for each p_i, its projection onto the y-axis lies in the Voronoi region of p_i then the values y_i are available in sorted order, and we can easily answer the question. Otherwise, there is a point p_i whose projection lies in the region of some other point p_j. Because of

$$|y_i - y_j| \leq d((0, y_i), p_j) < d((0, y_i), p_i) = \frac{i\,\varepsilon}{n} \leq \varepsilon,$$

in this case the answer is positive. □

On the other hand, sorting n *arbitrary* point sites by x-coordinates is not made easier by their Voronoi diagram, as Seidel [626] has shown.

With Definition 2.1 in mind one could think of computing each Voronoi region as the intersection of $n-1$ halfplanes. This would take time $\Theta(n \log n)$ per region, see [596]. In the following sections we describe various algorithms that compute the *entire* Voronoi diagram within this time; due to Theorem 3.1, these algorithms are (asymptotically) worst-case optimal.

3.2. Incremental construction

A natural idea first studied by Green and Sibson [378] is to construct the Voronoi diagram by *incremental insertion*, i.e., to obtain $V(S)$ from $V(S\backslash\{p\})$ by inserting the site p. In the beginning, when there are only two sites, their Voronoi diagram is just their bisector line.

Basically, the cyclic sequence of edges of the new Voronoi region $\mathrm{VR}(p, S)$ has to be constructed, and the invalidated parts of the diagram $V(S\backslash\{p\})$ inside $\mathrm{VR}(p, S)$ deleted, in a way quite similar to constructing *merge chains* of edges, as is detailed in the next section. Figure 3.2 illustrates the insertion process. As the region of p can have up to $n-1$ edges, for $n = |S|$, this leads to a runtime of $O(n^2)$.

(Symbol $|\cdot|$ denotes the *cardinality*, when applied to finite sets, that is, the number of elements in a set.)

Several authors fine-tuned the technique of inserting Voronoi regions, and efficient and numerically robust implementations are available nowadays; see Ohya *et al.* [570] and Sugihara and Iri [668]. In fact, overall runtimes of $O(n)$ can be expected for 'well-distributed' sets of sites.

The insertion process is, maybe, better described and implemented in the dual environment, for the Delaunay triangulation: Construct the triangulation $\mathrm{DT}_i = \mathrm{DT}(\{p_1, \ldots, p_{i-1}, p_i\})$ by inserting the site p_i into

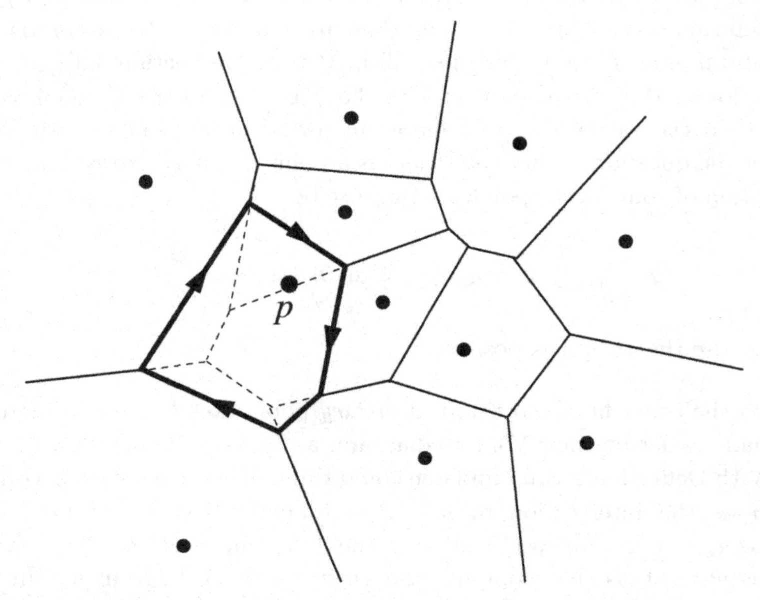

Figure 3.2. Inserting a Voronoi region. Invalidated portions of the diagram are drawn in dashed style.

DT_{i-1}. Clearly, DT_3 is a single triangle, and $\mathrm{DT}_n = \mathrm{DT}(S)$ is the final result. The advantage over a direct construction of $V(S)$ is that Voronoi vertices that appear in intermediate diagrams but not in the final one need not be constructed and stored. We follow Guibas and Stolfi [394] and construct DT_i by exchanging edges, using Lawson's [486] original *edge flipping* procedure, until all edges invalidated by p_i have been removed.

To this end, it is useful to extend the notion of triangle to the *unbounded face* of the Delaunay triangulation, which is the complement of the convex hull of S in the plane, $\mathbf{R}^2 \backslash \mathrm{conv}(S)$. If $\overline{pq}$ is an edge of $\mathrm{conv}(S)$, we call the supporting halfplane H not containing S an *infinite triangle* with edge $\overline{pq}$. Its 'circumcircle' is H itself, the limit of all circles through p and q whose centers tend to infinity within H; cf. Figure 2.2. As a consequence, each edge of a Delaunay triangulation is now incident to two triangles.

Those triangles of DT_{i-1} (finite or infinite) whose *circumcircles* contain the new site, p_i, are said to be *in conflict* with p_i. According to Theorem 2.1, they will no longer be Delaunay triangles.

Let $\overline{qr}$ be an edge of DT_{i-1}, and let $T(q,r,t)$ be the triangle incident to $\overline{qr}$ that lies on the other side of $\overline{qr}$ than p_i; see Figure 3.3. If its circumcircle $C(q,r,t)$ contains p_i then each circle through q,r contains at least one of p_i and t; see Figure 3.3 again. Consequently, $\overline{qr}$ cannot belong to DT_i, due to Definition 2.2. Instead, $\overline{p_i t}$ will be a new Delaunay edge, because there exists a circle contained in $C(q,r,t)$ that contains only p_i and t in its interior or boundary. This process of replacing edge $\overline{qr}$ by $\overline{p_i t}$ is called an *edge flip*.

The necessary edge flips can be carried out efficiently if we know the triangle $T(q,s,r)$ of DT_{i-1} that contains p_i, see Figure 3.4. (That is, we have

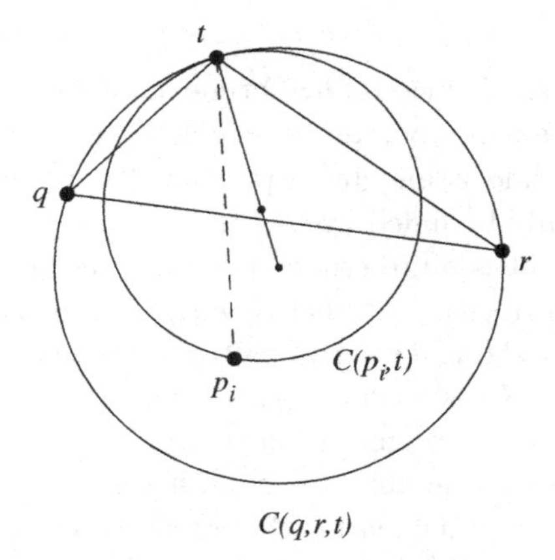

Figure 3.3. If triangle $T(q,r,t)$ is in conflict with p_i then the former Delaunay edge $\overline{qr}$ must be replaced by $\overline{p_i t}$.

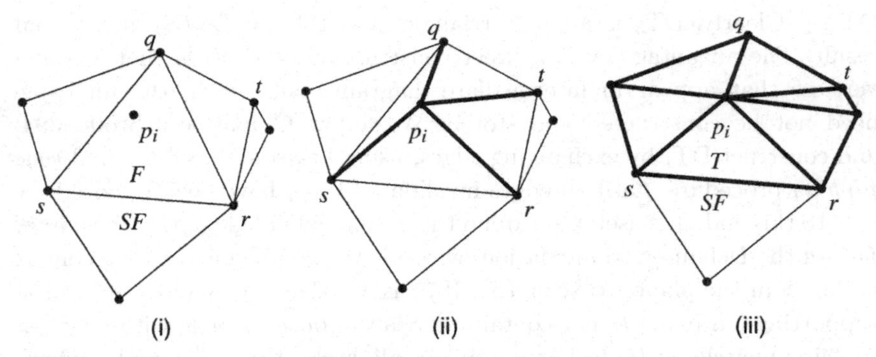

Figure 3.4. Updating DT_{i-1} after inserting the new site p_i. In (ii) the new Delaunay edges connecting p_i to q, r, s have been added, and edge $\overline{qr}$ has already been flipped. Two more flips are necessary before the final state shown in (iii) is reached.

to perform *point-location* of p_i in the temporary triangulation DT_{i-1}.) The line segments connecting p_i to q, r, and s will be new Delaunay edges, by the same argument as described above. Next we check if, e.g., edge $\overline{qr}$ must be flipped. If so, the edges $\overline{qt}$ and $\overline{tr}$ are tested, and so on. We continue until no further edge currently forming a triangle with, but not containing p_i, needs to be flipped, and obtain the triangulation DT_i.

Lemma 3.1. *If the triangle of* DT_{i-1} *containing* p_i *is known, the structural work needed for computing* DT_i *from* DT_{i-1} *is proportional to the* degree m *of* p_i *in* DT_i.

Proof. Continued edge flipping replaces $m - 2$ conflicting triangles of DT_{i-1} by m new triangles in DT_i that have p_i as a vertex; cf. Figure 3.4. □

Lemma 3.1 yields an obvious $O(n^2)$ time algorithm for constructing the Delaunay triangulation of n points: We can determine the triangle of DT_{i-1} containing p_i within linear time, by inspecting all candidates. Moreover, the degree of p_i is trivially bounded by $n - 1$.

The last argument is quite crude. There can be single vertices in DT_i that do have a high degree, but their *average* degree is bounded by 6, as Lemmas 2.3 and 2.4 show. This fact calls for *randomization.* Suppose we pick p_n at random in S, then choose p_{n-1} randomly from $S - \{p_n\}$, and so on. The result is a random permutation $(p_1, p_2, \ldots, p_n)$ of the set S of sites.

If we insert the sites in this order, each vertex of DT_i has the same chance of being p_i. Consequently, the *expected value* of the degree of p_i is $O(1)$, and the expected total number of structural changes in the construction of DT_n is only $O(n)$, due to Lemma 3.1.

In order to find the triangle that contains p_i it is sufficient to inspect all triangles that are in conflict with p_i. The following lemma shows that the expected total number of all *conflicting triangles* so far constructed is only logarithmic.

Lemma 3.2. *For each $k < i$, let t_k denote the expected number of triangles in DT_k but not in DT_{k-1} that are in conflict with p_i. Then,*

$$\sum_{k=1}^{i-1} t_k = O(\log i).$$

Proof. Let C denote the set of triangles of DT_k that are in conflict with p_i. A triangle $T \in C$ belongs to $\mathrm{DT}_k \backslash \mathrm{DT}_{k-1}$ iff it has p_k as a vertex. As p_k is randomly chosen in DT_k, this happens with probability $3/k$. Thus, the expected number of triangles in $C \backslash \mathrm{DT}_{k-1}$ equals $3 \cdot |C|/k$. Since the expected size of C is less than 6 we have $t_k < 18/k$, hence

$$\sum_{k=1}^{i-1} t_k < 18 \cdot \sum_{k=1}^{i-1} 1/k = \Theta(\log i). \qquad \square$$

Suppose that T is a triangle of DT_i incident to p_i, see Figure 3.4(iii). Its edge $\overline{sr}$ is in DT_{i-1} incident to two triangles: to its *father*, F, that has been in conflict with p_i; and to its *stepfather*, *SF*, who is still present in DT_i. Any further site in conflict with T must be in conflict with its father or with its stepfather, as illustrated by Figure 3.5.

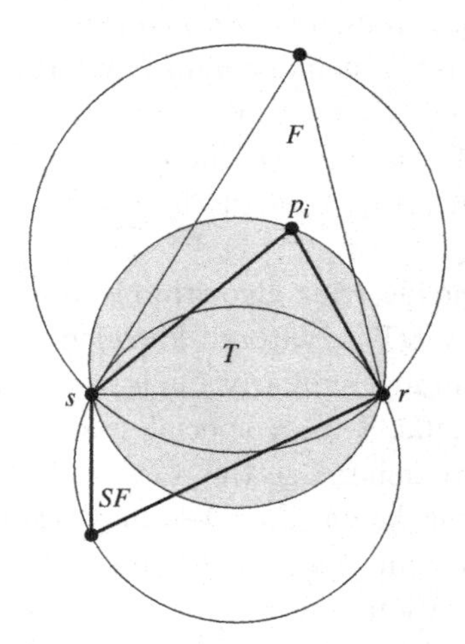

Figure 3.5. The circumcirle of T is contained in the union of the circumcircles of its father F and its stepfather SF.

This property can be exploited for quickly accessing all conflicting triangles. The *Delaunay tree* due to Boissonnat and Teillaud [148] is a *directed acyclic graph* that contains one node for each Delaunay triangle ever created during the incremental construction. (In other words, this graph reflects the *partial order* of the triangles imposed by the *construction history* of the current Delaunay triangulation.) Pointers run from fathers and stepfathers to their sons. The four triangles of DT_3 (three of which are infinite) are the sons of a dummy root node.

When p_i must be inserted, a Delaunay tree including all triangles up to DT_{i-1} is available. We start at its root and descend as long as the current triangle is in conflict with p_i. The above property guarantees that each conflicting triangle of DT_{i-1} will be found.

The expected number of steps this search requires is only $O(\log i)$, due to Lemma 3.2. Once DT_i has been computed, the Delaunay tree can easily be updated to include the new triangles. Thus, we have the following result.

Theorem 3.2. *The Delaunay triangulation of a set of n points in the plane can be constructed in expected time $O(n \log n)$, using expected linear space. The average is taken over the different orders of inserting the n sites.*

Note that we did not make any assumptions concerning the *distribution* of the sites in the plane; the incremental algorithm achieves its $O(n \log n)$ time bound for *every* possible input set. Only under a 'poor' insertion order can a quadratic number of structural changes occur, but this is unlikely.

The user annoyed by the uncertainty hidden in an *expected storage requirement* can apply a common trick to convert it into deterministic. Set some parameter c, and let a memory manager terminate the execution — and rerun the randomized construction — once the occupied space exceeds $c \cdot n$ (see Lemma 2.3 for a suitable choice of c). The expected runtime will still remain in $O(n \log n)$, though the constant in O will increase in dependency of c.

As a nice feature, the insertion algorithm is *online*. That is, it is capable of constructing DT_i from DT_{i-1} without knowledge of $p_{i+1}, \ldots, p_n$. In fact, *site deletions* can be handled similarly (as is sketched in Subsection 6.5.1, and also in Subsection 6.5.3 as a special case of an on-line algorithm which works in the more general setting of so-called *higher-order Voronoi diagrams*). This allows for *dynamizing* Delaunay triangulations and Voronoi diagrams, that is, maintaining these structures under insertion and deletion series of sites. Dynamization can also be based on the *divide & conquer* construction of Voronoi diagrams; see Section 3.3 for this earlier (and less efficient) approach, and Section 9.1.

Randomized geometric algorithms, though conceptually simple, tend to be tricky to analyze. Since Clarkson and Shor [229] introduced their technique, many researchers have been working on generalizing and simplifying the methods used. To mention but a few results, Boissonnat *et al.* [142] and Guibas *et al.* [390] have refined the methods of 'storing the past' in order to locate new conflicts quickly, Clarkson *et al.* [228] and Schwarzkopf [624] have generalized and simplified the analytic framework, and Seidel [631] systematically applied the technique of backward analysis first used by Chew [210]. A nice source is the monograph by Mulmuley [556]. The method in [390] for storing the past is applied in Section 6.5 for constructing order-k planar Voronoi diagrams.

If the set S of sites can be expected to be well distributed in the plane, *bucketing techniques* for accessing the triangle that contains a newly inserted site p_i have been used for speed-up. Joe [431], who implemented Sloan's algorithm [657], Su and Drysdale [664], who used a variant of Bentley *et al.*'s *spiral search* [124], and Lemaire and Moreau [502], who also gave probabilistic results for the higher-dimensional case, report on fast experimental runtimes. Snoeyink and van Kreveld [660] describe a simple way to precompute, in time $O(n \log n)$, an insertion order of the sites which then guarantees a *deterministic* $O(n)$-time incremental construction of the Delaunay triangulation — a task useful in the compression and transmission of triangular networks.

The arising issues of numerical stability have been addressed in Fortune [342], Sugihara [666], Jünger *et al.* [436], Shewchuk [648], and Avnaim *et al.* [108]. The main geometric primitive used by the algorithm is the *incircle test*, i.e., determining whether site p_i is enclosed by the circle defined by three other sites. As another practical issue, constructing Delaunay triangulations for '*imprecise point sets*' has been studied in Buchin *et al.* [172] and papers cited therein. In this setting, the sites are not known exactly but assumed to reside inside predefined regions (e.g., measurement tolerances), like small circles or squares.

Incremental insertion works well in the higher-dimensional case, and also for certain types of generalized Voronoi diagrams and their duals. We will see examples in Sections 6.1 and 6.5. Alternative methods for finding a *starting triangle*, i.e., a triangle of the current triangulation that contains the newly inserted site p_i, are discussed in Devillers *et al.* [266] and are reviewed, in the three-dimensional setting, in Section 6.1.

A technique similar to incremental insertion is *incremental search*. It starts with a single Delaunay triangle, and then incrementally discovers new ones, by growing triangles from edges of previously discovered triangles. This basic idea is used, e.g., in Maus [524] and in Dwyer [296]. It leads to efficient expected-time Delaunay algorithms, also in higher dimensions; see [296].

The paper [664] gives a thorough experimental comparison of available Delaunay triangulation algorithms.

3.3. Divide & conquer

The first deterministic worst-case optimal algorithm for computing the Voronoi diagram has been presented by Shamos and Hoey [637]. In their *divide & conquer approach*, the set of point sites, S, is split by a dividing line into subsets L and R of about the same size. Then, the Voronoi diagrams $V(L)$ and $V(R)$ are computed *recursively*, that is, the same strategy is applied to the (smaller) point sets L and R. If only three or two points are left in a set, their diagram is constructed directly, in $O(1)$ time.

The essential part is in finding the split line, and in *merging* $V(L)$ and $V(R)$, to obtain $V(S)$. If these tasks can be carried out in time $O(n)$ then the overall running time, $T(n)$, is only $O(n \log n)$, as we have the recurrence relation

$$T(n) = 2 \cdot T(n/2) + O(n).$$

During the recursion, vertical or horizontal split lines can be easily found if the sites in S are sorted by their x- and y-coordinates beforehand.

The merge step involves computing the so-called *merge chain* $B(L, R)$, that is, the set of all Voronoi edges of $V(S)$ that separate regions of sites in L from regions of sites in R.

Suppose that the split line is vertical, and that L lies to its left.

Lemma 3.3. *The edges of $B(L, R)$ form a single y-*monotone *polygonal chain. (That is, any line parallel to the x-axis intersects the chain in only one point.) In $V(S)$, the regions of all sites in L are to the left of $B(L, R)$, whereas the regions of the sites of R are to its right.*

Proof. Let b be an arbitrary edge of $B(L, R)$, and let $l \in L$ and $r \in R$ be the two sites whose regions are adjacent to b. Since l has a smaller x-coordinate than r, the edge b cannot be horizontal, and the region of l must be to its left. □

Thus, $V(S)$ can be obtained by gluing together $B(L, R)$, the part of $V(L)$ to the left of $B(L, R)$, and the part of $V(R)$ to its right; see Figure 3.6, where $V(R)$ is depicted by dashed lines.

The polygonal chain $B(L, R)$ is constructed by finding a starting edge at infinity, and by tracing $B(L, R)$ through $V(L)$ and $V(R)$.

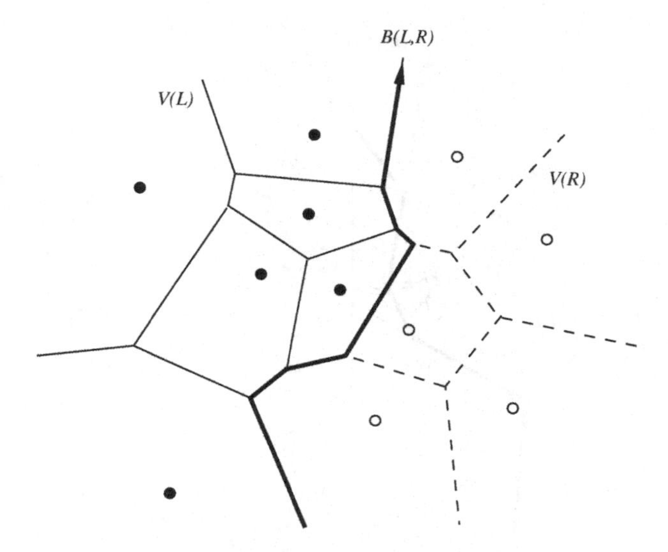

Figure 3.6. Merging $V(L)$ and $V(R)$ into $V(S)$.

Due to Shamos and Hoey [637], an unbounded starting edge of $B(L, R)$ can be found in $O(n)$ time by determining a line tangent to the *convex hulls* of L and R, respectively. Here we describe an alternative method by Lee [493], which was extended later by Chew and Drysdale [215], to be applicable for generalized Voronoi diagrams (Section 7.2). The unbounded regions of $V(L)$ and $V(R)$ are scanned simultaneously in cyclic order. For each non-empty intersection $\text{VR}(l, L) \cap \text{VR}(r, R)$, we test if it contains an unbounded piece of $B(l, r)$. If so, this must be an edge of $B(L, R)$, by Definition 2.1. Since $B(L, R)$ has two unbounded edges, by Lemma 3.3, this search will be successful. It takes time $|V(L)| + |V(R)| = O(n)$.

Now we describe how $B(L, R)$ is traced. Suppose that the current edge b of $B(L, R)$ has just entered the region $\text{VR}(l, L)$ at point v while running within $\text{VR}(r, R)$; see Figure 3.7. We determine the points v_L and v_R where b leaves the regions of l and of r, respectively. The point v_L is found by scanning the boundary of $\text{VR}(l, L)$ counterclockwise, starting from v. In our example, v_R is closer to v than v_L, so that it must be the endpoint of edge b.

From v_R, $B(L, R)$ continues with an edge b_2 separating l and r_2. Now we have to determine the points $v_{L,2}$ and $v_{R,2}$ where b_2 hits the boundaries of the regions of l and r_2. The crucial observation is that $v_{L,2}$ cannot be situated on the boundary segment of $\text{VR}(l, L)$ from v to v_L that we have just scanned; this can be inferred from the convexity of $\text{VR}(l, S)$. Therefore, we

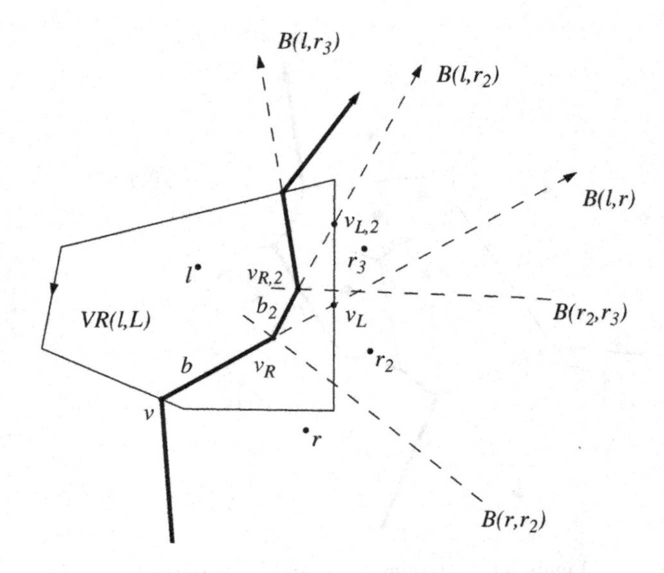

Figure 3.7. Computing the merge chain $B(L, R)$.

need to scan the boundary of $\mathrm{VR}(l, L)$ *only from* v_L *on*, in counterclockwise direction.

The same reasoning applies to $V(R)$; only here, region boundaries are scanned clockwise.

Even though the same region might be visited by $B(L, R)$ several times, no part of its boundary is scanned more than once. The edges of $V(L)$ that are scanned all lie to the right of $B(L, R)$. This part of $V(L)$, together with $B(L, R)$, forms a planar graph each of whose faces contains at least one edge of $B(L, R)$ in its boundary. As a consequence of Lemma 2.3, the size of this graph does not exceed the size of $B(L, R)$, times a constant. The same holds for $V(R)$. Therefore, the cost of constructing $B(L, R)$ is bounded by its size, once a starting edge is given. This leads to the following result.

Theorem 3.3. *The divide & conquer algorithm allows the Voronoi diagram of n point sites in the plane to be constructed within time $O(n \log n)$ and linear space, in the worst case. Both bounds are optimal.*

Of course, the divide & conquer paradigm can also be applied to the computation of the Delaunay triangulation $\mathrm{DT}(S)$. Guibas and Stolfi [394] give an implementation that uses the *quad-edge data structure* and only two geometric primitives: an orientation test and an *incircle test*. Fortune [342] showed how to perform these tests accurately with finite precision.

Dwyer's implementation [295] uses vertical and horizontal split lines in turn, and Katajainen and Koppinen's [447] merges square buckets in a *quad-tree* order. Both papers report on favorable results.

Using the '*history*' of constructing a Voronoi diagram by divide & conquer, a *dynamic Voronoi diagram* algorithm can be designed. Gowda *et al.* [376] pursue this approach. Based on a general dynamization paradigm in Overmars [573], they propose the so-called *Voronoi tree* as a data structure for supporting insertions and deletions of sites in $O(n)$ time, when n is the current size of the point set. The root of this binary tree stores the diagram $V(S)$ for the entire set S of sites, and its two children are the roots of the Voronoi trees for the subsets L and R that S got split into. Insertion or deletion of a site p amounts to an update in the Voronoi tree, which can be accomplished by traversing a path between its root and the leaf corresponding to p.

A related concept is the *Voronoi diagram for moving sites*, also called the *kinetic Voronoi diagram*, where the diagram is to be updated during a continuous movement of its sites along certain trajectories. We will elaborate on this topic to some extent in Section 9.1.

Divide & conquer algorithms are candidates allowing for *parallelization*. Efficient algorithms for computing in parallel the Voronoi diagram or the Delaunay triangulation have been proposed. We refer to the paper by Blelloch *et al.* [133] for references and for a practical parallel algorithm for computing DT(S). They highlight an algorithm by Edelsbrunner and Shi [314] that uses a *lifting map* for S (see Section 3.5) to construct a chain of Delaunay edges that divides S. They show experimentally that their implementation is comparable in *work* (i.e., product of runtime and processor number) to the best sequential algorithms.

In all the divide & conquer approaches described in this section, the emphasis is on the *bottom-up* phase — how to accomplish the merge step. This step gets considerably complicated for Voronoi diagrams in more general settings, because the merge chain may cycle and even heavily disconnect.

An alternative approach, with emphasis on the *top-down* phase — how to perform the divide step — has been recently considered in Aichholzer *et al.* [34, 27]. It is based on constructing the *medial axis* of a general planar shape, and exploits the tree structure of a medial axis to calibrate the divide step. The conquer step is trivial and consists of simply concatenating two partial medial axes. This algorithm, which also works for quite general planar objects as sites, is described in detail in Section 5.5.

3.4. Plane sweep

The well-known *plane sweep algorithm* (also called *sweep line* algorithm) by Bentley and Ottmann [121] computes the intersections of n line segments in the plane by moving a vertical line, H, from left to right across the plane. The line segments currently intersected by H are stored in y-order. This order must be updated whenever H reaches an endpoint of a line segment, or an intersection point. To discover the intersection points in time, it is sufficient to check, after each update of the y-order, those pairs of line segments that have just become neighbors on H.

It is tempting to apply the same approach to Voronoi diagrams, by keeping track of the Voronoi edges that are currently intersected by the vertical sweep line. The problem is in discovering new Voronoi regions in time. By the time the sweep line hits a new site, it has been intersecting Voronoi edges of its region for a while.

Fortune [344] was the first to find a way around this difficulty. He suggested a *planar transformation* under which each point site becomes the leftmost point of its Voronoi region, so that it will be the first point hit during a left-to-right sweep. His transformation does not change the combinatorial structure of the Voronoi diagram.

Later, Seidel [629] and Cole [230] have shown how to avoid this transformation altogether. They consider the Voronoi diagram of the point sites to the left of the sweep line H and of H itself, considered an *additional site* of straight-line shape; see Figure 3.8. Because the bisector of a line and a non-incident point is a parabola, the boundary of the Voronoi region of H is a connected chain of parabolic segments whose top- and bottommost edges tend to infinity. This chain is called the *wavefront*, W.

Let p be a point site to the left of H. Any point to the left of, or on, the parabola $B(p, H)$ is not farther from p than from H; hence, it is *a fortiori* closer to p than to any site to the right of H. Consequently, as the sweep line moves on to the right, the waves must follow because the sets to the left of $B(p_i, H)$ grow. On the other hand, each Voronoi edge to the left of W that currently separates the regions of two sites p_i, p_j will be (part of) a Voronoi edge in $V(S)$.

During the sweep, there are two types of events that cause the structure of the wavefront to change, namely when a new wave appears in W, or when an old wave disappears. The former one, called a *site event*, happens each time the sweep line hits a new site, e.g., p_6 in Figure 3.8. At that very

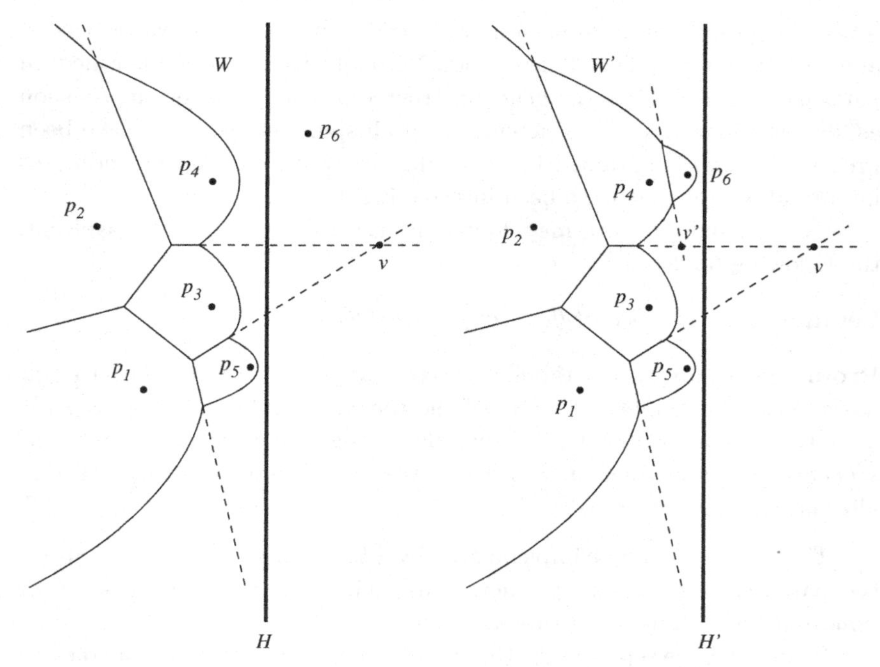

Figure 3.8. Voronoi diagrams of the sweep line H, and of the points to its left, which have been swept over already.

moment, $B(H, p_6)$ is a horizontal line through p_6 (as a degenerate case). A little later, its left halfline unfolds into a parabola that must be inserted into the wavefront by gluing it onto the wave of p_4 (which thereby is split into two waves of W.)

For describing the other type of event, let p, q be two point sites whose waves are neighbors in W. Their bisector, $B(p, q)$, gives rise to a Voronoi edge to the left of W. Its prolongation into the region of H is called a *spike*. In Figure 3.8 spikes are depicted as dashed lines; one can think of them as tracks along which the waves are moving. A wave disappears from W when it arrives at the point where its two bounding spikes intersect. Its former neighbors become now adjacent in the wavefront. This is called a *spike event*.

In Figure 3.8, the wave of p_3 would disappear at intersection point v, if the new site, p_6, that causes a site event prior to that did not exist. But after the wave of p_6 has been inserted, a spike event at point v' will take place, and the corresponding wave of p_4 will disappear.

While keeping track of the wavefront one can easily maintain the portion of the Voronoi diagram to the left of W. Spikes correspond to

Voronoi edges. At each event, a new adjacency between two waves occurs, and with it, a new spike. That is, a new Voronoi edge has been identified. In particular, a Voronoi vertex gets constructed at each spike event. As soon as all point sites have been detected and all spike intersections have been processed, $V(S)$ is obtained by removing the wavefront and extending to infinity all spikes that have been intersecting it.

Even though one site may contribute several waves to the wavefront, the following holds.

Lemma 3.4. *The size of the wavefront is $O(n)$.*

Proof. Since any two parabolic bisectors $B(p, H), B(q, H)$ can cross at most twice, the number of waves of the wavefront is bounded by $\lambda_2(n) = 2n - 1$, where $\lambda_s(n)$ denotes the maximum length of a *Davenport–Schinzel sequence* over n symbols in which no two symbols appear s times each in alternating positions; see [81]. □

The wavefront can be implemented by a *balanced search tree* that stores the parabola segments in y-order. This enables us to insert a wave, or remove a wave segment, in time $O(\log n)$.

Before the sweep starts, the point sites are sorted by increasing x-coordinates and inserted into an event *priority queue.* (See e.g. [234] or [538] for a description of basic data structures.) After each update of the wavefront, the newly adjacent spikes are tested for intersection. If they intersect at some point v, we insert into the event queue the time (i.e., the position x of the sweep line) when the wave segment between the two spikes arrives at v. Since the point v is a Voronoi vertex of $V(S)$, there are only $O(n)$ events caused by spike intersections. In addition, each of the n sites causes an event. For each active spike we need to store only its first intersection event. Thus, the size of the event queue never exceeds $O(n)$. We obtain the following result.

Theorem 3.4. *Plane sweep provides an alternative way of computing the Voronoi diagram of n points in the plane within $O(n \log n)$ time and $O(n)$ space.*

McAllister *et al.* [527] have pointed out a subtle difference between the sweep technique and the two methods mentioned before. The divide & conquer algorithm computes $\Theta(n \log n)$ many vertices, even though only a linear number of them appears in the final diagram. The randomized incremental construction method performs an expected $\Theta(n \log n)$ number of conflict tests. Both tasks, constructing a Voronoi vertex and testing a subset of sites for conflict, are usually handled by subroutines that deal directly with point coordinates, bisector equations, etc. They can become quite costly if we consider sites more general than points, and distance measures more general than the Euclidean distance; see Chapters 5 and 7.

The sweep algorithm, on the other hand, processes $O(n)$ spike events, and thus constructs only that many vertices.

The data structure to store the Voronoi diagram can be any standard data structure for planar subdivisions. See, for example, the *doubly connected edge list* [552] or the *quad-edge* data structure [394].

Note finally that the plane-sweep technique reduces a two-dimensional 'static' problem (for instance, the construction of a Voronoi diagram in the plane) to a one-dimensional 'dynamic' problem (here, the maintenance of the wavefront representation on the sweep line). This enables us to use efficient one-dimensional data structures, like balanced binary trees and priority queues. A kind of opposite philosophy underlies the algorithmic technique presented in the subsequent section: An interpretation of the planar problem in 3-space is sought, in order to gain insight into its structural properties, and to apply known but seemingly unrelated algorithms.

3.5. Lifting to 3-space

The following approach employs the powerful method of *geometric transformation*, which leads to an optimal algorithm for constructing the planar Delaunay triangulation.

Let $P = \{(x_1, x_2, x_3) \mid x_1^2 + x_2^2 = x_3\}$ denote the paraboloid depicted in Figure 3.9. For each point $x = (x_1, x_2)$ in the plane, let $x' = (x_1, x_2, x_1^2 + x_2^2)$ denote its lifted image on P.

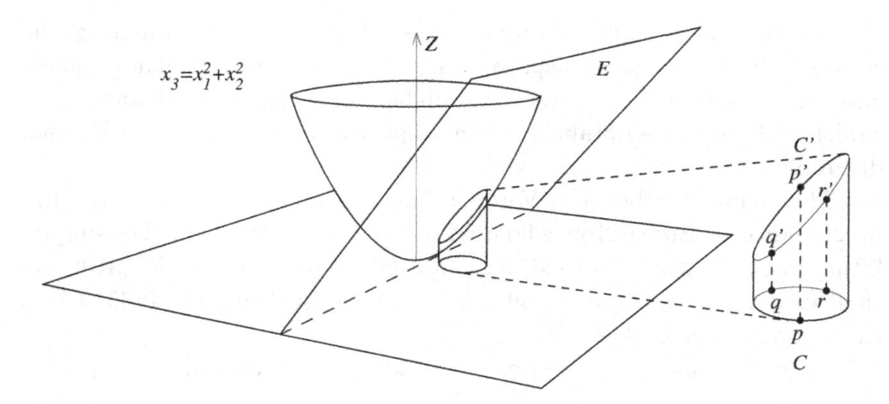

Figure 3.9. Lifting circles onto the paraboloid.

Lemma 3.5. *Let C be a circle in the plane. Then C' is a planar curve on the paraboloid P.*

Proof. Suppose that C is given by the equation

$$r^2 = (x_1 - c_1)^2 + (x_2 - c_2)^2 = x_1^2 + x_2^2 - 2x_1c_1 - 2x_2c_2 + c_1^2 + c_2^2.$$

By substituting $x_1^2 + x_2^2 = x_3$ we obtain

$$x_3 - 2x_1c_1 - 2x_2c_2 + c_1^2 + c_2^2 - r^2 = 0$$

for the points of C'. This equation defines a plane in 3-space. □

This lemma has an interesting consequence. Let S be a finite point set in the plane, and denote by S' its lifted image on P. The convex hull of S' is a *convex polyhedron* in 3-space. All points of S' appear as polyhedron vertices, because S' is in *convex position*, by construction. (Compare Figure 3.1 (i) in Section 3.1 for a picture in one dimension less.) By the *lower convex hull* of S' we mean that part of the convex hull which is visible from $x_3 = -\infty$.

Theorem 3.5. *The Delaunay triangulation* $\mathrm{DT}(S)$ *of S equals the vertical projection onto the x_1x_2-plane of the lower convex hull of S'.*

Proof. Let p, q, r denote three point sites of S. By Lemma 3.5, the lifted image, C', of their circumcircle C lies on a plane, E, that cannot be vertical. Under the lifting mapping, the points inside C correspond to the points on the paraboloid P that lie *below* the plane E.

By Theorem 2.1, p, q, r define a triangle of the Delaunay triangulation iff their circumcircle contains no further site. Equivalently, no lifted site s' is below the plane E that passes through p', q', r'. But this means that p', q', r' define a facet of the lower convex hull of S'. □

Because there exist $O(n \log n)$ time algorithms for computing the convex hull of n points in 3-space — see e.g. Preparata and Shamos [596], and for a survey of the various available construction methods — see Seidel [633], we have obtained another optimal algorithm for the Voronoi diagram.

The connection between Voronoi diagrams and convex hulls has first been studied by Brown [169] who used the *inversion transform.* The simpler lifting mapping has been used, e.g., in Edelsbrunner and Seidel [310]. We shall see several applications and generalizations in Chapter 6. In [310] also the following fact is observed.

For each point $s = (s_1, s_2)$ of S, consider the paraboloid

$$P_s = \{(x_1, x_2, x_3) \mid (x_1 - s_1)^2 + (x_2 - s_2)^2 = x_3\}.$$

If these paraboloids were opaque, and of pairwise different colors, an observer looking from $x_3 = -\infty$ upwards would see the Voronoi diagram $V(S)$. In fact, for $p, q \in S$, the projection $x = (x_1, x_2)$ of a point $(x_1, x_2, x_3) \in P_p \cap P_q$ belongs to their bisector $B(p, q)$; and there is no site s closer to x than p and q iff (x_1, x_2, x_3) lies below all paraboloids P_s. That is, $V(S)$ is the vertical projection onto the plane of the *lower envelope* of these paraboloids.

Instead of the paraboloids P_s one could use the surfaces

$$\{(x_1, x_2, x_3) = f((x_1 - s_1)^2 + (x_2 - s_2)^2)\}$$

generated by *any* function f that is strictly increasing. For example, $f(x) = \sqrt{x}$ gives rise to cones of slope 45° with apices at the sites. This setting illustrates the concept of circles expanding from the sites at equal speed, as mentioned after the proof of Lemma 2.1. Coordinate x_3 represents time.

In order to visualize a Voronoi diagram on a graphic screen one can feed the n surfaces to a z-buffer, and eliminate by brute force those parts not visible from below; cf. Section 11.1 for so-called *pixel Voronoi diagrams*.

Finally, we would like to mention a nice connection between the two ways of obtaining the Voronoi diagram by means of lower envelopes of paraboloids explained above; it goes back to [310]. For a point $w = (w_1, w_2, w_3)$, let $\hat{w}$ denote its mirror image $(w_1, w_2, -w_3)$. If we apply to 3-space the mapping which sends

$$x = (x_1, x_2, x_3) \text{ to } (x_1, x_2, x_3 - (x_1 - s_1)^2 - (x_2 - s_2)^2)$$

then, for each point s in the plane, the paraboloid P_s corresponds to the tangent plane of the paraboloid $\hat{P}$ at the point $\hat{s}'$, where s' denotes the lifted image of s; compare the plane equation derived in the proof of Lemma 3.5, letting $c = s$ and $r = 0$.

Lower envelope representations of generalized Voronoi diagrams will be briefly treated in Section 7.5. The corresponding surfaces and the resulting diagram, respectively, are sometimes called *Voronoi surfaces* and their *minimization diagram* in the literature. They are very useful for obtaining algorithms and, in addition, bounds on the combinatorial complexity of Voronoi diagrams.

Chapter 4

ADVANCED PROPERTIES

We have seen, in Chapters 2 and 3, quite a few basic properties of the Voronoi diagram for point sites in the plane. Many more shall be described in later chapters, along with generalizations of the Voronoi diagram concerning its sites and its underlying distance function, as certain properties extend to such settings in a natural way. The present chapter is devoted to some advanced features of the classical Voronoi diagram and its dual, the Delaunay triangulation.

4.1. Characterization of Voronoi diagrams

The process of constructing the Voronoi diagram for n point sites can be seen as an assignment of a planar convex region to each of the sites, according to the 'nearest neighbor rule'. We now address the following, in some sense inverse, question: Given a partition of the plane into n convex regions (which are then necessarily polygonal), do there exist sites, one for each region, such that the nearest neighbor rule is fulfilled? In other words, when is a given *convex partition* the Voronoi diagram of some set of sites?

Whether a *given* set of sites induces a given convex partition as its Voronoi diagram is, of course, easy to decide by exploiting symmetry properties among the sites. For the same reason, it is easy to check whether a given triangulation is Delaunay, by exploiting the *empty-circle property* of its triangles, stated in Theorem 2.1. Conditions for a given graph to be isomorphic to the Delaunay triangulation of *some* set of sites are mentioned, e.g., in the survey article by Fortune [343]. Below we concentrate on the recognition of Voronoi diagrams *without* knowing the sites. The process of restoring the sites, if they exist, is also called *Voronoi diagram inversion*; see Schoenberg *et al.* [619] for an overview.

Questions of this kind arise in *facility location* and in the recognition of biological *growth models* (as reported, e.g., in Suzuki and Iri [671]) and,

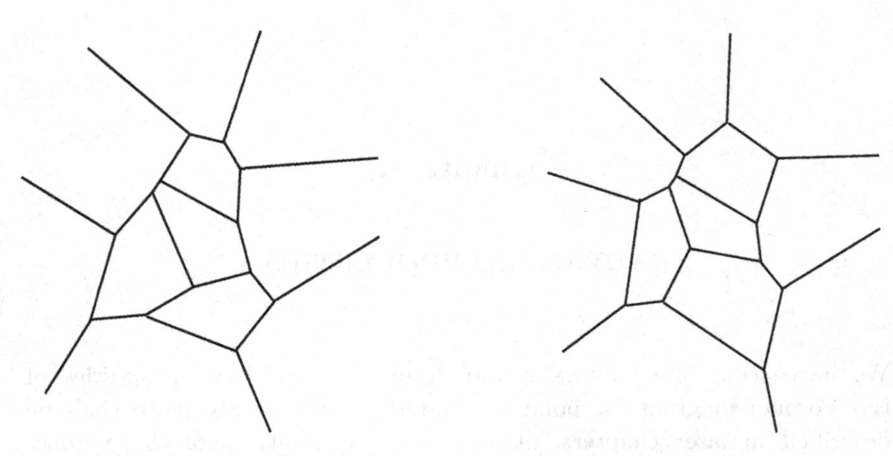

Figure 4.1. A Voronoi diagram? Figure 4.2. This one, yes.

in particular, in the so-called *gerrymander problem* mentioned in Ash and Bolker [79]: When the sites are regarded as polling places and election law requires each person to vote at the respective closest polling place, the election districts form a Voronoi diagram. If the legislature draws the district lines first, how can we tell whether election law is satisfied? (Figures 4.1 and 4.2).

Let R_i and R_j be two of the given regions. Assume that they share a common edge, and let h_{ij} be the straight line containing that edge. Further, let σ_{ij} denote the reflection at line h_{ij}.

Lemma 4.1. *A convex partition $R_1, \ldots, R_n$ of the plane defines a Voronoi diagram if and only if there exists a point p_i for each region R_i such that*

(1) $p_i \in R_i$ *(containment condition),*
(2) $\sigma_{ij}(p_i) = p_j$ *if R_j is adjacent to R_i (reflection condition).*

Proof. If we do have a Voronoi diagram then its defining sites exist and obviously fulfill (1) and (2). To prove the converse, assume that points $p_1, \ldots, p_n$ fulfilling both conditions exist. Take any region R_i and any point x therein. We show that $d(x, p_i)$ is minimum.

To get a contradiction, suppose p_j, $j \neq i$, is closest to x. Consider an edge of R_j that is intersected by the line segment $\overline{xp_j}$, and let R_k be the region adjacent to R_j at that edge; see Figure 4.3. Note that $k = i$ may happen. By convexity of R_j and by (1), the line h_{jk} separates p_j from x. Hence, by (2), we get $d(x, p_k) < d(x, p_j)$, which is a contradiction.

We conclude that p_i is closest to x among $p_1, \ldots, p_n$, which implies that R_i is the region of p_i in the Voronoi diagram $V(\{p_1, \ldots, p_n\})$. $\square$

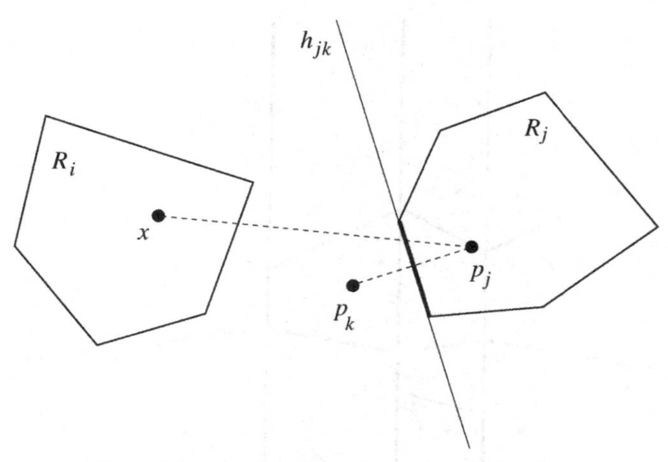

Figure 4.3. Region R_i must be closest to point x.

Based on Lemma 4.1, the recognition problem can now be formulated as a *linear programming problem*; see Hartvigsen [398]. We first exploit the reflection condition to get a system of linear equations.

Reflection at a line is an affine transformation, so we may write $\sigma_{ij}(x)$ as $A_{ij}x + b_{ij}$, for an appropriate matrix A_{ij} and vector b_{ij}. Consider a *depth-first search* order (see e.g. [363]) of the regions, in a way such that for each region R_i an adjacent region R_{i+1} is known. To get a linear system in x, put

$$p_1 = x,$$
$$p_2 = A_{12}x + b_{12} := C_2x + d_2,$$
$$p_3 = A_{23}(A_{12}x + b_{12}) + b_{23} := C_3x + d_3$$

and so on. This expresses all points p_i in terms of p_1 by using $n-1$ adjacencies among the regions. Each of the remaining adjacencies now gives an equation of the form

$$A_{ij}(C_ix + d_i) + b_{ij} = C_jx + d_j.$$

This system has at most $3n-3-(n-1) = 2n-2$ equations by Lemma 2.3. If it has no solution, or a unique solution, then we are done. In the former case, we cannot have a Voronoi diagram. In the latter, we get the coordinates of the first candidate site $p_1 = x$. The corresponding other sites are obtained simply by reflection. It remains to test these sites for containment in their regions.

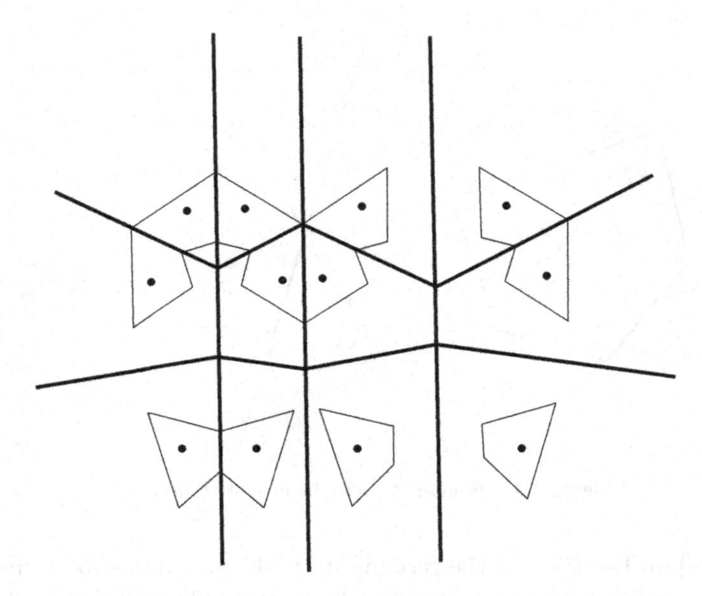

Figure 4.4. Sites might be chosen within the polygonal areas.

Setting up the system, solving it, and testing for containment can be accomplished in time $O(n)$ by standard methods. Note that only the equations of the lines bounding the regions and the adjacency information among the regions are needed. No coordinates of the region vertices are required. This is particularly interesting for the recognition problem in *higher dimensions*, to which the method above generalizes naturally.

The solution space of the linear system above may have dimension 1 or 2. Figure 4.4 reveals that certain symmetries among the regions lead to non-unique solutions. Now the containment condition is exploited to get, in addition, a set of inequalities for x. Consider each region R_i as the intersection of all halfplanes bounded by the lines h_{ij}. Then $p_i \in R_i$ gives a set of inequalities of the form

$$p_i^T t_{ij} \leq a_{ij},$$

which, by plugging in $p_i = C_i x + d_i$, yields

$$(C_i^T t_{ij})x \leq a_{ij} - d_i^T t_{ij}.$$

Finding a feasible solution of the corresponding linear program means finding a possible site $p_1 = x$ for R_1 which, by reflection, gives all the desired sites. Since we deal with a linear program with $O(n)$ constraints and of constant dimension (actually two), also only linear time (Megiddo [531]) is spent in this more complicated case.

Theorem 4.1. *Let $\mathcal{C}$ be a partition of the plane into n convex regions, given by halfplanes supporting the regions and by adjacencies among regions. $O(n)$ time suffices for deciding whether $\mathcal{C}$ is a Voronoi diagram, and also for restoring a suitable set of sites in case of its existence.*

This result is clearly optimal, and the underlying method easy to program. A generalization to higher dimensions is straightforward. Still, the method has to be used with care, as even a slight deviation from the correct Voronoi diagram (stemming from imprecise measurement or numerical errors) will cause the method to classify $\mathcal{C}$ as non-Voronoi. Suzuki and Iri [671] give a completely different method capable of *approximating* $\mathcal{C}$ by a Voronoi diagram.

Lemma 4.1 extends to more general Voronoi-like partitions. The characterizing configuration of points is then called a *reciprocal figure*, valuing Clerk Maxwell's work [525] of 1864. A nice survey on this subject is Ash *et al.* [80]. In particular, if the partition $\mathcal{C}$ is polygonal, then the containment and reflection conditions can be relaxed to

(1b) *orthogonality*: if polygons R_i and R_j share an edge e then the line segment $\overline{p_i p_j}$ is orthogonal to e, and
(2b) *orientation*: any ray which is parallel to the vector $(p_j - p_i)$ and intersects e meets R_i first.

Interestingly, the existence of a reciprocal figure $\{p_1, \ldots, p_n\}$ for $\mathcal{C}$, which is now also called an *orthogonal dual*, is equivalent to the property that $\mathcal{C}$ can be realized as the *equilibrium state* of a spider web: The edges of $\mathcal{C}$ allow positive tensions that balance out at all vertices of $\mathcal{C}$ — a property of use in the statics of *plane frameworks*; see e.g. Crapo and Whiteley [235].

Consult Figure 4.5. Suitable tensions for the edges $e_{ij} = R_i \cap R_j$ are the lengths of the vectors $(p_j - p_i)$, or constant multiples thereof. These tensions trivially add up to zero for the regions $R_1, \ldots, R_k$ around each vertex v, because the corresponding line segments $\overline{p_i p_{i+1}}$ border a face dual to v (a triangle in the non-degenerate case).

Moreover, by *Maxwell's theorem* [525], the existence of such tensions characterizes partitions of the plane that can be obtained as the *boundary projections* of convex polyhedra in 3-space; see also [235].

Clearly, for the Voronoi diagram $V(S)$ its set S of sites is a possible orthogonal dual. On the other hand, the *vertices* of $V(S)$ constitute an orthogonal dual for the Delaunay triangulation $\mathrm{DT}(S)$; the nonconvex

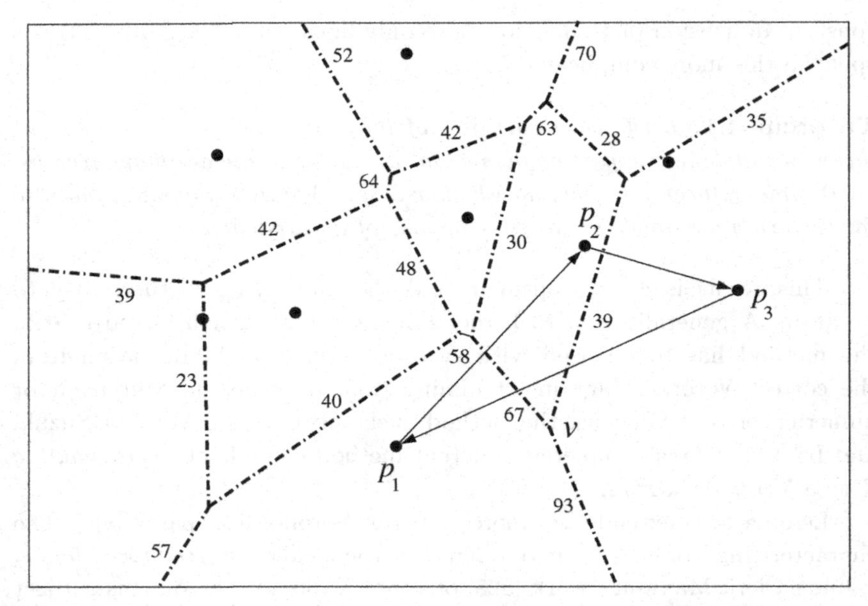

Figure 4.5. Spider web with orthogonal dual (•). Any scaled and translated copy is a valid orthogonal dual, too. Numbers at edges express tensions in equilibrium state at each vertex. This is witnessed for vertex v by the side lengths of its dual triangle $p_1p_2p_3$.

unbounded face of DT(S) can be treated consistently. Indeed, in Section 3.5 we have seen the existence of a projection polyhedron for DT(S), namely, the convex hull of the point set S lifted to $\mathbf{R}^3$.

The correspondences above still hold in higher dimensions; see Aurenhammer [88, 89]. By the relationship of Voronoi diagrams to convex polyhedra in the next dimension (Subsection 6.2.2), a partition $\mathcal{C}$ admits an orthogonal dual if and only if $\mathcal{C}$ is the *power diagram* of some set of spheres; the dual points p_i serve as sphere centers.

Various types of convex cell complexes can be identified as polyhedral projections in this way, including simple complexes, arrangements of hyperplanes, and higher-order Voronoi diagrams — structures we will encounter in Chapter 6. Moreover, projection polyhedra can be found efficiently [89].

The topic of lifting (not necessarily convex) cell complexes is briefly addressed in Subsection 6.3.3. For equilibrium states in nonconvex polygonal planar partitions, the *Maxwell–Cremona theorem* gives an elegant answer; negative tensions of edges (stresses) will now occur in the characterization. This theorem has surprising applications different from its direct use in *terrain recognition* and *scene analysis* — for instance to the long-standing open problem of *untangling simple polygons* in the plane, which has been solved in Connelli *et al.* [231].

4.2. Delaunay optimization properties

The Delaunay triangulation, DT(S), of a set S of n sites in the plane possesses a host of nice and useful properties many of which are well known and understood nowadays. As being the geometric dual of the Voronoi diagram $V(S)$, DT(S) comprises the proximity information inherent to S in a compact manner. Apart from the present section, various properties of DT(S) and their applications are described in Chapter 8 and, in particular, in Section 8.2. Here we look at DT(S) as a triangulation *per se* and concentrate on parameters which are optimized by DT(S) over all possible triangulations of the point set S.

Recall that a *triangulation* (or *triangular network*), T, of S is a maximal set of non-crossing line segments spanned by the sites in S. In other words, T defines a partition of the convex hull of S into triangles whose vertex set is exactly S. The number of different such partitions is large (in fact exponential, $\Omega(2.33^n)$) for *every* n-point set S in general position [43], which makes the optimality results to be presented even more valuable.

Let us call T *locally Delaunay* if, for each of its convex quadrilaterals Q, the corresponding two (adjacent) triangles have *circumcircles* empty of vertices of Q. Clearly, DT(S) is locally Delaunay because the circumcircles for its triangles are totally empty of sites; see Theorem 2.1. Interestingly, the local property also implies the global one.

Theorem 4.2. *A triangulation of S is locally Delaunay if and only if it equals the Delaunay triangulation,* DT(S).

Proof. Let T be a triangulation of S and assume that T is locally Delaunay. We show that, for each triangle Δ of T, its circumcircle $C(\Delta)$ is empty of sites in S.

Assuming the contrary, let s be inside $C(\Delta)$ for some $s \in S$ and some Δ in T. Observe $s \notin \Delta$ and let e be the edge of Δ closest to s. Suppose, without loss of generality, that (Δ, e, s) maximizes the angle spanned by e at s, for all such triples (triangle, edge, site).

See Figure 4.6. Because of s, edge e cannot be an edge of the convex hull of S. Let triangle Δ' be adjacent to Δ at e, and let s' be the third vertex of Δ'. As T is locally Delaunay, s' lies outside $C(\Delta)$, hence $s \neq s'$. Now, observe that s is enclosed by $C(\Delta')$, and let e' be the edge of Δ' closest to s. The angle at s for (Δ', e', s) is larger than that for (Δ, e, s), which gives a contradiction. □

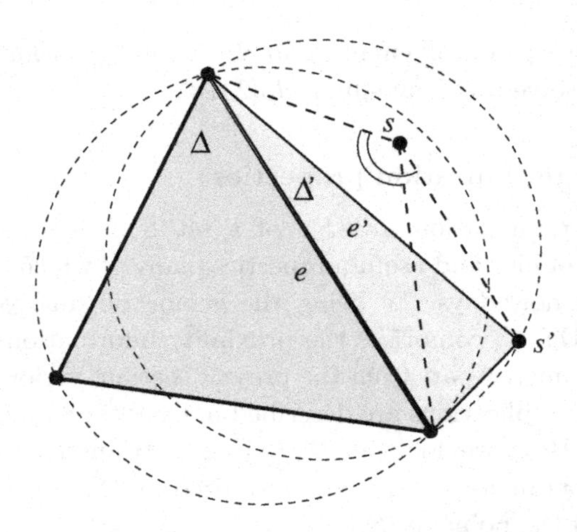

Figure 4.6. Illustration of the proof of Theorem 4.2.

An *edge flip* in a triangulation T of S is the exchange of the two diagonals in one of T's convex quadrilaterals; see Section 3.2. Call an edge flip *good* if, after the flip, the triangulation inside the quadrilateral (consisting of only 2 triangles) is locally Delaunay. Repeated exchange of diagonals in the same quadrilateral always produces an alternating sequence of good and not good flips. Theorem 4.2 now can be used to prove that DT(S) optimizes various quality measures, by showing that each good flip increases quality. Any sequence of good flips then terminates at the global optimum, the Delaunay triangulation. The length of such a sequence is bounded by $\binom{n}{2}$, as an edge that gets flipped away cannot reappear; see the argument on *flip distances* in planar triangulations given in Subsection 6.3.3.

One of the most prominent quality measures concerns the angles occurring in a triangulation. Recall that the number of edges (and thus of triangles) does not depend on the way of triangulating S, and let t be the number of triangles for S. The *angularity* of a triangulation is defined to be the sorted list of angles $(\alpha_1, \ldots, \alpha_{3t})$ of its triangles, in ascending order. A triangulation is called *equiangular* if it possesses lexicographically largest angularity among all possible triangulations for S.

As a matter of fact, every good flip increases angularity. Figure 4.7 gives evidence for this fact. Lawson [486] called a triangulation *locally equiangular* if no flip can increase angularity. Locally equiangular thus is equivalent to locally Delaunay. Sibson [652] and Lee [492] first proved Theorem 4.2, showing that locally equiangular triangulations are Delaunay and hence are unique. Edelsbrunner [300] observed that DT(S) is equiangular (in the global sense) as the global property implies the local one.

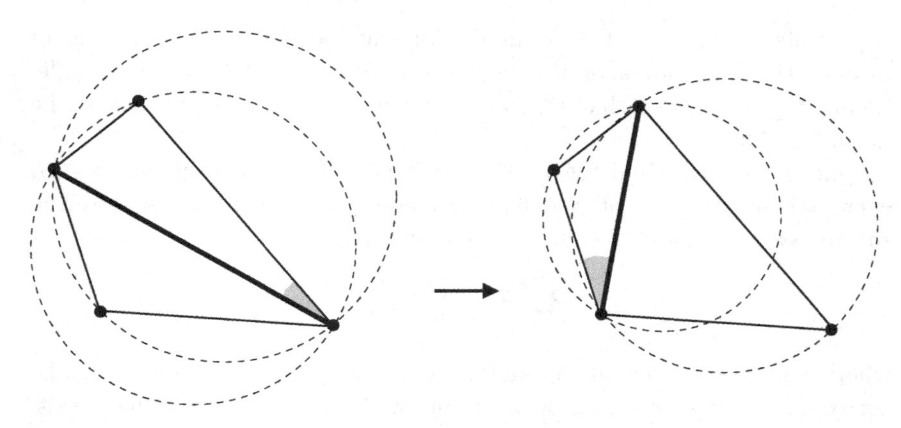

Figure 4.7. Equiangularity and the empty circle property.

In case of cocircularities among the sites, $\mathrm{DT}(S)$ is not a full triangulation; see Chapter 2. Mount and Saalfeld [550] showed that $\mathrm{DT}(S)$ can be completed by retaining local equiangularity, in $O(n \log n)$ time.

Theorem 4.3. *Let S be a finite set of sites in the plane. A triangulation of S is equiangular only if it is a completion of* $\mathrm{DT}(S)$.

The equiangular triangulation obviously maximizes the minimum angle that occurs in all the triangles. This property is desirable for *mesh generation* applications in *terrain modeling* or for the *finite element method*, as was first observed in Lawson [486] and McLain [528]. By Theorem 4.3, such triangulations can be computed in $O(n \log n)$ time with the Delaunay triangulation algorithms in Chapter 3.

More recently, it has been observed that several other parameters are optimized by $\mathrm{DT}(S)$. All the properties listed below can be proved by observing that every good edge flip locally optimizes the respective parameter.

Consider the *circumcircle* for each triangle in a triangulation, and measure *coarseness* by the largest such circle that arises; see Figure 4.7 again. As a matter of fact, $\mathrm{DT}(S)$ minimizes coarseness among all possible triangulations for S; see D'Azevedo and Simpson [240]. We may define coarseness also by taking *smallest enclosing circles* rather than circumcircles. (Note that the smallest enclosing circle differs from the circumcircle iff the triangle is obtuse.) D'Azevedo and Simpson proved that $\mathrm{DT}(S)$ minimizes coarseness in this sense, and Rajan [598] showed that this property of Delaunay triangulations — unlike others — generalizes to higher dimensions.

Similarly, *fatness* of a triangulation may be defined as the sum of *inradii*, that is, radius of the largest circle inscribed to each triangle. Lambert [480] showed that DT(S) maximizes fatness, or equivalently, the mean inradius.

Given an individual function value (height) $h(p)$ for each site $p \in S$, every triangulation T of S defines a *triangular surface* in 3-space. The *roughness* of such a surface may be measured by

$$\sum_{\Delta \in T} |\Delta| \cdot (\alpha^2 + \beta^2),$$

where $|\Delta|$ is the area of Δ, and α and β denote the slopes that the corresponding triangle in 3-space forms with the x-axis and the y-axis, respectively. In other words, roughness is the integral of the squared gradients. As has been shown by Rippa [605], roughness is minimum for the surface obtained from DT(S), for *any* fixed heights $h(p)$. For a simpler proof, see Powar [594]. See also Musin [558] for various functionals on triangular surfaces which are optimized by DT(S).

Let us mention that, in addition, DT(S) provides a means for *smoothing* the corresponding triangular surface. As was shown in Sibson [653], each point x within the convex hull of S can be expressed as the *weighted centroid* of its Delaunay neighbors p in DT($S \cup \{x\}$). Weights $w_p(x)$ can be computed from area properties of the corresponding diagram regions, and as functions of x, are continuously differentiable; see also Farin [336]. The corresponding interpolant to the spatial points $(p, h(p))$ is called the *natural neighbor interpolant*, and is given by

$$\phi(x) = \sum_{p \in S} w_p(x) h(p).$$

This useful property of DT(S) is shown to generalize to *regular triangulations* (duals of power diagrams for S, see Section 6.2), and to *higher-order Voronoi diagrams* (Section 6.5) in Aurenhammer [91]. For more material on natural neighbor interpolants and their applications, see, e.g., Boissonnat and Cazals [140].

On the negative side, DT(S) in general fails to fulfill optimization criteria similar to those mentioned above, such as minimizing the maximum angle, or minimizing the longest edge. Edelsbrunner *et al.* [315, 317] give near-quadratic time algorithms for computing triangulations optimal in that sense. DT(S) is not even *locally minimal*, in the sense that it does not always include the shorter diagonal for each of its convex quadrilaterals.

Kirkpatrick [455] proved that DT(S) may differ arbitrarily strongly from a *minimum-weight triangulation*, which is defined to have minimum total edge length. Computing a minimum-weight triangulation is an important and interesting problem, whose complexity remained unknown for a long time; see Garey and Johnson [352]. Recently, Mulzer and Rote [557] showed its NP-hardness. (For the class of so-called *NP-hard problems*, it is very unlikely that *polynomial-time algorithms* exist, i.e., algorithms with a running time of $O(n^k)$, with k being a constant.) Subsets of edges of DT(S) which always have to belong to a minimum-weight triangulation are exhibited in Subsection 8.2.3.

On the other hand, the widely used *greedy triangulation*, which is obtained by inserting non-crossing edges in increasing length order, can be constructed from DT(S) in $O(n)$ time, by a result in Levcopoulos and Krznaric [504].

Finally, let us mention that the Delaunay triangulation avoids an undesirable property that might be shared by other triangulations. Fix a point v in the plane, called the viewpoint. For two triangles Δ and Δ' in a given triangulation, write $\Delta < \Delta'$ if Δ fully or partially hides Δ' as seen from v. This defines a relation, called the *in-front/behind relation*, on the triangles. De Floriani *et al.* [245] observed that this relation is *acyclic* if the triangulation is Delaunay, no matter where the viewpoint is chosen. That is, the relation gives a *partial order*, also called a *shelling order* (with center v) of the triangles, see e.g. Brugesser and Mani [171] and Subsection 6.5.1.

An example of a triangulation which is cyclic in spite of being minimum-weight can be found in Aichholzer *et al.* [32].

Edelsbrunner [301] generalized the result in [245] to *regular simplicial complexes* in d dimensions, a class that includes higher-dimensional Delaunay triangulations as a special case; see Section 6.3. An application stems from a popular algorithm in computer graphics that eliminates hidden objects by first partially ordering the objects according to the in-front/behind relation and then displaying them from back to front, thereby overpainting invisible parts. In particular, this algorithm will work well for *α-shapes* (Subsection 8.2.2) and their weighted variants (Section 6.6). A rapid algorithm for displaying 3D Delaunay triangulations, or parts thereof, on a standard raster display is described in Karasick *et al.* [445]; cf. also Section 11.1 for *pixel Voronoi diagrams*.

For early structural discussions of planar triangulations, the reader is referred to the theses by Tan [675] and Lambert [481], respectively. A survey on optimization properties of triangulations is given in Aurenhammer and Xu [106].

Chapter 5

GENERALIZED SITES

In order to meet practical needs, the concept of Voronoi diagram has been modified and generalized in many ways, for example by changing the underlying space, the distance function used, or the shape of the sites. Chapters 5 to 7 give a systematic treatment of generalized Voronoi diagrams.

It is commonly agreed that most geometric scenarios can be modeled with sufficient accuracy by *polygonal objects*. Two typical and prominent examples are the description of the workspace of a robot moving in the plane, and the geometric information contained in a geographical map. In both applications, robot motion planning and geographical information systems, the availability of proximity information for the scenario is crucial. This is among the reasons why considerable attention has been paid to the study of Voronoi diagrams for polygonal objects.

Still, in some applications the scenario can be modeled more appropriately when *curved objects*, for instance, circular arcs are also allowed. Many Voronoi diagram algorithms working for line segments can be modified to work for curved objects as well.

5.1. Line segment Voronoi diagram

Let G be a *planar straight-line graph* on n points in the plane, that is, a set of non-crossing line segments spanned by these points. For instance, G might be a *tree* (a connected graph without edge cycles), or a collection of disjoint line segments or polygons (edge cycles), or a complete *triangulation* of the points (a maximal non-crossing set of edges). The number of segments of G is maximal, at most $3n-6$, in the last case. We will discuss several types of diagrams for planar straight-line graphs in the present and following sections.

The classical type is the (*closest site*) *Voronoi diagram*, $V(G)$, of G. It consists of all points in the plane which have more than one closest segment

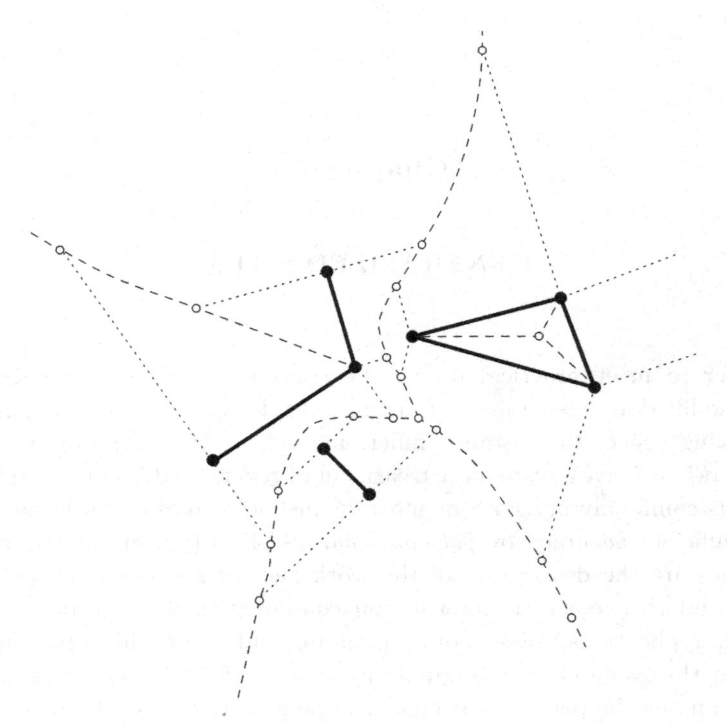

Figure 5.1. Line segment Voronoi diagram.

in G. $V(G)$ is known under different names in different areas, for example, as the *line Voronoi diagram* or *skeleton* of G, or as the *medial axis* or the *symmetric axis* when G is a simple polygon.

(A polygon is called *simple* if it is homeomorphic to a disk. Simple polygons have connected boundaries; in particular, they contain no 'holes'.)

Applications in such diverse areas as biology, geography, pattern recognition, computer graphics, and motion planning exist; see e.g. Kirkpatrick [454] and Lee [494] for early references, and the recent book by Siddiqi and Pizer [654].

See Figure 5.1. $V(G)$ is formed by straight-line edges and parabolically curved edges, both shown as dashed lines. Straight-line edges are part of either the perpendicular bisector of two segment endpoints, or of the angular bisector of two segments. Curved edges consist of points equidistant from a segment endpoint and a segment's interior. There are two types of vertices, namely of *type* 2 having degree two, and of *type* 3 having degree three (provided that the point set which spans G is in *general position*). Both are equidistant from a triple of objects (segment or segment endpoint), but for type-2 vertices the triple contains a segment along with one of its endpoints.

Together with G's segments, the edges of $V(G)$ partition the plane into regions. These can be refined by introducing certain normals through segment endpoints (shown dotted in Figure 5.1), in order to delineate *faces* each of which is closest to a particular segment or segment endpoint. Two such normals start at each segment endpoint where G forms a reflex angle, and also at each *terminal* of G which is an endpoint belonging to only one segment in G. A normal either ends at a type-2 vertex of $V(G)$ or extends to infinity.

Observe that each face f is *visible* from its defining component c in G; that is, for each point $x \in f$ there is a point $y \in c$ such that their straight-line connection $\overline{xy}$ stays entirely within f. (If c is a segment endpoint then $y = c$, and if c is a segment then take for y the point on c closest to x.) As a consequence, the union of the three faces for a segment of G plus its two endpoints is a *simply connected* set, i.e., it is connected and contains no holes.

It is well known that the number of faces, edges, and vertices of $V(G)$ is linear in n, the number of segment endpoints for G. The number of vertices is shown to be at most $4n - 3$ in Lee and Drysdale [496]. An exact bound, which also counts the 'infinite' vertices at unbounded edges and segment normals, is given below.

Lemma 5.1. *Let G be a planar straight-line graph on n points in the plane, and let G realize t terminals and r reflex angles. The number of (finite and infinite) vertices of $V(G)$ is exactly $2n + t + r - 2$.*

Proof. Suppose first that G consists of e disjoint segments (that do not touch at their endpoints). Then there are e regions, and each type-3 vertex belongs to three of them. By *Euler's formula* for planar graphs, there are exactly $2e - 2$ such vertices, if we also count those at infinity. To count the number of type-2 vertices, observe that each segment endpoint is a terminal and gives rise to two segment normals each of which, in turn, yields one (finite or infinite) vertex of type 2. Hence there are $4e$ such vertices, and $6e - 2$ vertices in total.

Now let G be a general planar straight-line graph with e segments. We simulate G by disjoint segments, by shortening each segment slightly such that the segment endpoints are in *general position.* Then we subtract from $6e - 2$ the number of vertices which have been generated by this simulation.

Consider an endpoint p that is incident to $d \geq 2$ segments of G. Obviously, p gives rise to d copies in the simulation.

The Voronoi diagram of these copies has $d - 2$ finite vertices, which are new vertices of type 3. As the sum of the degrees $d \geq 2$ in G is $2e - t$, we get $2e - t - 2(n - t)$ new vertices in this way.

Each convex angle at p gives rise to two new normals emanating at the respective copies of p, and thus to two (finite) type-2 vertices. A possible reflex angle at p gives rise to one (finite or infinite) type-3 vertex, on the perpendicular bisector of the corresponding copies of p. There are r reflex angles in G, and thus $2e - t - r$ convex angles. This gives $r + 2(2e - t - r)$ new vertices in addition.

The lemma follows by simple arithmetic. □

Surprisingly, the number of *edges* of G does not influence the bound in Lemma 5.1. The maximum number of vertices, $3n - 2$, is achieved, for example, if G is a set of disjoint segments ($t = n$ and $r = 0$), or if G is a *simple polygon* P ($t = 0$ and $r = n$). In the latter case, the majority of applications concerns the part of $V(P)$ interior to P. This part is commonly called the *medial axis* of P. The medial axis of an n-gon with r reflex interior angles has a *tree-like* structure and realizes exactly $n + r - 2$ vertices and $2(n + r) - 3$ edges, a bound first mentioned in Lee [494].

Several algorithms for computing $V(G)$, for general or restricted planar straight-line graphs G, have been proposed and tested for practical efficiency. $V(G)$ can be computed in $O(n \log n)$ time and $O(n)$ space by *divide & conquer algorithms* (Kirkpatrick [454], Lee [494], and Yap [712]), by using the *plane sweep technique* (Fortune [344]), and by *randomized incremental insertion* (Boissonnat *et al.* [142] and Klein *et al.* [466]); cf. Chapter 3 on basic algorithms.

Burnikel *et al.* [173] give an early overview of existing methods for computing $V(G)$. They also discuss implementation details of an algorithm in Sugihara *et al.* [669] that first inserts all segment endpoints, and then all the segments, of G in random order. An enhanced fast implementation of this algorithm is given in Held [399]. A method of comparable simplicity and practical efficiency (though with a worst-case running time of $\Theta(n^2)$) is proposed in Gold *et al.* [366, 367]. They first construct a Voronoi diagram for point sites by selecting one endpoint for each segment, and then maintain the diagram while expanding the endpoints, one by one, to their corresponding segments. During an expansion, the resulting topological updates in the diagram can be carried out efficiently.

In fact, *Voronoi diagrams for moving sites*, which are sometimes also called *kinetic Voronoi diagrams* in the literature, are well-studied concepts; they will be discussed in Section 9.1.

Recently, a practical and efficient divide & conquer algorithm, which works for general curved sites, and constructs the diagram $V(G)$ as a special straight-line case, has been devised in Aichholzer *et al.* [25]. We will detail this method in Section 5.5.

If G is a connected graph then $V(G)$ can be computed in *randomized* time $O(n \log^* n)$; see Devillers [261]. (Here $\log^* n$ denotes the *iterated logarithm*, that is, the number of times the (dual) logarithm has to be applied to n in order to obtain a number smaller than 1.) Moreover, $O(n)$ time randomized, and deterministic, algorithms for the medial axis of a simple polygon have been designed by Klein and Lingas [465] and Chin *et al.* [218], settling open questions of long standing. The case of a convex polygon is considerably easier; see Section 5.2.

An efficient $O(n \log^2 n)$ work *parallel algorithm* for computing $V(G)$ is given in Goodrich *et al.* [373]. This is improved to $O(\log n)$ parallel (randomized) time using $O(n)$ processors in Rajesekaran and Ramaswami [599]. The latter result also implies an optimal parallel construction method for the Voronoi diagram for point sites.

The line segment Voronoi diagram $V(G)$ is a *planar graph* embedded in the plane, and as such has a well-defined combinatorial *dual graph*. However, this dual reflects the adjacencies between the regions of $V(G)$ in a way not always desirable in applications. More specifically, the dual is a 'triangulation' of the segments in G that may contain multiple edges, due to multiple adjacencies between Voronoi regions. This shortcoming is remedied in Chew and Kedem [216], who define a so-called *segment Delaunay triangulation* $\mathrm{DT}(G)$ for G, by placing certain additional points on G's segments. Based on this idea, Brévilliers *et al.* [165] define general segment triangulations, and show that *local optimality* that characterizes the classical Delaunay triangulation (Section 4.2) extends to these structures. Moreover, they prove in [166] that $\mathrm{DT}(G)$, and thus, also the primal structure, $V(G)$, can be constructed by improving flips in segment triangulations.

Along with the study of closest-site Voronoi diagrams go their *farthest-site* counterparts. In that model, each site s gets allotted the region of all points for which s is the farthest site (rather than the closest) in the input set. See Section 6.5 for the case of point sites, where the diagram has a simple *tree structure* — just the sites on the *convex hull* of the input point set contribute nonempty regions.

For *line segments* as sites, their farthest-site Voronoi diagram is investigated in Aurenhammer *et al.* [95]. The properties of this diagram deviate from the obvious. For example, regions may be disconnected, and region emptiness cannot be characterized by convex hull properties; see Figure 5.2. Moreover, and unlike the closest line segment case, the number of edges and vertices of the diagram remains $O(n)$ in the worst case when the segments in the input set G do *not* form a planar graph (i.e., define crossings).

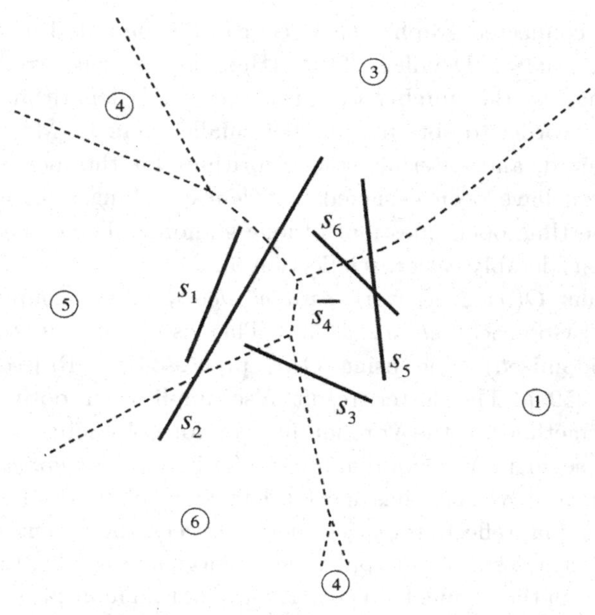

Figure 5.2. Farthest-segment Voronoi diagram for six sites. Encirculated numbers indicate affiliation of regions to segments (from [95]).

The diagram can be computed in time $O(n \log n)$, by applying a generalized (and primal) version of the ear-clipping algorithm for the farthest-site Delaunay triangulation in Subsection 6.5.1.

Basically, a convex projection surface in 3-space is constructed, by using, for each point x in the plane, its distance to the farthest segment in G as a third coordinate for x — an algorithmic idea that also works for the medial axis or the straight skeleton of a *convex* polygon; cf. Sections 5.2 and 5.3. For nonconvex simple polygons such an algorithm would however fail, due to the nonconvexity of the resulting projection surface.

The farthest-segment Voronoi diagram is still a tree; all its connected faces are unbounded. The upper bound of $8n + 4$ in [95] on their number has been improved recently to $6n - 6$ in Papadopoulou and Dey [580].

The farthest-site Voronoi diagram for *simple polygons* as sites is studied in Cheong *et al.* [208]. Its size is still $O(n)$, though its structure is more complicated, leading to an $O(n \log^3 n)$ time algorithm.

Line segment Voronoi diagrams of *order* k (cf. Section 6.5) are another natural generalization, covering the closest line segment Voronoi diagram ($k = 1$) and the farthest-segment Voronoi diagram ($k = |G| - 1$). Surprisingly, they did not receive much attention for general k until recently, when Papadopoulou and Zavershynskyi [584] showed that their combinatorial complexity remains in $O(k(n-k))$, as in the point site case.

In dimensions more than two, the known results on Voronoi diagrams for generalized sites are sparse. Exceptions are power diagrams (Section 6.2), Voronoi diagrams for spheres, and the medial axis in three dimensions (both Section 6.6).

5.2. Convex polygons

Voronoi diagrams for a single *convex polygon* have a particularly simple structure. Tailor-made algorithms for their construction have been designed.

Let C be a convex n-gon in the plane. The part of the Voronoi diagram of C's sides which lies inside C is called the *medial axis*, $M(C)$, of C. It is a *tree* whose edges, by convexity of C, are pieces of *angular bisectors* of the sides. In fact, $M(C)$ coincides with the *straight skeleton* SK(C) of C (discussed in Section 5.3) in this case. See Figure 5.3 for an illustration. $M(C)$ realizes exactly n faces, $n-2$ vertices, and $2n-3$ edges.

There is a simple *randomized incremental algorithm* by Chew [210] that computes $M(C)$ in $O(n)$ time. The algorithm first removes, in random order, the halfplanes whose intersection is C. Removing a halfplane means removing the corresponding side e of C, and extending the two sides e' and e'' adjacent to e so that they become adjacent in the new convex polygon. This can be done in constant time per side. (If the extensions of e' and e'' do not intersect then the obtained polygon is unbounded but still convex; we omit the easy modifications necessary for that case.)

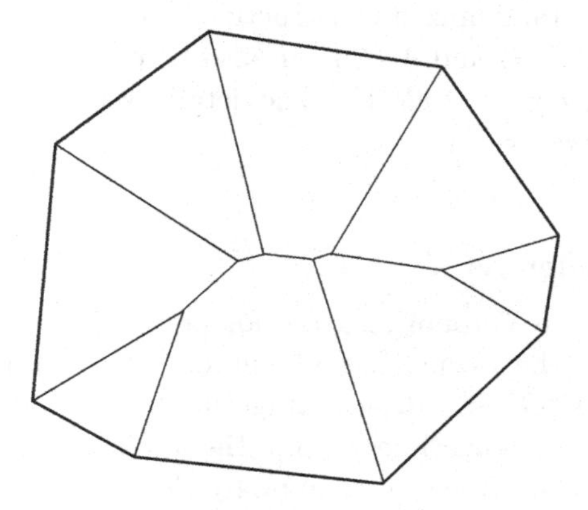

Figure 5.3. Medial axis of a convex polygon.

The *adjacency history* of C is stored. That is, for each removed side e, one of its formerly adjacent sides is recorded. In a second stage, the sides are put back in reverse (still randomized) order, and the medial axis is maintained during these insertions.

Let us focus on the insertion of the ith side, e_i. We have to integrate the face, $f(e_i)$, of e_i into the medial axis of the $i-1$ sides that have been inserted before e_i. From the adjacency history, we already know a side e' of the current polygon that will be adjacent to e_i after its insertion. Hence we know that the angular bisector of e_i and e' will contribute an edge to $f(e_i)$.

Having a starting edge available in $O(1)$ time, the face $f(e_i)$ now can be constructed in time proportional to its size. We construct $f(e_i)$ edge by edge, by simply tracing and deleting parts of the old medial axis interior to $f(e_i)$. As the medial axis of an i-gon has $2i-3$ edges, and each edge belongs to two faces, the expected number of edges of a randomly chosen face is less than 4. Thus $f(e_i)$ can be constructed in constant expected time, which gives an $O(n)$ randomized algorithm for computing $M(C)$.

The same technique also applies to the *Voronoi diagram* for the vertices *of a convex* n*-gon* C, that is, to the Voronoi diagram of a set S of n point sites in *convex position*. By Lemma 2.2, all regions in $V(S)$ are unbounded, such that the edges of $V(S)$ have to form a *tree*. Hence $V(S)$ has the same number of edges and (finite) vertices as the medial axis of C.

For each $p \in S$, its region $\mathrm{VR}(p, S)$ shares an unbounded edge with the regions $\mathrm{VR}(p', S)$ and $\mathrm{VR}(p'', S)$, respectively, where p' and p'' are the neighbors of p on the boundary of the convex hull of S (which is the polygon C). An adjacency history can be computed in $O(n)$ time, by removing the sites in random order and maintaining their convex hull. For each site that is re-inserted, the expected number of edges is less than 4, as before. So an $O(n)$ randomized construction algorithm is obtained.

The diagrams $V(S)$ and $M(C)$ can also be computed in *deterministic* linear time; see Aggarwal *et al.* [17]. The details of this algorithm are much more involved, however.

5.3. Straight skeletons

In comparison to the Voronoi diagram for point sites, which is composed of straight edges, the occurrence of curved edges in the line segment Voronoi diagram $V(G)$ is a disadvantage in the computer representation and construction, and sometimes also in the application, of $V(G)$.

There have been several attempts to *linearize* and simplify $V(G)$, mainly for the sake of efficient point location and motion planning; see

Canny and Donald [181], Kao and Mount [439], de Berg *et al.* [242], and McAllister *et al.* [527]. The *compact Voronoi diagram* in [527] is particularly suited to these applications. It is defined for the case where G is a set of k disjoint convex polygons. Its size is only $O(k)$, rather than $O(n)$, and it can be computed in time $O(k \log n)$; see Subsection 8.1.1 for more details.

As a different alternative to $V(G)$, we now describe the *straight skeleton*, $\mathrm{SK}(G)$, of a planar straight-line graph G. This structure is introduced, and discussed in more detail, in Aichholzer and Aurenhammer [29]. $\mathrm{SK}(G)$ is composed of *angular bisectors* and thus does not contain curved edges. In general, its size is even less than that of $V(G)$.

Its use as a type of *skeleton* for G partially stems from the fact that G can be reconstructed from $\mathrm{SK}(G)$ in a simple manner. This fact is considered important, for example, in picture processing; see Pfaltz and Rosenfeld [590], who observe this property for the medial axis of a simple polygon. The recently introduced *triangulation axis* of a polygon P also shares this feature. This skeletal axis is piecewise linear, too, and is defined via some optimal triangulation of P. Its size is (sometimes drastically) smaller than the size of the medial axis or the straight skeleton of P; see Aigner *et al.* [44].

The relevance of the straight skeleton $\mathrm{SK}(G)$ in shape and object recognition is discussed, e.g., in Tanase and Veltkamp [677] and Demuth *et al* . [256]. Applications in robotics and to origami design have been described in Barequet *et al.* [113], and in Lang [483] and Demaine *et al.* [255], respectively. A lifted 3D model of $\mathrm{SK}(G)$ turns out to be of use in architecture and geographic information systems, as will be sketched later. (Interestingly, and seemingly having faded into oblivion, utilizing straight skeletons as a means for designing roof constructions dates back more than a hundred years, as is documented in a monograph for engineers by Peschka [589], and later in a textbook on descriptive geometry by Müller [553].) Straight skeletons also appear implicitly in a certain Voronoi diagram for time distances, the so-called *city Voronoi diagram* for point sites under a rectilinear transportation network; see Aichholzer *et al.* [38]. We will give details on the city Voronoi diagram in Section 7.6.

$\mathrm{SK}(G)$ is defined as the *interference pattern* of certain *wavefronts* propagated from the segments and segment endpoints of G. Let F be a connected component (called a *figure*) of G. Imagine F as being surrounded by a belt of (infinitesimally small) width ε. For example, a single segment s gives rise to a rectangle of length $|s| + 2\varepsilon$ and width 2ε, and a simple polygon P gives rise to two offset copies of P with minimum distance 2ε. In general, if F partitions the plane into c connected domains then F gives rise to c simple polygons, called *wavefronts* for F.

The wavefronts arising from all the figures of G are now propagated simultaneously, at the same speed, and in a self-parallel manner. Wavefront vertices move on angular bisectors of wavefront edges which, in turn, may increase or decrease in length during the propagation. This situation continues as long as the wavefronts do not change combinatorially. Basically, there are two types of changes, called events.

(1) *Edge event*: A wavefront edge collapses to length zero. (The wavefront may vanish altogether, due to three simultaneous edge events.)
(2) *Split event*: A wavefront edge splits due to interference or self-interference. In the former case, two wavefronts merge into one, whereas in the latter case a wavefront splits into two.

After either type of event, we are left with a new set of wavefronts which are propagated recursively.

The *edges* of SK(G) now are the pieces of angular bisectors traced out by wavefront vertices. Each *vertex* of SK(G) corresponds to some point where an edge event or a split event takes place. SK(G) is a unique structure defining a polygonal partition of the plane; see Figure 5.4.

During the propagation, each wavefront edge e sweeps across a certain area which we call the *face* of e. Each segment of G gives rise to two wavefront edges and thus to two faces, one on each side of the segment.

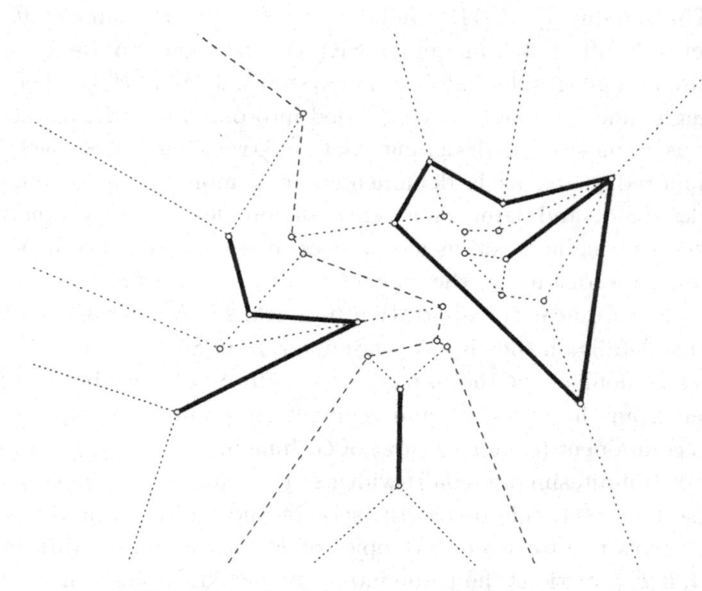

Figure 5.4. Straight skeleton for three figures.

Each terminal of G (endpoint of degree one) gives rise to one face. Faces can be shown to be *monotone polygons*, i.e., each intersection of a face with lines having some fixed direction is connected. In our case, this direction is normal to the generating wavefront edge. As a consequence, faces are *simply connected* sets. This implies a total of $2m + t = O(n)$ faces, if G realizes m edges and t terminals. There is also an exact bound on the number of vertices of SK(G).

Lemma 5.2. *Let G be a planar straight-line graph on n points, and t of which are terminals. The number of (finite and infinite) vertices of* SK(G) *is exactly $2n + t - 2$.*

From Lemma 5.1 it is apparent that SK(G) has r vertices less than $V(G)$, if G defines r reflex angles. In particular, if G is a *simple polygon* with r reflex interior angles, then the part of SK(G) interior to G is a *tree* with only $n - 2$ vertices, whereas the medial axis of G has $n + r - 2$ vertices.

By construction, SK(G) encodes all self-parallel inward and outward offsets for G, also called *mitered offsets* in the polygon case; see e.g. Devadoss and O'Rourke [260]. However, this offsetting process is different from the classical offsetting process, which is based on a wavefront model (or *growth model*) for the Voronoi diagram or the medial axis $V(G)$ of G; cf. the end of Section 5.5, and the 'expanding waves' view in Chapter 2. The propagation speed of all points on the wavefront is the same in that model, whereas, in the model for SK(G), the speed of each wavefront vertex, u, is controlled by the interior angle, $\beta(u)$, between its incident wavefront edges. The sharper the angle, the faster is the movement of the vertex. More precisely, the speed $s(u)$ of vertex u is given by the formula

$$s(u) = 1/\sin\frac{\beta(u)}{2}$$

and tends to infinity if the angle $\beta(u)$ approaches 0. This behavior makes SK(G) completely different from the Voronoi diagram of G. It can be shown that, without prior knowledge of its structure, SK(G) *cannot* be defined by means of distances from G, by lack of its *non-procedural definition*; see [28]. Also, SK(G) does not fit into the framework of *abstract Voronoi diagrams* described in Section 7.5: The bisector of two segments of G would be the interference pattern of the rectangular wavefronts they send out, but these curves do not fulfill condition (ii) in Definition 7.7. As a consequence, the well-developed machinery for constructing Voronoi diagrams (see Chapter 3) does not apply to SK(G).

An algorithm that simulates the wavefront propagation is given in [29]. It maintains a triangulation, T, of the wavefront vertices in 'free space' —

the part of the plane not yet swept over by any wavefront. Collision of offsetting figures can be predicted by the collapse of triangles in T. Though the known upper bounds are much higher, T typically undergoes only $O(n)$ updates (of several kinds, including collapses). A runtime of $O(n \log n)$ is then achieved, so the method is simple and practically efficient. We conjecture that the techniques for kinetic *collision detection* in Kirkpatrick *et al.* [459], which are based on a so-called *pseudo-triangulation* (Section 6.3) of the free space, are capable of theoretically speeding up this approach.

More sophisticated though more involved construction methods running in subquadratic time are given in Eppstein and Erickson [326], Cheng and Vigneron [205], and recently in Vigneron and Yan [691] who achieve a running time of $O(n^{4/3+\varepsilon})$. These works also describe additional structural properties of the straight skeleton. The so-called *motorcycle graph* is introduced and utilized, which captures an important aspect of straight skeletons. This planar graph resolves the interplay of the reflex vertices' trajectories, and thus is of substantial help in predicting the (by nature non-local) split events. The paper by Vyatkina [695] shows that SK(G) can be decomposed into (pruned) medial axes of certain convex polygons. Recently, practical implementations of straight skeleton algorithms have been developed in Huber and Held [410] and in Palfrader *et al.* [577].

Whether SK(G) can be computed in $O(n \log n)$ time, and possibly even faster inside a simple polygon, is still an open problem. In fact, only the trivial lower bound, $\Omega(n)$, is known in the polygon case.

In view of the notoriously difficult task of computing the straight skeleton SK(G) efficiently, simpler instances for G have been studied as well. Clearly, if G is a *convex polygon*, then SK(G) is just the *medial axis* of G, which can be constructed in $O(n)$ time (Section 5.2).

More interesting is the case of a *monotone polygon*, which is solvable in $O(n \log n)$ time; see Das *et al.* [239]. Basically, the polygon's boundary is decomposed into two monotone chains of edges. The straight skeleton of each chain can be constructed without the trouble of split events, and the two partial skeletons merged in a simple way, similar as in the Voronoi diagram case; see Section 3.3.

SK(G) has a three-dimensional interpretation, obtained by defining the height of a point x in the plane as the unique time when x is reached by a wavefront. In this way, SK(G) lifts up to a *polygonal surface* Σ_G, where points on G have height zero. In a geographical application (e.g., *terrain modeling*), G may delineate rivers, lakes, and coasts, and Σ_G represents a corresponding terrain with fixed slope; see Figures 5.5 and 5.6 (taken from [29]). Σ_G has the nice property that every raindrop that hits a terrain facet f drains off to the segment or terminal of G defining f; see [28]. This

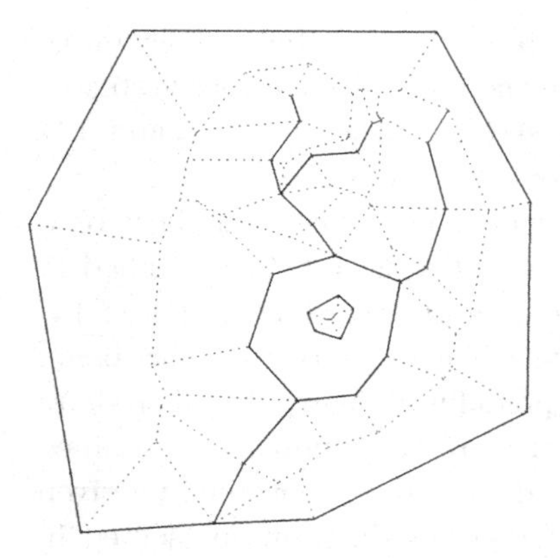

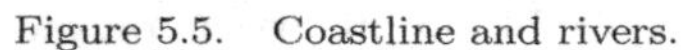

Figure 5.5. Coastline and rivers.

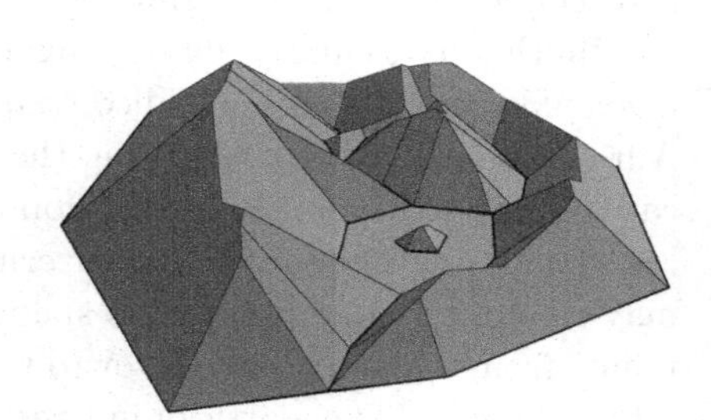

Figure 5.6. Reconstructed terrain.

may have applications in the study of rain water fall and its impact on the flooding caused by rivers in a given geographic area, where currently the Voronoi diagram of G is used; see, for example, Barett [116].

When restricted to the interior of a simple polygon P, the surface Σ_P can be used as a canonical construction of a *roof* of given slope above P. For *rectilinear* (and axis-aligned) *polygons* P, the medial axis of P in the L_∞-*metric* will do the job. SK(P) coincides with this medial axis for such polygons, and thus generalizes this roof construction technique to general shapes of P. Applications arise, e.g., in the automated generation of building models; see Brenner [164], Kelly and Wonka [451], and references therein.

Sometimes, the volatile behavior of the straight skeleton SK(P), which stems from the high propagation speed of sharp reflex vertices of the polygon P, is considered a disadvantage. A possible way out is to 'cut off' such vertices in a controlled way, an idea sketched in [29] and applied systematically in Tanase and Veltkamp [676]. The resulting (appropriately pruned) straight skeleton is called the *linear axis* of the polygon P in the latter paper, and is shown to approximate the medial axis of P in a certain sense. See also Bookstein [153] for an early study of the straight skeleton for polygons with large interior angles, who termed it the *line skeleton.*

The concept of straight skeleton SK(G) of a planar straight line graph G can be modified by tuning the propagation speed of the individual wavefront edges. That is, edges can be given a weight that expresses their velocity. For the 3D model of SK(G) described above, this means putting prescribed facet slopes for the corresponding surface. The maximal size of the skeleton, and its procedural definition, remain unaffected. (The exceptional case where two parallel edges of different weights become adjacent after an edge event can be treated consistently, by considering the created wavefront vertex as a convex vertex.) The *monotonicity* of the skeleton faces, however, and if G is disconnected even their simple connectedness, may be lost.

In the particularly interesting case where G is the boundary of a *convex polygon* P, this so-called *weighted skeleton*, $\mathrm{SK}_W(P)$, is studied in Aurenhammer [94]. By adjusting the edge weights accordingly, $\mathrm{SK}_W(P)$ is capable of partitioning a given n-gon P into n convex parts, each part being based on a single edge of P and covering a predefined 'share' of P. The share may relate, for example, to the spanned area, to the number of contained points from a given point set, or to the total edge length covered in a given set of curves. (The partitioning results for power diagrams presented in Section 6.4 are related but not equivalent.) Possible applications of such *fixed-share decompositions* include priority-based or fair facility allocation, which may concern real estate or access to power lines, aquifers, or oil wells.

Area decomposition may be done recursively, yielding a new instance of a structure called *Voronoi tree map*, of use for visualizing attributed hierarchical data; see Andrews *et al.* [62], Granitzer *et al.* [377], and Balzer and Deussen [111].

Straight skeletons can be generalized to higher dimensions. For example, in 3D the input could be a nonconvex and boundary-connected polyhedron P with n vertices. (For convex polyhedra, the straight skeleton coincides with the medial axis.) The shrinking process now involves offsetting inwards the boundary facets of P in a self-parallel way and at unit speed. Thereby, facets change their shape and may disconnect (as do polygon edges in the 2D case), and the shrinking process continues with each part. The volumes traced out by the facets are polyhedra again. Hence, the structure retains its piecewise-linear shape and offers an alternative to the medial axis of P whose geometric description is rather complex; see Section 6.6.

However, and unlike the planar case, there is no unique (mitered) offsetting process for P, in general. This was observed in Demaine *et al.* [254] and Barequet *et al.* [115]. In fact, as there possibly exist several different 'skeleton-like' partitions of a nonconvex polygon by means of angular bisectors [28], even such having super-linear size, there may be different ways of locally offsetting a vertex v of P of high degree. In certain cases, the planar weighted straight skeleton can be used to compute a canonical offset structure [115]. In the general case (when the edges incident to v positively span 3-space), a generalization of the straight skeleton to the sphere will work; see Aurenhammer and Walzl [105].

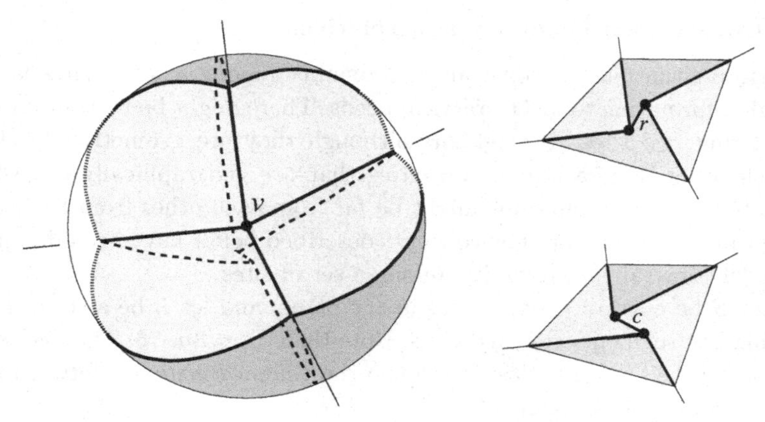

Figure 5.7. Two local offset surfaces for a saddle point (right). The spherical skeleton (left, dashed style) produces the lower solution. The short skeleton arc corresponds to a convex edge c in the offset polytope. It connects the two vertices which the saddle point is split into. In the upper solution, the offset edge r is reflex.

Figure 5.7 shows two possible offsets for a saddle point v of degree 4. Either choice leads to a different though valid straight skeleton for the underlying polyhedron P.

Concerning size, $O(n^2\alpha(n)^2)$ is a lower bound on the combinatorial complexity of a 3D skeleton [115]. (Here, $\alpha(n)$ denotes the inverse of the rapidly growing *Ackermann function*; see e.g. [642]. We have $\alpha(n) \leq 5$ even for 'astronomically large' arguments n, such that $\alpha(n)$ can be considered a constant for all practical purposes.) A trivial upper bound is $O(n^4)$, because each 4-tuple of facet planes for P can yield only one common point of intersection during the entire offsetting process.

If P is an *orthogonal polyhedron* (all edges of P are parallel to the coordinate axes) then the skeleton size reduces to $O(n^2)$, and an $O(n^2 \log n)$ construction algorithm exists [115]. A recent implementation using the plane sweep technique is given in Martinez *et al.* [516]. The straight skeleton is the L_∞-*medial axis* of P in this case, which also enables its efficient computation in *voxel representation.* Yet, by its non-interpretability in terms of distances in general [28], one cannot resort to powerful distance transforms (like the EDT for the medial axis; see Section 11.1), which makes the accurate computation of *pixel straight skeletons* an interesting task already in 2D; see Demuth *et al.* [256].

An efficient algorithm for the straight skeleton of general nonconvex 3D polyhedra (in either representation) is still missing, as are non-trivial upper bounds on its combinatorial complexity.

5.4. Constrained Delaunay and relatives

In certain situations, unconstrained proximity among a set of sites is not enough information to meet practical needs. There might be reasons for not considering two sites as neighbors although they are geometrically close to each other. For example, two cities that are geographically close but separated by high mountains might be far from each other from the point of view of a truck driver. The concepts described below have been designed to model constrained proximity among a set of sites.

Let S be a set of n point sites in the plane, and let L be a set of non-crossing line segments spanned by S. Note that $|L| \leq 3n-6$. The segments in L are viewed as obstacles: We define the *bounded distance* between two points x and y in the plane as

$$b(x,y) = \begin{cases} d(x,y) & \text{if } \overline{xy} \cap L = \emptyset, \\ \infty & \text{otherwise,} \end{cases}$$

where d stands for the Euclidean distance. In the resulting *bounded Voronoi diagram* $V(S,L)$, regions of sites that are close but not visible from each other are clipped by segments in L. Regions of sites being segment endpoints are *nonconvex* near the corresponding segment; see Figure 5.8 (left).

The dual of $V(S,L)$ is not a full triangulation of S, even if the segments in L are included. However, $V(S,L)$ can be modified to dualize into a triangulation which includes L and, under this restriction, is as much 'Delaunay' as possible.

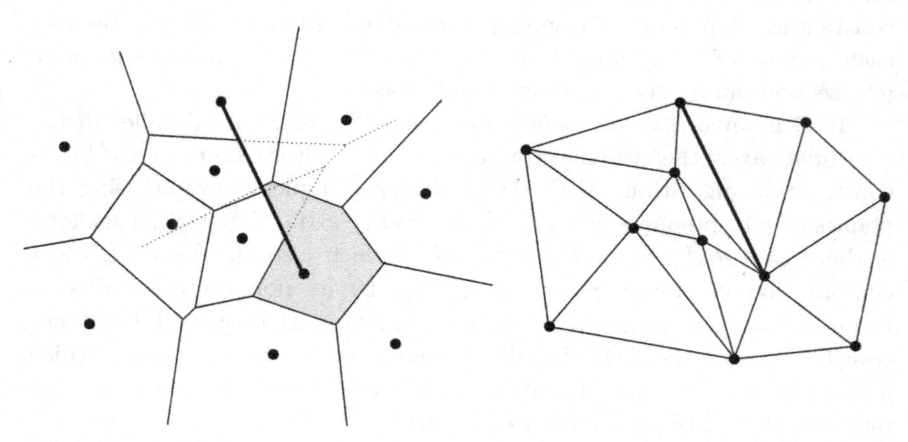

Figure 5.8. Bounded Voronoi diagram extended, and its dual, the constrained Delaunay triangulation. The single constraining segment is drawn in bold style.

The modification is simple but nice. For each segment $\ell \in L$, the regions clipped by ℓ from the right are extended to the left of ℓ, as if only the sites of these regions were present. The regions clipped by ℓ from the left are extended similarly. See Figure 5.8 again, where these modifications are shown in dotted lines. Of course, extended regions may overlap now, so they fail to define a partition of the plane. If we dualize now by connecting sites of regions that share an edge, a full triangulation that includes L is obtained: the *constrained Delaunay triangulation*, $\mathrm{CDT}(S, L)$. It is clear that the number of edges of $\mathrm{CDT}(S, L)$ is at most $3n - 6$, and that, in general, the number of edges of $V(S, L)$ is even less. Hence both structures have a linear size.

The original definition of $\mathrm{CDT}(S, L)$ in Lee and Lin [497] is based on a modification of the *empty-circle property*: $\mathrm{CDT}(S, L)$ contains L and, in addition, all edges between sites $p, q \in S$ that have $b(p, q) < \infty$ and that lie on some circle enclosing only sites $r \in S$ with at least one of $b(r, p), b(r, q) = \infty$.

Algorithms for computing $V(S, L)$ and $\mathrm{CDT}(S, L)$ have been proposed in Lee and Lin [497], Chew [211], Wang and Schubert [697], Wang [696], Seidel [629], and Kao and Mount [440]. The last two methods seem best suited to implementation. For an application of $\mathrm{CDT}(S, L)$ to quality *mesh generation* see Chew [213] and Shewchuk [647]; the latter paper carefully discusses the implementation details.

We sketch the $O(n \log n)$ time *plane sweep* approach in [629]. If only $V(S, L)$ is required then the plane sweep algorithm described in Section 3.4 can be applied without much modification. If $\mathrm{CDT}(S, L)$ is desired then the extensions of $V(S, L)$ as described above are computed in addition. To this end, an additional sweep is carried out for each segment $\ell \in L$. The sweep starts from the line through ℓ in both directions. It constructs, on the left side of this line, the (usual) Voronoi diagram of the sites whose regions in $V(S, L)$ are clipped by ℓ from the right, and vice versa.

The special case where S and L are the sets of vertices, and sides, of a *simple polygon* has received special attention, mainly because of its applications to visibility problems in polygons. The bounded Voronoi diagram $V(S, L)$ is constructible in $O(n)$ randomized time in this case; see Klein and Lingas [463]. If the L_1-*metric* instead of the Euclidean metric is used to measure distances, the same authors [465] give a deterministic linear time algorithm. Both algorithms, as well as the linear time medial axis algorithms in [464] and in [218], first decompose the polygon into smaller parts called *histograms*. These are polygons whose vertices, when considered in cyclic order, appear sorted in some direction.

An alternative concept that forces a set L of line segments spanned by S into $\mathrm{DT}(S)$ is the *conforming Delaunay triangulation*. For each segment $\ell \in L$ that does not appear in $\mathrm{DT}(S)$, new sites on ℓ are added such that ℓ becomes expressible as the union of Delaunay edges in $\mathrm{DT}(S \cup C)$, where C is the total set of added sites. For several site adding algorithms, $|C|$ depends on the size as well as on the geometry of L. See, e.g., the survey article by Bern and Eppstein [126] and references therein. Edelsbrunner and Tan [316] show that $|C| = O(k^2 n)$ is always sufficient, and construct a set of sites with this size in time $O(k^2 n + n^2)$, for $k = |L|$.

Conforming triangulations should not be confused with the concept of *compatible triangulations*; see Aichholzer *et al.* [37]. These are triangulations T_1 and T_2 of two planar point sets S_1 and S_2, respectively, such that there is a triangle-preserving bijection $S_1 \rightarrow S_2$. Whether a compatible triangulation T_2 always exists, given S_1, S_2, and T_1 (under the obvious necessary size and convex hull restrictions for the point set S_2), is still an unsettled question.

Things get remarkably different if the constraining line segments (obstacles) in L do *not* have their endpoints included in the set S of sites. In this setting, the obstacles will just block visibility, without exerting proximity influence — a scenario useful in visibility, motion planning, and guarding problems. The resulting *visibility-constrained Voronoi diagram* is a piecewise linear structure of superlinear combinatorial complexity in $n = |S|$. Its regions are disconnected in general, being the union of polygons bounded by bisectors and visibility rays. Already for $|L| = 2$, when visibility is constrained to a 'window' between two segments, and the so-called *peeper's Voronoi diagram* is obtained, the size may increase to $\Theta(n^2)$; see Aurenhammer and Stöckl [103]. This bound remains valid if general *visibility angles* (of less than π) are attached to the sites, as in Figure 5.9; see Fan *et al.* [335]. $O(n^2)$ and $O(n^2 \log n)$ time algorithms, respectively, exist for these scenarios.

For the general case of n sites among m line segment obstacles, where the combinatorial complexity may be as large as $\Theta(n^2 m^2)$, Wang and Tsin [698] gave an $O(n^4 + n^2 m^2)$ algorithm, worst-case optimal for $m = \Omega(n)$.

A different, and geometrically even more complicated, type of constrained Voronoi diagram is the *geodesic Voronoi diagram*, also called the *Voronoi diagram for obstacles*. Here, the *geodesic distance* between a site p and a point x in the plane is the length of the *shortest obstacle-avoiding path* between p and x. It leads to a diagram with (path-)connected regions and linear size; see Aronov [66]. Once the geodesic Voronoi diagram is available, several proximity questions that respect the obstacles can be

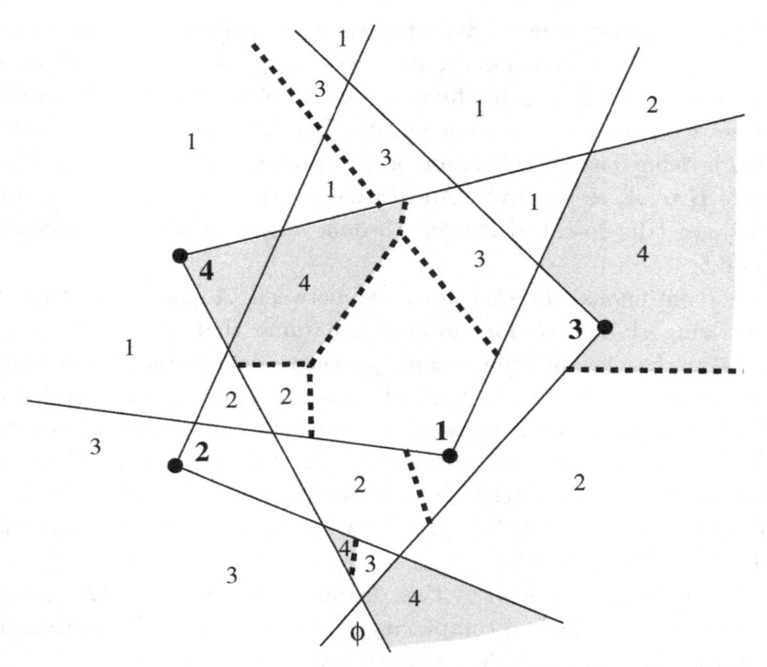

Figure 5.9. Voronoi diagram of four sites, constrained by angular visibility. The plane is dissected by visibility rays into cells $A(T)$, exclusively seen by subsets $T \subseteq S$, and partitioned by their (classical) Voronoi diagram $V(T)$. The region of a site p can be expressed as $\bigcup_{T \ni p}(A(T) \cap \mathrm{VR}(p,T))$, where $\mathrm{VR}(p,T)$ is the region of p in $V(T)$. The shaded areas above constitute the region of site 4. Cells not visible from S, like the bottommost cell $A(\emptyset)$, belong to no site.

answered efficiently, like nearest neighbor queries or the construction of a geodesic minimum spanning tree; cf. Chapter 8. The obstacles are usually modeled by a set L of non-crossing line segments. If all their endpoints are sites then the *bounded Voronoi diagram* is obtained.

Computing geodesic distances is not a constant-time operation, which complicates matters. Bisectors may consist of various linear and hyperbolic pieces. A subquadratic construction algorithm is given by Mitchell [546], and Hershberger and Suri [404] speeded up his *wavefront propagation* method (based on the *continuous Dijkstra technique*) to $O((n+m)\log(n+m))$ time and space, for n sites and m segments. The storage is reduced to optimal, $O(n+m)$, for the geodesic Voronoi diagram inside a simple m-gon, in Papadopoulou and Lee [582].

Voronoi diagrams induced by geodesic distances (shortest paths) on *surfaces in 3-space* are discussed in Section 7.1.

There are many other ways of measuring distances in the plane in the presence of line segments, or more generally, of polygonal objects or straight-line graphs. Let us briefly mention two of them at this place. Let A and B be two planar straight-line graphs. The *Hausdorff distance* between A and B is defined as the maximum of the minimum distances from A to B, and from B to A, respectively. An instance of the Voronoi diagram under this distance (the so-called *cluster Voronoi diagram*) will be discussed in Section 6.5.

The (continuous) *Fréchet distance* between A and B is based on parametrizing A and B; for simplicity, assume that A and B are two polygonal paths. As an intuitive interpretation, imagine a person walking all along A and his/her dog walking all along B. Then the Fréchet distance between the paths A and B corresponds to the minimum possible length of a leash that enables both, the person and the dog, to complete their walk. Computing Fréchet distances is quite elaborate. See Alt and Godau [51] for an efficient algorithm, which is based on the *parametric search* paradigm introduced in Megiddo [530].

Using a discrete version of this distance, Bereg *et al.* [125] recently analyzed the combinatorial complexity of the respective Voronoi diagram for a set of polygonal paths in two (and higher) dimensions.

5.5. Voronoi diagrams for curved objects

Some of the algorithms mentioned in Section 5.1 that compute the Voronoi diagram for line segments will also work for curved objects.

The *plane sweep algorithm* in Fortune [344] elegantly handles arbitrary sets of *circles* (i.e., the additively weighted Voronoi diagram, or Johnson–Mehl model; see Section 7.4) without modification from the point site case. Circular arcs, however, cannot be treated easily with that approach. Yap's *divide & conquer algorithm* [712] allows sets of disjoint segments of arbitrary *degree-two curves*. A *randomized incremental algorithm* for general *curved objects* is given by Alt *et al.* in [50]. They show that complicated curved objects can be partitioned into 'harmless' ones by introducing new points. All these algorithms achieve an optimal running time, $O(n \log n)$, but are not easy to implement.

The topic of this section is a simple and practical algorithm for computing the Voronoi diagram of a set of general (not necessarily polygonal) sites, developed in Aichholzer *et al.* [25]. In fact, such diagrams may have all kinds of artifacts. Their edge graph may be disconnected, and their bisectors may be closed curves. In particular, the abstract Voronoi diagram setting in Klein [461] and Klein *et al.* [466] (see Section 7.5) does not apply. Moreover, divide & conquer is usually involved when emphasis is on the *bottom-up phase* (i.e. the merge step; cf. Section 3.3), even if the sites are of relatively simple shape. This is due to the missing separability condition for the sites, which would prevent the merge chain from cycling and breaking into several components.

The idea is to put emphasis on the *top-down phase*, namely, the divide step. The edge graph of the Voronoi diagram can be divided into a certain *tree* that corresponds to the *medial axis* of a (generalized) planar domain, as will be described in Subsection 5.5.1. (Usually, it is the other way round: Voronoi diagram algorithms are applied to compute the medial axis; cf. Section 6.6.) Division into base cases is then possible, which, in the bottom-up phase, can be merged by trivial concatenation. The resulting construction algorithm, in Subsection 5.5.2, is *not bisector-based* and merely computes dual links between the sites — similar to Delaunay triangulation methods. This guarantees computational simplicity and numerical stability. No part of the Voronoi diagram, once constructed, has to be discarded again.

5.5.1. *Splitting the Voronoi edge graph*

In the Voronoi diagram for general objects to be considered now, sites are allowed to be two-dimensional objects (any topological disks), one-dimensional objects (topological line segments), or simply points.

Let S denote the given set of sites. The distance of a point x to a site $s \in S$ is given by $\min_{y \in s} d(x, y)$, where d denotes the Euclidean distance function. As done e.g. in [50, 712], we define the Voronoi diagram, $V(S)$, of S via its *edge graph*, G_S, which is the set of all points having more than one closest point on the union of all sites. Our aim is to relate the graph G_S to the medial axis of a generalized planar domain. In this way, we will be able to construct the Voronoi diagram $V(S)$ by means of the medial axis algorithm presented in Subsection 5.5.2.

An edge of G_S containing points equidistant from two or more different points on the same site s is called a *self-edge* for s. The *regions* of $V(S)$ are the maximal connected subsets of the complement (of the closure) of G_S in the plane. The differences to a bisector-based definition of the Voronoi diagram should be noticed. Self-edges are ignored in such a definition unless the sites are split into suitable pieces. Such pieces, however, share boundaries — a fact that, if not treated with care, may give rise to unpleasant phenomena like two-dimensional bisectors.

To get rid of the unbounded components of the diagram, we include a surrounding circle, Γ, into the set S of sites, in a way such that each vertex of $V(S\backslash\{\Gamma\})$ is also a vertex of $V(S)$.

Removal of certain points from the edge graph G_S of $V(S)$ will now break all its cycles and make it a *tree*. Finding such points is non-trivial, in view of the possible presence of self-edges. For a site $s \neq \Gamma$, let $p(s)$ be a point on s with smallest y-coordinate, and denote with $q(s)$ the closest point on G_S vertically below $p(s)$. Then the geometric graph

$$T_S = G_S\backslash\{q(s) \mid s \in S\backslash\{\Gamma\}\}$$

can be shown to be a tree.

Our next goal is to interpret the tree T_S as the medial axis of a generalized planar domain. We first consider $V(S)$ to be the medial axis of a planar shape B, by taking the surrounding circle Γ as part of the shape boundary, and considering each remaining site $s \in S$ as a (possibly degenerate) 'hole'. That is, we define

$$B = B_0\backslash\{s \in S \mid s \neq \Gamma\},$$

where B_0 denotes the disk bounded by Γ. The medial axis $M(B)$ is just the (closure of the) edge graph G_S of $V(S)$. We now want to combinatorially disconnect the shape B at appropriate positions, such that the medial axis of the resulting domain corresponds to the tree decomposition T_S of $V(S)$ above. Observe from Lemma 5.4 (in the next subsection, 5.5.2) that any *maximal inscribed disk* can be used to split the medial axis of a shape into two (or more) components which share a point at the disk's center. In order to adapt this fact to our situation, the notion of a so-called augmented domain is introduced. Its definition is recursive, as follows.

An *augmented domain* is a set A together with a projection $\pi_A : A \to \mathbf{R}^2$. Initially, A is the original shape B, and the associated projection π_B is the identity. Now, consider a maximal inscribed disk D of an augmented domain A, which touches the boundary ∂A of A at exactly two points u and v. Denote by $\widehat{uv}$ and $\widehat{vu}$ the two circular arcs which the boundary of D is split into. The new augmented shape, A', which is obtained from A by splitting it with D, is defined as

$$A' = A^0 \cup D^1 \cup D^2,$$

where $A^0 = \{(x, 0) \mid x \in A\backslash D\}$, $D^1 = \{(x, 1) \mid x \in D\}$, and $D^2 = \{(x, 2) \mid x \in D\}$. The associated projection is

$$\pi_{A'} : A' \to \mathbf{R}^2, \quad (x, i) \mapsto \pi_A(x).$$

We say that the line segment in A between points (x, i) and (y, j) is *contained* in A' if one of the following conditions is satisfied:

(1) $i = j$ and the line segment $\overline{xy}$ avoids ∂D,
(2) $\{i, j\} = \{0, 1\}$ and $\overline{xy}$ intersects the arc $\widehat{uv}$, or
(3) $\{i, j\} = \{0, 2\}$ and $\overline{xy}$ intersects the arc $\widehat{vu}$.

For any two points (x, i) and (y, j) in A', their *distance* now can be defined. It equals the distance of $\pi_A(x)$ and $\pi_A(y)$ in the plane, provided the connecting line segment is contained in A', and is ∞, otherwise. An (open) disk in A' with center (m, i) and radius ϱ is the set of all points in A' whose distance to (m, i) is less than ϱ. Such a disk is said to be *inscribed* in A' if its projection into the plane is again an open disk.

Having specified inscribed disks for A', the boundary of A' and the medial axis of A' can be defined as in the case of (usual) planar shapes. In particular, $\partial A'$ derives from ∂A by disconnecting the latter boundary at the contact points u and v of the splitting disk D, and reconnecting it with the circular arcs $\widehat{uv}$ and $\widehat{vu}$. See Figure 5.10, which shows the boundary of a domain augmented with two disks, one for each of its two holes (i.e., the sites s_1 and s_2).

Concerning the medial axis, every maximal inscribed disk in A different from D corresponds to exactly one maximal inscribed disk in A'. The medial axis of A' therefore is the same geometric graph as $M(A)$, except that the edge of $M(A)$ containing the center of D is split into two disconnected edges which both have the center of D as one of their endpoints. These two points are two leaves of $M(A')$.

To draw the connection to the edge graph G_S of $V(S)$, the initial shape B from above is augmented with $|S| - 1$ maximal inscribed disks,

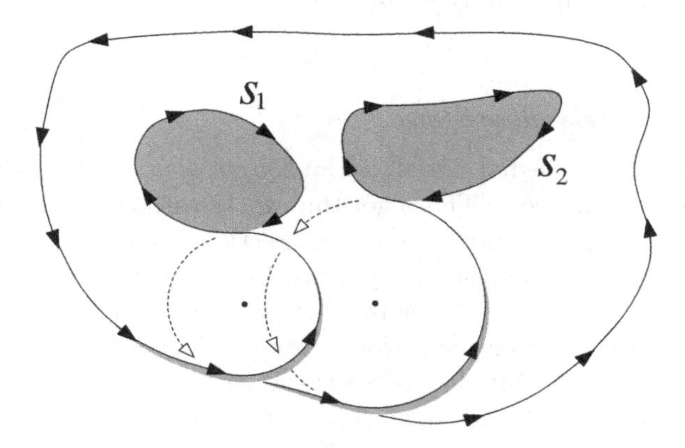

Figure 5.10. Oriented boundary of an augmented domain (from [25]).

namely, the ones centered at the points $q(s) \in G_S$, where $q(s)$ was the vertical projection onto G_S of a point with smallest y-coordinate on the site s. Denote by A_S the resulting domain after these $|S| - 1$ augmentation steps. We may conclude the main finding of this subsection as follows.

Lemma 5.3. *The tree* $T_S = G_S \backslash \{q(s) \mid s \in S \backslash \{\Gamma\}\}$ *is the medial axis of the augmented domain* A_S.

From the algorithmic point of view, augmenting B amounts to connecting its boundary ∂B to a *single* cyclic sequence, ∂A_S, that consists of pieces from ∂B and from circles bounding the splitting disks. (One-dimensional sites contribute to ∂B with two curves, one for either orientation, and the special case of point sites can be handled consistently.) Each such boundary piece is used exactly once on ∂A_S, and traversing ∂A_S corresponds to tracing the medial axis tree $M(A_S)$ in *preorder* (i.e., root first, then the subtrees for its children recursively, in radial order).

Of course, the construction of ∂A_S is trivial once the *splitting disks* are available. The non-trivial task is to find these disks D_i, one for each site $s_i \in S \backslash \{\Gamma\}$. Recall that D_i is horizontally tangent to s_i at a lowest point $p(s_i)$ of s_i. The center $q(s_i)$ of D_i lies on the edge graph G_S of $V(S)$ but, of course, D_i needs to be found without knowledge of G_S. Indeed, a simple and efficient *plane sweep* can be applied in $O(n \log n)$ time if the sites in S are described by a total of n objects, each being manageable in constant time.

Note that ∂A_S then consists of $\Theta(n)$ pieces, that either bound splitting disks, or pieces that stem from site boundaries. The former pieces are used only to link the site segments in the correct cyclic order. They do not play any geometric role, and can be ignored when the medial axis algorithm of Subsection 5.5.2 is applied to A_S in order to compute the desired Voronoi diagram $V(S)$.

5.5.2. *Medial axis algorithm*

We now describe the medial axis algorithm for *circular arc shapes* in [27, 34] that is suited to our needs. This algorithm works without modifications for the augmented domains introduced in Subsection 5.5.1.

Let A be some planar shape bounded by n circular arcs. We restrict attention to circular curves, because they are well suited to approximate more general objects accurately (for example, objects bounded by *spline curves*), and the case of line segments is trivially covered. Moreover, if the approximation by circular arcs (in particular, *biarcs*, i.e., smooth concatenations of two circular arcs) is done carefully, stability of the medial axis can be guaranteed [34]. For a recent survey on medial axes for general objects, the interested reader is referred to Attali *et al.* [83].

For the shape A, call a disk $D \subseteq A$ *maximal* if there exists no disk D' different from D such that $D' \supset D$ and $D' \subseteq A$ holds. Then the *medial axis*, $M(A)$, of A can be defined as the (infinite) set of all centers of maximal disks for A. Observe that this definition is equivalent to the distance-based definition given in Subsection 5.5.1. The corresponding infinite set of disks (its union gives A) is commonly called the *medial axis transform* of A. Sometimes also the term *symmetric axis* is used to denote the medial axis (transform) of a shape.

$M(A)$ is connected and cycle-free and thus forms a tree. It consists of $O(n)$ *edges*, which are maximal pieces of straight lines and (possibly all four types of) conics. Endpoints of edges will be called *vertices* of $M(A)$. Compared to polygonal shapes, the medial axis for circular arc shapes is not more complicated, as both structures contain edges of algebraic degree 2 in general; see Figure 5.11.

We give a simple and practical *divide & conquer randomized* algorithm for computing $M(A)$. The costly part is delegated to the divide step, which basically will consist of *inclusion tests* for arcs in circles. The merge step is trivial; it just concatenates two partial medial axes. The expected runtime is bounded by $O(n^{3/2})$, and can be proven to be $O(n \text{ polylog } n)$ for several types of shape.

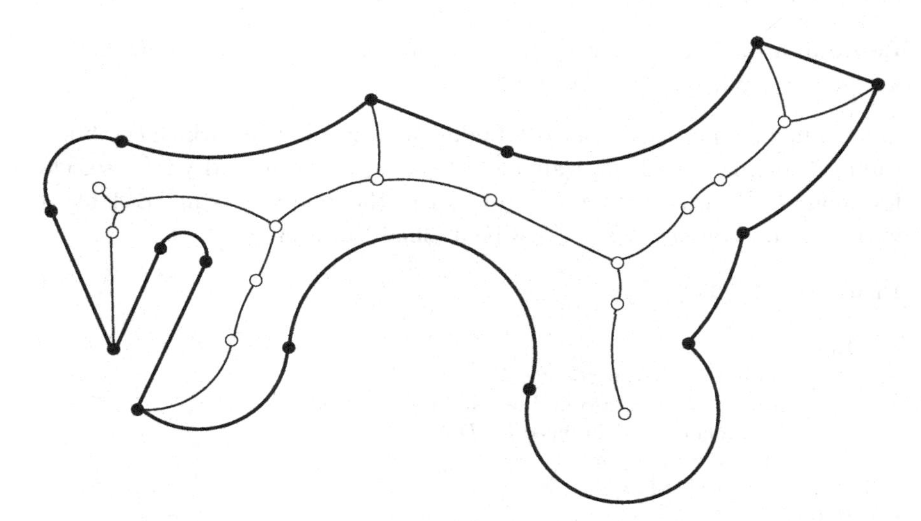

Figure 5.11. Medial axis of a shape bounded by circular arcs.

A qualitative difference to existing medial axis algorithms is that a *combinatorial* description of $M(A)$ is extracted first, which can then be directly (and robustly) converted into a geometric representation. The algorithm is based on the following simple though elegant *decomposition lemma* in Choi *et al.* [219].

Lemma 5.4. *Consider any maximal disk D for A. Let $A_1, \ldots, A_t$ be the connected components of $A \backslash D$, and denote by p the center of D.*

$$(1) \quad M(A) = \bigcup_{i=1}^{t} M(A_i \cup D),$$

$$(2) \quad \{p\} = \bigcap_{i=1}^{t} M(A_i \cup D).$$

That is, using any maximal disk one can compute the medial axes for the resulting components recursively, and then glue them together at the disk's center. By a moving argument, a *balanced* decomposition always exists.

Lemma 5.5. *There exists a maximal disk D for A such that at most $\frac{n}{2}$ arcs from ∂A are (completely) contained in each component of $A \backslash D$.*

For an edge e of $M(A)$, define Walk(e) as the path length in $M(A)$ from e to p^*, the center of a balanced disk as in Lemma 5.5. Further, define Cut(e) as the size of the smaller one between the two subtrees which constitute $M(A) \backslash \{e\}$. Any tree with small 'cuts' tends to have short 'walks', in the following respect.

Lemma 5.6. *If an edge e of $M(A)$ is chosen uniformly at random, then we have $E[\text{Walk}(e)] = \Theta(E[\text{Cut}(e)])$.*

Lemma 5.6 motivates the disk finding algorithm below, which combines random cutting with local walking. Its main subroutine, MAX(b), selects for an arc $b \subset \partial A$ its midpoint x and returns the unique maximal disk for A with x on its boundary. Let $c \geq 3$ be a (small) integer constant.

Procedure CUT(A)
  Put $A' = A$
  Repeat
    Choose a random arc b of $\partial A'$
    Compute D=MAX(b) and let A_0 be the larger
      component of A induced by D
    Assign $A' = A' \cap A_0$
  Until A_0 contains less than $n - \frac{n}{c}$ arcs
  Report D

Procedure WALK(A)
Choose a random arc b of ∂A
Compute D=MAX(b)
Let A_0 be the larger component induced by D
While A_0 contains more than $n - \frac{n}{c}$ arcs do
Let b_1 (b_2) be the first (last) complete arc of ∂A in A_0
Compute D_1=MAX(b_1) and D_2=MAX(b_2)
Assign to A_0 the smaller one of the respective larger components of A for D_1 and D_2
Memorize the corresponding disk $D \in \{D_1, D_2\}$
Report D

The disk finding algorithm now combines the CUT procedure and the WALK procedure as follows. The repeat loop of CUT and the while loop of WALK are executed by turns. Whenever CUT is closer to the goal (i.e., yields a smaller largest component than does WALK), we readjust the current disk for WALK to be that of CUT. Termination takes place in either WALK or CUT. Using Lemma 5.6, a bound of $O(\sqrt{n})$ on the expected number of total loop executions in CUT and WALK can be proven.

The costly part in both procedures is their subroutine MAX, whose expected number of calls obeys the same bound, $O(\sqrt{n})$. Computing D = MAX(b) has a trivial implementation which runs in $O(n)$ time: We initialize the disk D as the (appropriately oriented) halfplane that supports b at its midpoint x. Then, for all remaining arcs $b_i \subset \partial A$ that intersect D, we shrink D so as to touch b_i while still being tangent to b at x. The most complex operation for shrinking D is computing the intersection of two circles. In particular, and unlike previous medial axis algorithms, *no conics* take part in geometric operations.

The randomized complexity for computing the medial axis is thus given by the recurrence relation

$$T(n) = T\left(\frac{1}{c} \cdot n\right) + T\left(\left(1 - \frac{1}{c}\right) \cdot n\right) + O(n^{3/2}) = O(n^{3/2}).$$

In many cases, however, the algorithm will perform substantially better. Let Δ be the *graph diameter* of $M(A)$ (i.e., the maximal number of edges on a path in this tree). If $\Delta = \Theta(\log n)$, which is the smallest value possible, then an overall runtime of $O(n \log^2 n)$ is met. For the other extreme case, $\Delta = \Theta(n)$, our strategy is even faster, $O(n \log n)$. The latter situation is quite relevant in practice, because an input shape, even if not branching much, is typically approximated by a large number of circular arcs.

Notice that the algorithm works exclusively on the *boundary* ∂A of A, except for a final step, where the conic edges of $M(A)$ are explicitly

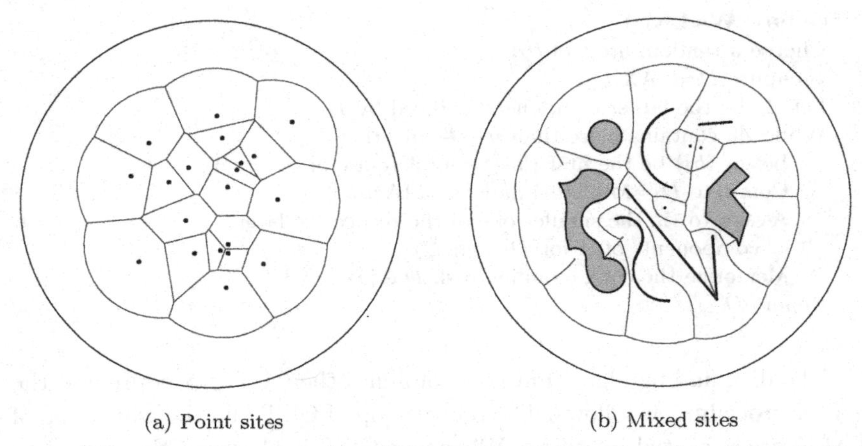

Figure 5.12. Voronoi diagrams for different kinds of sites, computed by the generalized medial axis algorithm (from [25]).

calculated and reassembled. This gives rise to increased numeric stability in comparison to other existing approaches. The necessary implementation details, including a treatment of degenerate situations, and a discussion of the base cases occurring in this divide & conquer algorithm, are given in the paper [27]. Figures 5.12(a) and 5.12(b) give two examples of the output.

We remark at this place that approximating *general objects* by circular arcs — instead of line segments — drastically reduces the input volume, namely from N line segments to $O(N^{\frac{2}{3}}) = n$ circular arcs, for fixed accuracy ε; see e.g. [34]. This is particularly desirable in applications concerning *motion planning* and *offset calculation* (a standard operation in image processing).

The δ-offset of a shape A is the *Minkowski sum* of A and a disk with radius δ. Whereas the class of polygons is not stable under taking offsets (because the δ-offset of a polygon will also contain circular arcs in its boundary), the class of circular arc shapes does have this property. This is useful, for example, in computer-aided manufactoring; see Held *et al.* [400] and Seong *et al.* [635]. In fact, the medial axis algorithm presented in this subsection delivers a link structure for ∂A that allows for offset computations without constructing $M(A)$ explicitly; see [25].

A different possibility for (polygon) offsetting, the so-called *mitered offset*, results from the straight skeleton of a polygon, discussed in Section 5.3.

Chapter 6

HIGHER DIMENSIONS

An important generalization of the Voronoi diagram for point sites is to Euclidean d-space $\mathbf{R}^d$, for $d \geq 3$. Naturally the three-dimensional space — the space we live in — holds a predominant role, especially concerning applications in the natural sciences. But also higher dimensions are strongly relevant, either in direct applications like clustering multivariate data, or indirectly, like embedding a generalized Voronoi diagram into a polyhedral cell complex of larger dimension, for analytic or algorithmic purposes.

Several nice properties are retained in higher dimensions (e.g., the convexity of the regions, and with it, the piecewise linear structure), while others are lost (e.g., the linear worst-case size of the diagram).

Voronoi diagrams in d-space are closely related to geometric objects in different dimensions. These relationships, and their structural and algorithmic implications, are also discussed in this chapter.

6.1. Voronoi and Delaunay tessellations in 3-space

6.1.1. *Structure and size*

Let us consider the classical Voronoi diagram in $\mathbf{R}^3$ first. Let S be a set of n point sites in 3-space. The *bisector* of two sites $p, q \in S$ is the perpendicular plane through the midpoint of the line segment $\overline{pq}$. The region $\mathrm{VR}(p, S)$ of a site $p \in S$ is the intersection of halfspaces bounded by bisectors, and thus is a three-dimensional convex polyhedron. The boundary of $\mathrm{VR}(p, S)$ consists of *facets* (maximal subsets within the same bisector), of *edges* (maximal line segments in the boundary of facets), and of *vertices* (endpoints of edges).

The regions, facets, edges, and vertices of $V(S)$ define a *polyhedral cell complex* in 3-space. This cell complex is *face-to-face*: If two regions have a non-empty intersection f, then f is a face (facet, edge, or vertex) of both regions.

As an appropriate data structure for storing a three-dimensional cell complex we mention the *facet-edge data structure* in Dobkin and Laszlo [284].

The number of facets of $\mathrm{VR}(p, S)$ is at most $n-1$; at most one for each site $q \in S \setminus \{p\}$. Hence, by the Eulerian polyhedron formula, the number of edges and vertices of $\mathrm{VR}(p, S)$ is $O(n)$, too. This shows that the total number of components of the diagram $V(S)$ in 3-space is $O(n^2)$. In fact, there are configurations S that force each pair of regions of $V(S)$ to share a facet, thus achieving their maximum possible number of $\binom{n}{2}$; see, e.g., Dewdney and Vranch [268].

This fact sometimes makes Voronoi diagrams in $\mathbf{R}^3$ less useful compared to those in the plane. On the other hand, Dwyer [296] showed that the *expected size* of $V(S)$ in $\mathbf{R}^d$ is only $O(n)$ for constant dimensions d, provided S is drawn uniformly at random in the unit ball. The analogous result was shown by Bienkowski *et al.* [131] for points drawn from a d-dimensional cube. These results indicate that high-dimensional Voronoi diagrams will be small in many practical situations (Figure 6.1).

In analogy to the two-dimensional case, the *Delaunay tessellation* $\mathrm{DT}(S)$ in 3-space is defined as the geometric dual of $V(S)$. It contains a tetrahedron for each vertex, a triangle for each edge, and an edge for each facet, of $V(S)$. Equivalently, $\mathrm{DT}(S)$ may be defined using the *empty sphere property*, by declaring a tetrahedron spanned by S as Delaunay iff its circumsphere is empty of sites in S. The centers of these empty spheres are just the vertices of $V(S)$.

$\mathrm{DT}(S)$ is a partition of the convex hull of S into tetrahedra, provided S is in *general position* (no four points coplanar, no five points co-spherical), which will be assumed in the sequel. Note that, for special configurations of point sites in 3-space, the edges of $\mathrm{DT}(S)$ may already form the *complete graph* on S, i.e., the graph where all $\binom{n}{2}$ vertex pairs are matched with an edge. The cell complex $\mathrm{DT}(S)$ is also called the *Delaunay tetrahedrization* or the *Delaunay simplicial complex* for S.

From the many optimization properties of planar Delaunay triangulations (described in detail in Section 4.2), only a few find their generalizations in three and higher dimensions. As shown in Rajan [598], one of them is *coarseness*, that is, $\mathrm{DT}(S)$ minimizes the largest circumsphere of its tetrahedra, among all possible tetrahedrizations of the point set S. (Simultaneously, also the largest *minimum enclosing sphere* is minimized.) Both concepts of coarseness are equivalent for so-called *self-centered* tetrahedrizations (or more generally, simplicial complexes in d-space), where each tetrahedron (d-simplex) is required to contain its circumcenter.

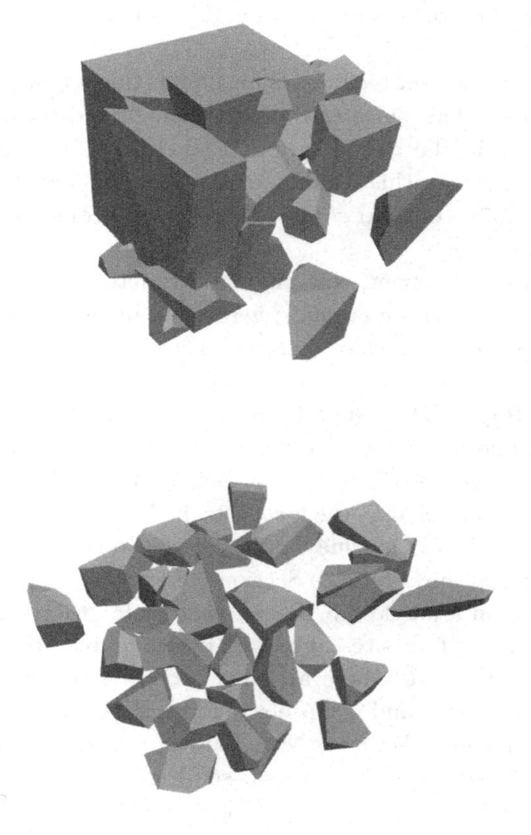

Figure 6.1. A cube is split into various Voronoi regions.

As an interesting fact, each self-centered simplicial complex is necessarily a Delaunay tessellation [598, 618].

Regrettably, the important and useful *equiangularity* property is lost in dimensions $d \geq 3$; see Schmitt and Spehner [618]. $\mathrm{DT}(S)$ does not maximize the smallest occurring angle in its tetrahedra, for none of the direct three-dimensional generalizations of an angle. Still, a certain dual optimization property for angles is preserved [618].

6.1.2. *Insertion algorithm*

Various methods for constructing the Voronoi diagram $V(S)$ in 3-space have been proposed; they are surveyed to some extent in [93]. Among them, *incremental insertion* of sites (cf. Section 3.2) is most intuitive and easy to implement.

Basically, two different techniques for integrating a new site p into $V(S)$ have been applied. The more obvious method first determines all facets of the region of p in the new diagram, $V(S \cup \{p\})$, and then deletes the parts of $V(S)$ interior to this region; see e.g. Watson [699], Field [340], and Tanemura *et al.* [678]. Inagaki *et al.* [422] describe a robust implementation of this method.

In the dual environment, this amounts to detecting and removing all tetrahedra of $\mathrm{DT}(S)$ whose circumspheres contain p, and then filling the 'hole' with *empty circumsphere* tetrahedra with p as an apex, to obtain $\mathrm{DT}(S \cup \{p\})$.

Joe [430], Rajan [598], and Edelsbrunner and Shah [313] follow a different and numerically more stable approach. Like in the planar case, after having added a site to the current Delaunay tessellation, certain *flips* (described below) that change the local tetrahedral structure are performed in order to restore local Delaunayhood. The existence of such a sequence of flips is less trivial, however; see Section 6.3.

Let us focus on a single flip. Recall that in $\mathbf{R}^2$ there are exactly two ways of triangulating four sites in convex position, and that a flip changes one into the other. In $\mathbf{R}^3$ there are also two ways of tetrahedrizing five sites in convex position, and a flip per definition exchanges them. (Such a flip is also called a *bistellar flip.*) Note, however, that the flip will replace two tetrahedra by three or vice versa; see Figure 6.2. This indicates an important difference between triangulations and tetrahedral tessellations: The number of tetrahedra *does* depend on the way of tetrahedrizing S. It may vary from $\Theta(n)$ to $\Theta(n^2)$; see Section 6.3 for more details.

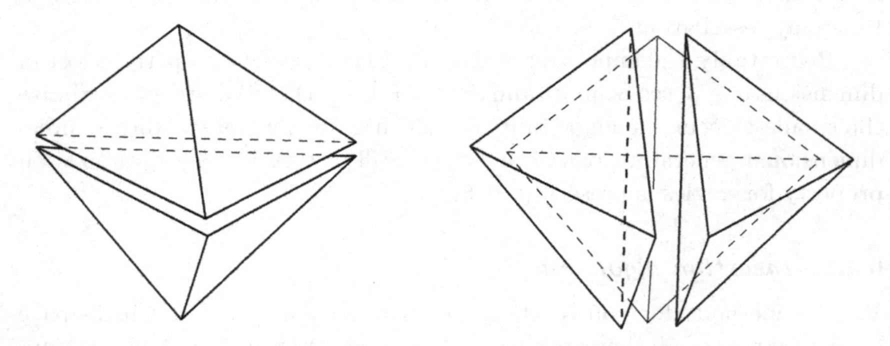

Figure 6.2. The 2-to-3 tetrahedra flip in 3-space.

Each triangle Δ of the tessellation (except for convex hull triangles) is shared by two tetrahedra T and T' which, in turn, are spanned by five sites. Three of them span Δ, and the remaining two sites q and q' belong to T and T', respectively. The triangle Δ is called *locally Delaunay* if the circumsphere of T does not enclose q' (or, equivalently, if the circumsphere of T' does not enclose q). A third tetrahedron might be spanned by these five sites. Δ is called *flippable* if the union of these two or three tetrahedra is convex.

After having added a site $p \in S$ to the current Delaunay tessellation, the algorithm first splits the tetrahedron containing p into four tetrahedra with apex p, in the obvious way. The four triangles opposite to p are then considered, which are the bases of the tetrahedra with apex p. For each such triangle Δ, a flip is performed that involves the five sites corresponding to Δ, provided Δ is flippable and not locally Delaunay. Thereby, new triangles become opposite to p and possibly have to be flipped, too. This sequence of flips terminates with the Delaunay tessellation that includes p.

The algorithm works locally and thus is elegant and relatively easy to understand. It runs in worst-case optimal $O(n^2)$ time. Still, the incremental approach has its drawbacks. It might construct intermediate tessellations of quadratic size, in spite of a possible linear size of the final tessellation. This unpleasant phenomenon even cannot be overcome by inserting the sites in random order. Proving the existence of an — in this sense — efficient insertion order is an open problem. See Snoeyink and van Kreveld [660] for a related result in the plane.

The algorithm can be found in more detail in Edelsbrunner and Shah [313], including correctness proofs and a data structure for storing tetrahedral tessellations. Joe [431] provides an efficient implementation in 3-space, and Cignoni *et al.* [223] propose a hybrid method that works in general d-space and efficiently combines insertion, divide & conquer, and *bucketing*. Rajan [598] and Edelsbrunner and Shah [313] discuss the d-dimensional variant of the incremental insertion algorithm, and the latter paper also provides a generalization to so-called *regular simplicial tessellations*. Flips and other material on such regular simplicial cell complexes will be discussed in some detail in Section 6.3.

6.1.3. *Starting tetrahedron*

An important issue we have ignored so far in our discussion: Finding a *starting tetrahedron* that contains a newly inserted site. Such a tetrahedron has to be identified before the process of restructuring the tessellations can take place. Interestingly, an efficient search may be based on the *construction history* of the tessellation. See Edelsbrunner and Shah [313], who describe a d-dimensional generalization of the strategy for planar Delaunay triangulations in Guibas *et al.* [390]. This strategy is similar in spirit to the method of *conflicting triangles* described in Sections 3.2 and 6.5.

From the practical point of view, conceptually simpler techniques may be preferable, even though they are theoretically inferior. Bucketing techniques like the one implemented in Joe [431] may be used. As another alternative, we may simply 'walk' in the tessellation from some chosen point q to the target point p (the site to be inserted). Walking strategies work for arbitrary tetrahedral tessellations and have been systematically studied in Devillers *et al.* [266]. We briefly review their findings below.

The *straight walk* passes through all the tetrahedra along the line segment $\overline{qp}$, whereas the *orthogonal walk* visits the tetrahedra along an *isothetic* (i.e., axis-parallel) path, moving from q to p by changing only one coordinate at a time. *Visibility walk*, on the other hand, moves from the current tetrahedron t to some tetrahedron adjacent to t, through a facet (triangle) Δ if the plane supporting Δ separates t from p. More than one facet may fulfill this condition, and choice may be done randomly (stochastic visibility walk). For Delaunay tessellations and regular tessellations (Section 6.3), visibility walk always reaches the correct target. This is a consequence of the face *acyclicity property* with respect to the *in-front/behind relation* of these structures, described in De Floriani *et al.* [245] and in Edelsbrunner [301]; see Section 4.2. In the case of an arbitrary tessellations, visibility walk may in fact loop, though stochastic visibility walk will terminate with high probability.

The expected number of triangles visited by straight walk in a planar Delaunay triangulation of uniformly distributed points is $O(d(q,p)\sqrt{n})$, where $d(q,p)$ denotes the distance between q and p; see Bose and Devroye [158] and Devroye *et al.* [267]. In the worst case in 3-space, however, this walk may have to pass through $\Theta(n^2)$ tetrahedra. The best choice is the stochastic visibility walk, in spite of its extreme behavior in the worst case. It performs experimentally slightly better than the straight walk and the orthogonal walk, is easier to code, and does not encounter problems with geometric degeneracies where, for example, the line segment $\overline{qp}$ intersects an edge or vertex of a tetrahedron.

Of course, choosing a 'good' starting point q is an issue, no matter which kind of walk is used. An efficient technique for selecting q in two- and three-dimensional Delaunay tessellations is described by Mücke *et al.* [551]. Their *point-location* method for walking to the target point p also does not require any preprocessing of the tessellation.

6.2. Power diagrams

Voronoi diagrams are intimately related to geometric objects in higher dimensions. This fact, along with one of its algorithmic applications, has already been addressed in Section 3.5. Here, we base the discussion on a generalization of Voronoi diagrams called *power diagrams*; the geometric correspondences to be described extend to that type in a natural manner. We refer to d dimensions in order to point out the general validity of the results.

6.2.1. *Basic properties*

Consider a set S of n point sites in d-space $\mathbf{R}^d$. Assume that each point in S has assigned an individual *weight* $w(p)$. In some sense, $w(p)$ measures the capability of p to influence its neighborhood. This is expressed by the so-called *power function* of a point $x \in \mathbf{R}^d$ with respect to a site $p \in S$,

$$\mathrm{pow}(x,p) = (x-p)^T(x-p) - w(p).$$

A nice geometric interpretation is the following. For positive weights, a weighted site p can be viewed as a sphere with center p and radius $\sqrt{w(p)}$; for all points x outside this sphere, $\mathrm{pow}(x,p) > 0$, and $\sqrt{\mathrm{pow}(x,p)}$ expresses the length of a line segment starting from x and tangent to the sphere.

The locus of equal power with respect to two weighted sites p and q is a hyperplane, called the power hyperplane of p and q. (In the plane, the power line of two circles is also called their *chordale* or *radical axis*. It passes through their points of intersection, if any, because the power function vanishes there for both circles.) Let $h(p,q)$ denote the closed halfspace bounded by this hyperplane and containing the points of less power with respect to p. Then the *power cell* of p is given by

$$\mathrm{cell}(p) = \bigcap_{q \in S \setminus \{p\}} h(p,q).$$

In analogy to the classical Voronoi regions, the power cells define a partition of $\mathbf{R}^d$ into convex polyhedra, the so-called *power diagram*, $\mathrm{PD}(S)$, of S. Also the names *Laguerre-Voronoi diagram* or *generalized Dirichlet tessellation* are used in the literature; see e.g. Imai *et al.* [420]. Figure 6.3 depicts a planar example.

$\mathrm{PD}(S)$ coincides with the Voronoi diagram of S if all weights are the same, or even are zero. In contrast to Voronoi regions, however, power cells might be empty if general weights are used; for instance, $\mathrm{cell}(p) = \emptyset$ in Figure 6.3. A condition on weights sufficient for making each site remain within its power cell is stated in Section 10.5. Note that adding a fixed

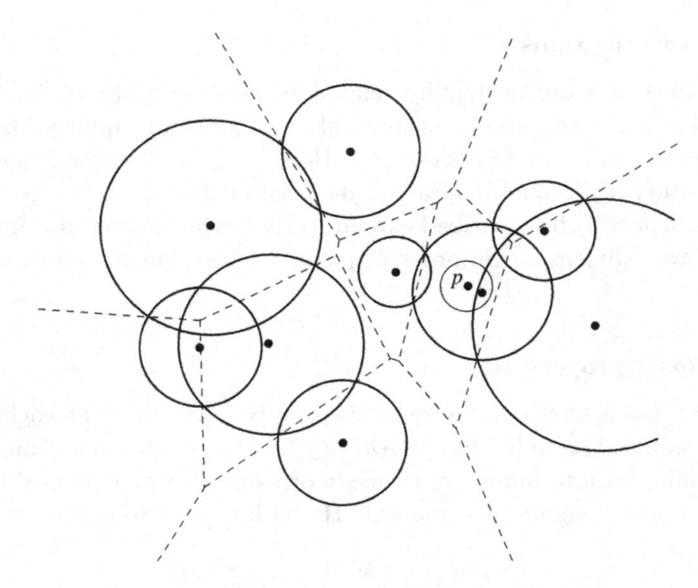

Figure 6.3. Power diagram for 10 circles in the plane.

constant Δ to all weights leaves the power diagram unchanged. So insisting on $w(p) \geq 0$ for all $p \in S$ would cause no loss of generality.

PD(S) is a *face-to-face cell complex* in d-space that consists of polyhedral faces of various dimensions j, for $0 \leq j \leq d$. In the non-degenerate case, exactly $d+1$ *edges*, $\binom{d+1}{2}$ *facets* (faces of dimension $d-1$), and $d+1$ *cells* (the power cells) meet at each vertex of PD(S). Such complexes are commonly called *simple cell complexes* in $\mathbf{R}^d$. For storing a d-dimensional cell complex, the *cell-tuple data structure* in Brisson [168] seems appropriate. This data structure represents the incidence and ordering information in a cell complex in a simple uniform way.

When each (positively) weighted site $p \in S$ is interpreted as the sphere $\sigma_p = (p, \sqrt{w(p)})$, we can make the following nice observation. The part of σ_p that contributes to the union of all these spheres, $\bigcup_{p \in S} \sigma_p$, is just the part of σ_p within cell(p). This means that PD(S) defines a partition of this union into simply-shaped and algorithmically tractable pieces. Several algorithms concerning the *union* (and also the *intersection*) *of balls* are based on this partition; see Avis *et al.* [107], Aurenhammer [90], and Edelsbrunner [302]. For the intersection of balls, their *order-*$(n-1)$ *power diagram* (Subsection 6.5.2) is used.

6.2.2. *Polyhedra and convex hulls*

Power diagrams (and thus Voronoi diagrams) are, in a strong sense, equivalent to the boundaries of *convex polyhedra* in one dimension higher. This is a fact with far-ranging implications and has been observed in Brown [169], Klee [460], Edelsbrunner and Seidel [310], Paschinger [586], and Aurenhammer [87]. The power function $\mathrm{pow}(x, p)$ in $\mathbf{R}^d$ can be expressed by the hyperplane

$$\pi(p) : x_{d+1} = 2x^T p - p^T p + w(p)$$

in $\mathbf{R}^{d+1}$, in the sense that a point x lies in $\mathrm{cell}(p)$ of $\mathrm{PD}(S)$ iff, at x, $\pi(p)$ is vertically above all hyperplanes $\pi(q)$ for $q \in S \setminus \{p\}$. Hence $\mathrm{PD}(S)$ corresponds, by *vertical projection*, to the *upper envelope* of these hyperplanes, which is the surface of an unbounded convex polyhedron, $\Pi(S)$, in $\mathbf{R}^{d+1}$. Conversely, it is not difficult to see that every upper envelope of n non-vertical hyperplanes in $\mathbf{R}^{d+1}$ corresponds to the power diagram of n suitably weighted sites in $\mathbf{R}^d$.

Cell complexes that can be obtained from projecting the boundary of a convex polyhedron (onto one of its facets) are also called *Schlegel diagrams* in convex geometry. We conclude that power diagrams and Schlegel diagrams are the same structures, up to clipping the unbounded faces of the power diagram. For a wealth of material on convex polytopes and cell complexes, and on convex geometry in general, the interested reader is referred to the monograph by Grünbaum [384] and to the handbook by Gruber and Wills [383].

The following upper bound is a direct consequence of the *upper bound theorem* for convex polyhedra proved in McMullen [529]. The bound is trivially sharp for power diagrams, but is achieved also for Voronoi diagrams, as was shown in Seidel [625]. A respective construction of a point configuration S in d-space is sketched at the beginning of Chapter 10.

Theorem 6.1. *Let S be a set of n point sites in $\mathbf{R}^d$. Any power diagram for S, and in particular, the Voronoi diagram for S, realizes at most f_j faces of dimension j, for*

$$f_j = \sum_{i=0}^{a} \binom{i}{j} \binom{n-d+i-2}{i} + \sum_{i=0}^{b} \binom{d-i+1}{j} \binom{n-d+i-2}{i},$$

where $a = \lceil \frac{d}{2} \rceil$ and $b = \lfloor \frac{d}{2} \rfloor$. The numbers f_j are $O(n^{\lceil \frac{d}{2} \rceil})$, for $0 \le j \le d-1$.

For algorithmic issues, power diagrams in $\mathbf{R}^d$ can be brought in connection to *convex hulls* in $\mathbf{R}^{d+1}$, by exploiting a *duality transform* (actually, *polarity*) between upper envelopes of hyperplanes (or intersections of upper halfspaces) and convex hulls of points. This connection is best described by generalizing the *lifting map* in Section 3.5 to weighted points. A site $p \in S$ with weight $w(p)$ is transformed into the point

$$\lambda(p) = \begin{pmatrix} p \\ p^T p - w(p) \end{pmatrix}$$

in $\mathbf{R}^{d+1}$. There is a interrelation called *polarity* between the transforms λ and π. The point $\lambda(p)$ is called the pole of the hyperplane $\pi(p)$ which, in turn, is called the polar hyperplane of $\lambda(p)$. Polarity defines a one-to-one correspondence between arbitrary points in $\mathbf{R}^{d+1}$ and non-vertical hyperplanes in $\mathbf{R}^{d+1}$. It is well known that polarity preserves the *relative position* of points and hyperplanes.

In particular, if $w(p) = 0$, then point $\lambda(p)$ lies on the surface of a paraboloid P in $\mathbf{R}^{d+1}$ whose axis of revolution is the x_{d+1}-axis, and the hyperplane $\pi(p)$ is tangent to P at $\lambda(p)$; cf. Section 3.5 for the three-dimensional case.

To show the connection to convex hulls, consider an arbitrary face f of the unbounded polyhedron $\Pi(S)$ defined above. Let f be the intersection of $m = d - j + 1$ hyperplanes $\pi(p_1), \dots, \pi(p_m)$, such that f is of dimension j. Each point $x \in f$ lies on these but above all other hyperplanes $\pi(q)$ defined by S. Hence, the polar hyperplane of x has the points $\lambda(p_1), \dots, \lambda(p_m)$ on it and the remaining points $\lambda(q)$ above it. This shows that the points $\lambda(p_1), \dots, \lambda(p_m)$ span a face of dimension $d - j$ of the *lower convex hull* of the point set $\{\lambda(p) \mid p \in S\}$.

We conclude that each j-dimensional face of $\Pi(S)$, and thus of $\mathrm{PD}(S)$, is represented by a $(d - j)$-dimensional face of this convex hull. This implies a duality between power diagrams in $\mathbf{R}^d$ and convex hulls (i.e., *convex polytopes*) in $\mathbf{R}^{d+1}$.

In the special case of an unweighted point set S in the plane, the parts of the convex hull that are visible from the plane project to the vertices, edges, and triangles of the Delaunay triangulation of S, and we obtain Theorem 3.5 of Section 3.5. A triangulation which can be obtained by (*vertically*) *projecting* the boundary faces of some convex hull is called a *regular triangulation*; see, e.g. Lee [491] and Edelsbrunner and Shah [313]. Regular triangulations are just those being dual to planar power diagrams. We will provide some material on regular simplicial complexes in general dimensions in Section 6.3.

Once the convex hull of the point set $\{\lambda(p) \mid p \in S\}$ has been computed, the faces of PD(S), as well as their incidence and ordering relations, can be obtained in time proportional to the size of PD(S).

Theorem 6.2. *Let $C_{d+1}(n)$ be the time needed to compute a convex hull of n points in $\mathbf{R}^{d+1}$. A power diagram (and in particular, the Voronoi diagram) of a given n-point set in $\mathbf{R}^d$ can be computed in $C_{d+1}(n)$ time.*

Worst-case optimal *convex hull algorithms* working in general dimensions have been designed by Clarkson and Shor [229], Seidel [630], and Chazelle [194], yielding $C_{d+1}(n) = O(n \log n + n^{\lceil \frac{d}{2} \rceil})$. So Theorem 6.2 is asymptotically optimal in the worst case. Note, however, that power diagrams in higher dimensions may as well have a fairly small size, $O(n)$, which emphasizes the use of *output-sensitive* convex hull algorithms. The algorithm in Seidel [627] achieves $C_{d+1}(n) = O(n^2 + f \log f)$, where f is the total number of faces of the convex hull constructed. The latest achievements are $C_4(n) = O((n + f) \log^2 f)$ in Chan *et al.* [191] and $C_5(n) = O((n + f) \log^3 f)$ in Amato and Ramos [55]. Mehlhorn *et al.* [535] provide a kernel for higher-dimensional geometric computation and describe its application to convex hulls and Delaunay triangulations.

A thorough and detailed overview of the existing convex hull algorithms for various dimensions is provided in Seidel [633].

Among the possible direct methods that construct the power diagram in $\mathbf{R}^2$ without resorting to one more dimension are *randomized incremental insertion* (see Section 3.2) and *divide & conquer* (see Imai *et al.* [420]). Both run in optimal time, $O(n \log n)$. The linear-time insertion method for power cells in Aurenhammer and Edelsbrunner [96] yields a worst-case optimal $O(n^2)$ algorithm in $\mathbf{R}^3$.

6.2.3. *Related diagrams*

Space constraints preclude our discussion of power diagrams. Still, some remarks are in order to point out their central role within the context of Voronoi diagrams. For a more detailed discussion of the following material, and references, see Aurenhammer and Imai [98].

The regions of a Voronoi diagram are defined by a set of sites and a distance function. If the regions are polyhedral, then *any* Voronoi diagram defined in this way can be shown to be the power diagram of some suitable set of weighted point sites. For instance, this is the case for the *farthest site Voronoi diagram*, whose regions consist of all points having the same farthest site in the given set. In fact, any *Voronoi diagram of higher order* (see Section 6.5) is a power diagram.

Note that the feature of regions being polyhedral is equivalent to the feature that all *bisectors* are hyperplanes, and thus the latter is a characterizing property of power diagrams as well. This fact turns out useful, for example, in the study of Voronoi diagrams in parametric statistical spaces, where distances between points measure divergences between distributions. We refer to Sadakane *et al.* [614] for the *Kullback–Leibler divergence*, and to Boissonnat *et al.* [145] for a study of general *Bregman divergences*. In particular, certain *hyperbolic Voronoi diagrams* (see Subsection 7.1.1) that arise in this context can be interpreted as power diagrams.

Voronoi diagrams all of whose separators are hyperplanes are also called *affine Voronoi diagrams*. Such diagrams can always be generated by general quadratic-form distances; see Subsection 7.4.5 for a short discussion of some interesting instances.

A polyhedral cell complex in $\mathbf{R}^d$ is called *simple* if exactly $d + 1$ cells meet at each vertex. For example, the Voronoi diagram of a set of point sites in $\mathbf{R}^d$ is simple if no $d + 2$ sites are co-spherical. If $d \geq 3$, *any* simple cell complex can be shown to be a power diagram. Note that, interestingly, for $d = 2$ this property is lost. An intuitive reason is that polygonal partitions of the plane allow plenty of freedom, whereas forming a valid cell complex in higher dimensions comes with several geometric restrictions.

The class of power diagrams is closed under taking *cross-sections* with a hyperplane. That is, the diagram obtained from intersecting a power diagram in $\mathbf{R}^d$ with some hyperplane is again a power diagram, in $\mathbf{R}^{d-1}$. As a seemingly unrelated application of this fact, certain *linear combinations* between a weighted site and its neighbors in the power diagram can be obtained, which are of use in smoothing surfaces based on Delaunay triangulations and regular triangulations; see Section 4.2.

Several generalized Voronoi diagrams in $\mathbf{R}^d$ have an *embedding* in a power diagram in $\mathbf{R}^{d+1}$, in the sense that they can be obtained by intersecting a power diagram with simple geometric objects (like cones, spheres, or paraboloids), and then projecting this intersection.

For example, the *additively weighted Voronoi diagram* (i.e., the Voronoi diagram for spheres, or the *Johnson–Mehl model* [433]), and the *multiplicatively weighted Voronoi diagram* [96, 615] (or the *Apollonius model*) have this property. Weighted Voronoi diagrams tend to have complex geometric and combinatorial properties, especially in higher dimensions, and their embeddings in power diagrams can be viewed as *linearizations* at the expense of adding one more dimension. We will come back to weighted models of Voronoi diagrams and their applications in

Section 7.4. The Voronoi diagram for spheres is also briefly treated at the end of Section 6.6.

In all situations mentioned above, a set of weighted sites for the corresponding power diagram can be computed easily. Thus general methods of handling Voronoi diagrams and cell complexes in d-space become available. For example, the Voronoi diagram for spheres in 3-space, and the multiplicatively weighted Voronoi diagram in the plane, can both be computed in $O(n^2)$ time, which is asymptotically optimal in the worst-case.

6.3. Regular simplicial complexes

A *polyhedral cell complex* in d-space $\mathbf{R}^d$ is called *simplicial* if all its cells are simplices. A k*-simplex* in $\mathbf{R}^d$ (for $0 \leq k \leq d$) is the convex hull of $k+1$ affinely independent points. For example, 3-simplices are tetrahedra, 2-simplices are triangles, 1-simplices are line segments, and 0-simplices are just points. The boundary of a simplex consists of various simplices of lower dimensions. Cell complexes that are dual to simplicial complexes are necessarily *simple cell complexes*, and vice versa.

For instance, the classical Voronoi diagram (respectively, the power diagram) in $\mathbf{R}^d$ are simple cell complexes, if sites (respectively, weights) are chosen sufficiently general, implying that the Delaunay complex is then simplicial.

While the numbers of edges and triangles in any *triangulation* are fixed with a given configuration S of its n vertices in the plane (see Section 2.2), the size of simplicial complexes in higher-dimensional space $\mathbf{R}^d$ may vary largely, from $\Theta(n)$ to $\Theta(n^{\lceil \frac{d}{2} \rceil})$. See Rothschild and Straus [610] for the upper bound, which is already attained by Delaunay simplicial complexes; cf. Section 10.1. Note that this is also the asymptotic maximum size of a convex hull in $\mathbf{R}^{d+1}$, by the upper bound theorem in Subsection 6.2.2.

Edelsbrunner *et al.* [312] showed that, for any n-point set in $\mathbf{R}^3$, there is a tetrahedrization consisting of only $3n - 11$ tetrahedra or less. Very recently, this was generalized to $d \geq 4$ by Aichholzer *et al.* [40], who showed a number of d-simplices of at most $dn + \Omega(\log n)$. On the other hand, if we ask for the *largest-sized* tetrahedrization every n-point set in $\mathbf{R}^3$ (in general position) has to admit, the best known bound is $O(n^{\frac{5}{3}})$; see Brass [163].

The fact that every point set in $\mathbf{R}^d$ has a small simplicial complex suggests the idea of 'slimming' the Delaunay simplicial complex DT(S), by adding point sites that lie in many Delaunay spheres. As a surprising result, Bern *et al.* [128] showed that only $O(n)$ extraneous points suffice to achieve a size of $O(n)$ in general dimensions d, improving the bound given in Chazelle *et al.* [198].

6.3.1. *Characterization*

So-called *regular simplicial complexes* are a subclass well studied in convex geometry. As mentioned in Section 6.1, these complexes are intimately related to convex hulls and power diagrams, by projection and duality, respectively. Algorithmic applications arise, for example, from certain subcomplexes introduced as *weighted α-shapes* in Edelsbrunner [302], a concept useful for describing finite *unions of balls*, and thus, for the approximate shape and medial axis description of solids; see Section 6.6. Also, certain optimization properties of Delaunay triangulations described in Sections 4.2 and 6.1 generalize to regular simplicial complexes. We review some more of their properties in the sequel.

Let S be a set of n point sites in $\mathbf{R}^d$, and assume each site $p \in S$ has assigned some real weight $w(p)$. Like the Delaunay tessellation can be defined by the empty-circle or empty-sphere property (see Sections 4.2 and 6.1), we can define a regular simplicial complex for weighted sites by generalizing this property.

Consider two weighted sites p and q in $\mathbf{R}^d$, not necessarily in S, with weights $w(p)$ and $w(q)$. We call p and q *orthogonal* (*obtuse*, *acute*) if $d(p,q)^2$ is equal, larger, or smaller than $w(p) + w(q)$, respectively. Notation stems from interpreting these two sites as spheres, with radii $\sqrt{w(p)}$ and $\sqrt{w(p)}$ if $w(p), w(q) \geq 0$ (see Section 6.2), which intersect at a respective angle.

Now, consider $d+1$ weighted sites in the set S above. Their convex hull is a d-simplex Δ. It is easy to show that there exists a unique sphere σ orthogonal to each of these $d+1$ weighted sites. (If their weights are zero then σ is just the circumsphere of Δ.) Simplex Δ is called *regular* (in S) if all other weighted sites in S are obtuse to σ. See Figure 6.4 for an example in the plane.

The collection of all regular d-simplices for the weighted point set S defines a simplicial complex which resides in the convex hull conv(S) of S, and which is called a *regular tessellation* of S.

Not every simplicial complex is regular. Already for dimension $d = 2$, well-known examples of triangulations exist which are not regular; see e.g. [88, 300].

By the geometric relations presented in Section 6.2, we have the following characterization.

Theorem 6.3. *Let S be a set of n point sites in $\mathbf{R}^d$, and let Σ be a simplicial complex in* conv(S) *whose vertex set is a subset of S. The following statements are equivalent:*

(a) *Σ is the regular tessellation of S, for some weights for S.*
(b) *Σ is dual to the power diagram of S, for some weights for S.*

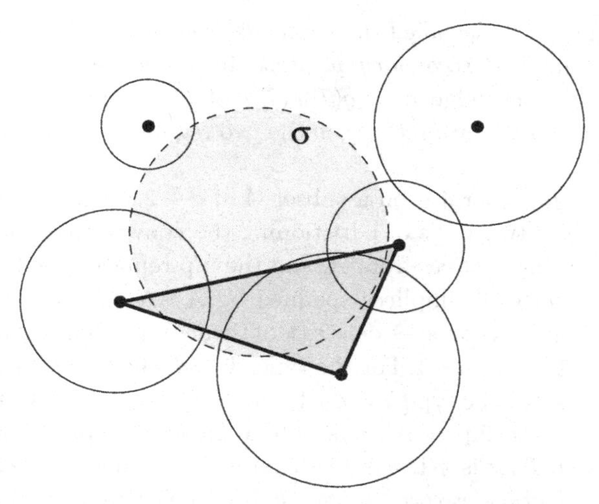

Figure 6.4. A regular 2-simplex (triangle). The dashed circle σ orthogonally intersects the three weighted sites (circles) centered at the triangle vertices, and is obtuse to the remaining two weighted sites.

(c) *Σ is the vertical projection of the lower boundary part of the convex hull of a point set S' in $\mathbf{R}^{(d+1)}$, obtained from S by assigning some $(d+1)$th coordinates.*

6.3.2. *Polytope representation in weight space*

Given a set S of sites, a variety of regular tessellations exists, depending on the weight vector W we assign to S. Certain sites in S may be missing as vertices in the tessellation, however, due to the fact that their dual power cells are empty in the power diagram for the chosen weights. Interestingly, there is structure in this family of regular tessellations for S. This becomes transparent when interpreting such a tessellation, T, in the two different ways below.

From a combinatorial point of view, T can be associated with the point $q(T) = W$ in the high-dimensional *configuration space* $\mathbf{R}^n$ of weights, for $n = |S|$. From a geometric point of view, T corresponds to the lower part of a convex hull in $\mathbf{R}^{d+1}$ (Theorem 6.3(c)), that is, to a convex simplicial surface $\varphi(T)$ in $(d+1)$-space. By its convexity, $\varphi(T)$ minimizes the volume below it, over all possible (in general, nonconvex) simplicial surfaces based on the same vertex set in $(d+1)$-space.

Volume minimality for $\varphi(T)$ translates to extremality of $q(T)$ in the n-space of weights, meaning that the points $q(T)$ are the vertices of a *convex polytope*, $Q(S)$, in $\mathbf{R}^n$, which is just their convex hull. This polytope of regular triangulations is called the *associahedron* in Lee [490] and Gel'fand *et al.* [357], and the *secondary polytope* in Billera *et al.* [132]. As an important fact, each edge $e = q(T)q(T')$ of $Q(S)$ corresponds to a so-called bistellar flip operation between the two regular tessellations T and T' for S.

A *bistellar flip* operates on a subset A of $d+2$ weighted sites from S. There are exactly two ways of partitioning the convex hull, conv(A), of A into simplices using vertices from A, and the flip replaces one by the other. The total number of d-simplices spanned by A is $\binom{d+2}{d+1} = d+2$, and the flip — depending on its type — deletes k of them and constructs $d+2-k$ new ones, for some $1 \leq k \leq d+1$. For the types $k \leq d$, the subset A is in *convex position*, whereas for the type $k = d+1$, one site $p \in A$ is interior to conv(A). In the last case, the flip constructs only a single d-simplex, conv(A), and leaves p isolated. That is, p is *not* a vertex of the new simplicial complex any more, and we talk of a *(vertex-)removing flip.* (In the dual environment, this means that the cell of p is now empty in the power diagram for the current weights.) By contrast, the former flip types may be called *exchanging flips*, since they just exchange certain d-simplices but leave unchanged the vertex set of the simplicial complex.

For $d=3$, the exchanging flip of type 2-to-3 (for tetrahedra) is described and illustrated in Section 6.1.

The existence of the secondary polytope $Q(S)$ has an important implication. Since the edge graph of a convex polytope is connected, there exists for any two regular tessellations T_1 and T_2 of S, a sequence of flips that takes T_1 into T_2.

In fact, a stronger result holds, as has been proven recently in Pournin and Liebling [595]: T_1 and T_2 are still connected in the edge graph of $Q(S)$ if all edges that correspond to removing flips are ignored. This means that any two regular tessellations of S can be transformed into each other by using *exchanging flips* only, that is, without ever removing a vertex.

6.3.3. *Flipping and lifting cell complexes*

If we leave the class of *regular* simplicial complexes, flipping becomes difficult. The main reason behind is that an interpretation in terms of convex surfaces (Theorem 6.3(c)) is missing.

Even the classical result in $\mathbf{R}^2$, that an arbitrary triangulation of S can be flipped into the Delaunay triangulation DT(S) by *Delaunay flips* (Section 4.2), does not generalize if a specific regular triangulation is to be reached: We can define *locally regular* faces in a simplicial complex on S, similar to locally Delaunay faces — based on the notions of *obtuse* and *acute* pairs of weighted sites introduced in the beginning of this section — but global regularity will not be reached if only *regularizing flips* are to be used; see e.g. [30]. This is due to the fact that not every interior edge of a triangulation might be flippable, see below. Still, the class of triangulations is connected under arbitrary, in general non-regularizing, edge-exchanging flips.

In particular, $O(n^2)$ is an upper bound on the *flip distance* between *any* two triangulations T_1 and T_2 of a given n-point set S in the plane: T_1 as well as T_2 can be flipped into $\mathrm{DT}(S)$, and by reversing the latter flip sequence, T_1 transforms into T_2. On the way to $\mathrm{DT}(S)$, any fixed edge — once flipped away — will never reappear, because the corresponding lifted surface (see Subsection 6.3.2), which now lives in 3-space, will decrease in height until the convex triangular surface $\varphi(\mathrm{DT}(S))$ is attained. As $\binom{n}{2}$ edges are spanned by S, the above bound follows. It is asymptotically tight, as an $\Omega(n^2)$ lower bound exists; see e.g. Fortune [343]. The problem of computing a *shortest* flip sequence that transforms two triangulations into each other has recently been shown to be *NP-hard* in Lubiw and Pathak [513] and Pilz [591].

We observe that not all edges in a triangulation of S can be flipped. This is trivially true for each convex hull edge e, but also if e borders two triangles whose union is nonconvex. Hurtado *et al.* [411] have shown that every triangulation contains at least $\frac{(n-4)}{2}$ flippable edges. When transforming triangulations, these can be flipped *in parallel*, as long as they are *independent*, i.e., they do not share a vertex (or equivalently, they do not belong to the same triangle). In fact, linearly many edges always are both flippable *and* independent, leading to a flip distance of only $O(n)$ for this more powerful parallel operation; see Galtier et al. [348], and recently Souvaine *et al.* [661] for an improved bound of $\frac{(n-4)}{5}$ which is tight.

In 3 dimensions already, it is an open problem whether every tetrahedral complex can be turned into $\mathrm{DT}(S)$ by a sequence of exchanging flips. An affirmative answer would imply that any two given tetrahedral complexes of a finite point set in $\mathbf{R}^3$ can be flipped into each other. In fact, for dimensions 5 (and higher), a counterexample has been found recently by Santos [616]. On the other hand, flipping works in general dimensions for the *incremental construction algorithm* in Edelsbrunner and Shah [313], because there site insertions take place already in a regular tessellation for the sites processed so far, which guarantees a special structure.

The practical relevance of the flipping problem above is given by a result in Guibas and Russel [393]. They have shown by experiments that flipping will convert to $\mathrm{DT}(S)$ certain 'near'-Delaunay tetrahedrizations (stemming from small vertex movements, for example) up to three times

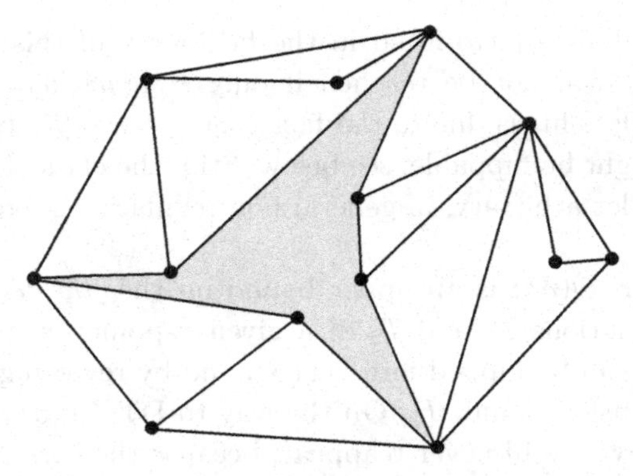

Figure 6.5. Pseudo-triangulation with 13 vertices. A pseudo-triangle with many vertices is highlighted.

faster than DT(S) can be constructed from scratch, provided flipping does not get stuck.

Flippability can be retained if we extend the class of simplicial complexes to so-called *pseudo-simplicial complexes.*

Of particular interest are *pseudo-triangulations* in the plane, which contain, in addition to triangles, general polygonal faces with exactly three convex interior angles. See Figure 6.5 for an illustration. Pseudo-triangulations have many properties useful in diverse applications and are investigated, concerning their *flipping properties* and embedding properties into *polygonal surfaces*, in Aichholzer *et al.* [30]. Flip distances of $O(n \log n)$ between arbitrary triangulations are achieved, using 'shortcuts' via (less edge-consuming hence more flexible) pseudo-triangulations. Pseudo-simplicial complexes in general dimensions are introduced and studied in Aurenhammer and Krasser [101]. Both papers also give short summaries of earlier literature on pseudo-triangulations; a comprehensive survey is provided in Rote *et al.* [609]. An even further relaxation of triangulations, called *pre-triangulations* in Aichholzer *et al.* [33], allows the definition and characterization of various polygonal partitions with special flipping and *lifting properties.*

In fact, deciding the liftability to 3-space of a given polygonal partition of the plane is an intriguing topic. Various geometric and combinatorial results exist, see e.g. [30, 33] and references given there. A classical result is known as *Steinitz's theorem* [663], presented nicely in Ziegler's textbook [717]: Every triangulation in the plane, as being a 3-connected planar graph, is combinatorially equivalent to the edge graph of some convex polyhedron in 3-space.

Finally, let us consider so-called *constrained simplicial complexes*, which are restricted to the interior of *nonconvex polytopes.* A most popular representative we have encountered in Section 5.4 is the *constrained Delaunay triangulation* of a polygon (or, more generally, of a given set of line segments). While every polygon can be triangulated, *constrained regular triangulations* either do not always exist, or are not unique in general, depending on their definition; see [30]. In dimensions higher than two, another difficulty arises: Not every nonconvex polytope admits a simplicial complex without extraneous vertices. The *Schönhard polytope* [621, 717] is a well-known counterexample in $\mathbf{R}^3$. In fact, it is *NP-hard* to decide whether a nonconvex polytope in $\mathbf{R}^3$ can be tetrahedrized; see Ruppert and Seidel [613].

Using a different construction, Grislain and Shewchuk [381] showed that deciding the existence of a *constrained Delaunay tetrahedrization* in $\mathbf{R}^3$ remains NP-hard if co-spherical degeneracies in the vertex set of the underlying polytope are allowed, and that the problem has a *polynomial solution* in the non-degenerate case.

A thorough treatment of constrained Delaunay (and regular) tessellations in general dimensions is given in Shewchuk [651]. For example, a sufficient existence condition is that each constraining $(d-2)$-simplex is separable by a sphere from the remaining vertices in the tessellation. Shewchuk's papers also describe implementations and applications of constrained Delaunay and regular simplicial complexes, in the fields of *mesh generation* and *finite element methods.*

When generalizing to pseudo-simplicial complexes, unique existence and flippability of constrained regular tessellations is guaranteed; see [30, 33]. This is due to the existence of respective convex polytope representations in the high-dimensional configuration space $\mathbf{R}^n$, which constitute generalizations of the *secondary polytope* discussed in Subsection 6.3.2.

6.4. Partitioning theorems

Voronoi diagrams partition their underlying space according to a given set of sites and a distance function. In this sense, they naturally induce a *clustering* of space (or of possible objects situated therein). We will discuss this fact now, in the context of power diagrams, because the possibility of weighting the sites gives room for interesting existence and optimality results; see Aurenhammer *et al.* [97]. The general role of Voronoi diagrams in geometric clustering is described in Section 8.4.

6.4.1. *Least-squares clustering*

Consider a set S of n point sites, and another set X of m points, in d-space $\mathbf{R}^d$. The (closest-site Euclidean) Voronoi diagram of S defines an *assignment function* $A : X \to S$, given by

$$A(x) = s \quad \text{if and only if} \quad x \in \text{VR}(s).$$

Here $\text{VR}(s)$ denotes the Voronoi region of site $s \in S$. Equivalently, $A^{-1}(s) = X \cap \text{VR}(s)$, for all $s \in S$. (Points of X that have more than one closest site in S are not uniquely covered by this definition; by convention, A assigns each such point to an arbitrary but fixed closest site.) The total number of points assigned to a particular site s, $|A^{-1}(s)|$, is called the *capacity* of s. The capacities of all sites add up to $m = |X|$. The assignment A has an obvious optimization property: It minimizes the sum of the distances between sites and their assigned points, over all possible assignments of X to S.

Given S and X, we would like to be able to change the assignment by varying the distance function that underlies the Voronoi diagram of S. To this end, we attach a set $W = \{w(s) \mid s \in S\}$ of weights to the sites, and replace the Euclidean distance $d(x, s)$ between a point x and a site s by the *power function*

$$\text{pow}_W(x, s) = d(x, s)^2 - w(s).$$

In the resulting *power diagram* (see Section 6.2), each region is still a convex polyhedron, and has the property of shrinking (or expanding) when the weight of its defining site is decreased (respectively, increased). As above, we obtain an assignment function $A_W : X \to S$ which now depends on the choice of weights. In particular, the site capacities depend on W.

A *least-squares clustering* of a set X of points with respect to a set S of sites (or cluster centers) in d-space is defined to minimize the total squared Euclidean distance between the sites and their associated points. Our interest is in *constrained* least-squares clusterings, where the number of associated points per site (i.e., its capacity) is prescribed. Interestingly, assignments defined by power diagrams are constrained least-squares clusterings.

Lemma 6.1. *Let S and X be finite sets of sites and points in $\mathbf{R}^d$, respectively, and fix a set W of weights for S. The assignment A_W minimizes*

$$\sum_{x \in X} d(x, A(x))^2$$

over all assignments $A : X \to S$ *with capacity constraints* $|A^{-1}(s)| = |A_W^{-1}(s)|$ *for all* $s \in S$.

Proof. From the definition of A_W it is evident that A_W minimizes the expression

$$\sum_{x \in X} \mathrm{pow}_W(x, A(x)) = \sum_{x \in X} d(x, A(x))^2 - \sum_{x \in X} w(A(x))$$

over all possible assignments $A : X \to S$, regardless of the capacity constraints. The last sum, being equal to $\sum_{s \in S} |A^{-1}(s)| \cdot w(s)$, is a fixed constant for all assignments A with capacities $|A^{-1}(s)| = |A_W^{-1}(s)|$. The lemma follows. □

Power diagrams also give rise, in the obvious way, to mappings of the *entire* d-space to the set of sites. Let $A_W : \mathbf{R}^d \to S$ be the assignment induced by the power diagram of S with weights W. That is, $A_W^{-1}(s) = \mathrm{reg}_W(s)$ is the power cell of site s in the diagram. The capacity of a site now can be defined, for example, as the fraction of the *unit hypercube* $[0,1]^d$ contained in its power cell.

Formally, let ϱ be a continuous and non-vanishing probability distribution on $[0,1]^d$, and let $\mu(X) = \int_X \varrho(x)dx$ denote the measure of a set $X \subset \mathbf{R}^d$ with respect to ϱ. Then $\mu(A_W^{-1}(s))$ is the capacity of s that results from A_W. The capacities of all sites add up to 1. A continuous version of Lemma 6.1 can be proved easily. In fact, we have the following general result [97].

Theorem 6.4. *Let* S *be a finite set of sites in* $\mathbf{R}^d$. *Any (discrete or continuous) assignment induced by a power diagram of* S *is a least-squares clustering, subject to the resulting capacities. Conversely, a least-squares clustering for* S, *subject to any given capacities (whose sum is the total number of assigned points in the discrete case, and* 1 *in the continuous case) exists and can be realized by a power diagram of* S.

For finite point sets X, the existence of a constrained least-squares clustering $L : X \to S$ is trivial. Its realizability by power diagrams follows from an algorithm that constructs such a power diagram, given in Subsection 6.4.2. Existence and realizability in the continuous case, $L : [0,1]^d \to S$, can be proved by using the fact that, for any two sites $s, t \in S$, the sets $L^{-1}(s)$ and $L^{-1}(t)$ are separable by a hyperplane. An alternative proof is given in Cuesta-Albertos and Tuero-Diaz [236], in the more general context of the so-called *Monge–Kantorovich mass transference problem*; see Gangbo and McCann [350], Caffarelli *et al.* [175], and Subsection 7.4.1.

The existence of least-squares clusterings, together with their realizability by power diagrams, immediately implies several *partitioning results* for power diagrams. The discrete variant reads as follows.

Theorem 6.5. *Let S and X be a set of n sites and m points in Euclidean d-space $\mathbf{R}^d$, respectively. For any choice of integer site capacities $c(s)$ with $\sum_{s \in S} c(s) = m$, there exists a set W of weights such that $|A_W^{-1}(s)| = c(s)$, for all sites $s \in S$.*

In other words, there always exists a power diagram whose regions partition a given d-dimensional finite point set X into clusters of prescribed sizes, no matter where the sites of the power diagram are chosen. Moreover, we have the following continuous version of Theorem 6.5.

Theorem 6.6. *Let S be a set of n sites in $\mathbf{R}^d$, let ϱ be a continuous and non-vanishing probability distribution on $[0,1]^d$, and let μ denote the measure with respect to ϱ. For any capacity function $c : S \to [0,1]$ satisfying $\sum_{s \in S} c(s) = 1$, there is a set W of weights such that $\mu(\mathrm{reg}_W(s)) = c(s)$, for all sites $s \in S$.*

By taking, for instance, ϱ to be the uniform distribution in $[0,1]^d$ we get the following corollary.

Corollary 6.1. *For any set of n sites in $\mathbf{R}^d$ there exists a power diagram that partitions the unit hypercube into n polyhedral regions of prescribed volumes.*

This seems surprising, as the placement of the sites determines the normals of the facets separating the power regions. Corollary 6.1 is, in a close sense, related to *Minkowski's theorem* for *convex polytopes* (see e.g., Grünbaum [384]): Let V be any collection of n non-zero non-parallel vectors that span $\mathbf{R}^{d+1}$ and sum up to zero. Then there exists a convex $(d+1)$-polytope with n facets in one-to-one correspondence with vectors of V, such that each facet is normal to its corresponding vector and has d-dimensional volume equal to the vector length. Though each power diagram is a projection of a convex polytope (Section 6.2), the statements are not equivalent, however, due to the presence of unbounded components in the latter polytope. Similar is also the partitioning result in Aurenhammer [94] for convex polygons in the plane, where the so-called *weighted skeleton* of a polygon (a weighted variant of the straight skeleton discussed in Section 5.3) is used to optimally partition various sets of objects.

A recent result on *equipartitions* into convex sets (i.e., partitions into sets of equal volume), which takes surface areas into account, is based on power diagrams as well; see Hubard and Aronov [409]. For further partitioning results, applications, and references; see e.g., Carlsson *et al.* [184], and Subsection 7.4.1.

6.4.2. *Two algorithms*

It is of practical interest to compute the partitions in Theorems 6.5 and 6.6 efficiently. In the discrete case, the following point insertion algorithm can be used. The continuous case can be treated with a gradient method, described later in this subsection.

The *incremental insertion algorithm* computes, for a set S of n sites and a set X of m points in $\mathbf{R}^d$, a least-squares clustering $L : X \to S$ subject to a given integer capacity vector c with $\sum_{s \in S} c(s) = m$. By Lemma 6.1, it suffices to compute a weight vector $W = (w(s))_{s \in S}$ such that we have $|X \cap \text{reg}_W(s)| = c(s)$ for all $s \in S$.

We start with $W = 0$, for which the power diagram is just the classical Voronoi diagram of S, and proceed in m phases, each inserting one point $x \in X$ into the current diagram. The vector W, and with it the power diagram, is then recomputed such that the invariant $b(s) \leq c(s)$ is maintained for all $s \in S$, where $b(s)$ denotes the current number of points in $\text{reg}_W(s)$.

(1) Add point x to the region $\text{reg}_W(s)$ containing x in the current power diagram. If now $b(s) \leq c(s)$ the phase ends, as there is no need to change W. Otherwise, let $D = \{s\}$. Intuitively, the set D will contain the sites whose regions are too large and must be shrunk.
(2) Repeat the following two steps (consult Figure 6.6):
    (a) Shrink all D-regions by simultaneously decreasing their weights. More formally, find the smallest positive number Δ so that decreasing the weights of all D-sites simultaneously by more than Δ causes one of the shrinking regions to lose a point, say p'. Notice that in this process a site in $S \setminus D$ cannot lose a point to a D-site, and that no point can move between two D-regions or between two non-D-regions.
    (b) Decrease the weights of all D-sites by Δ. Consider the region $\text{reg}(s')$ where p' would end up, had we shrunk the weights by more than Δ. If $b(s') < c(s')$ then go to Step 3, as we have found a region which is not full. Else add s' to D and repeat (a).
(3) We have found a region $\text{reg}(s')$ that is not full and a point p' on its boundary. Assign p' to s'. This makes some region $\text{reg}(s'')$ with $s'' \in D$ less than full. But s'' was added to D because of some point p'' that it shared with a site s''' that had already been in D. So assign p'' to s''

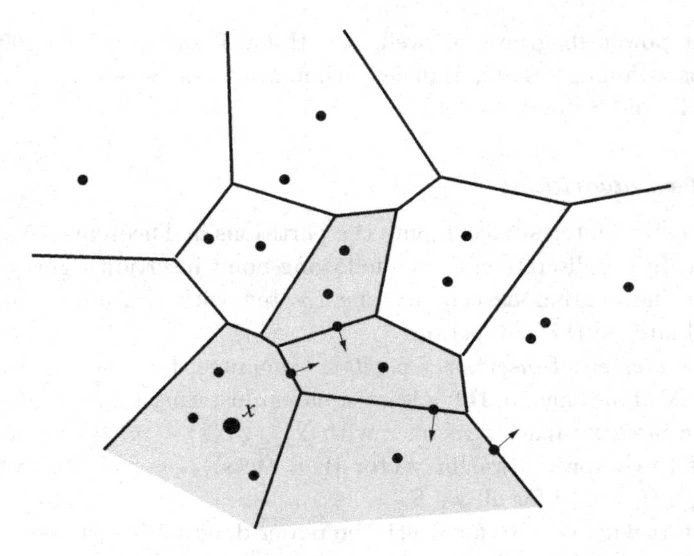

Figure 6.6. A set of D-regions (shaded) losing a point. After inserting point x in the lower left corner, reassignment takes place as indicated by arrows.

and follow the chain back, reassigning points to regions and adjusting b-values accordingly, until the original site s is encountered and relieved of one point.

This procedure can be implemented in $O(n^2 m \log m + nm \log^2 m)$ time and $O(m)$ space, for the (probably most interesting) planar case [97]. This improves over the $O(nm^2 + m^2 \log m)$ time and $O(nm)$ space algorithm that results from transforming the problem into a *minimum-cost flow problem*; see Fredman and Tarjan [345]. Note that, in the terminology of network flows, the chain-like process of reassigning points to sites at the end of a phase corresponds exactly to an *augmenting path*.

For the continuous problem, a *gradient method* for iteratively improving the weight vector W can be used. In fact, for a fixed set of sites, the value of the assignment induced by the power diagram with weights W is a concave function of W. Finding a weight vector such that the resulting assignment fulfills the capacity requirements is then equivalent to finding the maximum of a related function whose domain is the *configuration space* of weights.

Let ϱ be a continuous and non-vanishing probability distribution in $[0,1]^d$. For an arbitrary but fixed assignment $A : [0,1]^d \to S$, define the function $f_A : \mathbf{R}^n \to \mathbf{R}$ by

$$f_A(W) = \int_{[0,1]^d} \varrho(x) \cdot \mathrm{pow}_W(x, A(x))dx.$$

Let $B(A) = (\mu(A^{-1}(s)))_{s \in S}$ be the vector of capacities resulting from the assignment A, and put

$$Q(A) = \int_{[0,1]^d} \varrho(x) \cdot \delta(x, A(x))^2 dx,$$

the value of A. With this notation, f_A can be written as

$$f_A(W) = -B(A) \cdot W + Q(A).$$

Hence f_A is a linear function of W. Now consider the function $f : W \mapsto f_{A_W}(W)$; recall that A_W is the assignment induced by the power diagram with weights W. We claim that f is the pointwise minimum of the class of functions f_A. Indeed, for fixed W, the assignment A_W minimizes the value $f_A(W)$ by definition of the power diagram of S with weights W. In other words, the graph of f is the *lower envelope* of an (infinite) set of hyperplanes in $\mathbf{R}^{n+1}$, showing that f is a concave function.

By the choice of properties of ϱ, $B(A_W)$ and $Q(A_W)$ depend continuously on W. Hence, for each $W = W'$, the graph of f has at point $(W', f(W'))$ a unique tangent hyperplane $x_{n+1} = -B(A'_W) \cdot W + Q(A'_W)$ that changes continuously with W. That is, f describes a smooth surface. Note that the gradient $\bigtriangledown f(W)$ of f at W is given by $-B(A_W)$.

Recall that we aim to find a weight vector W^* such that $B(A_{W^*}) = C$, the given capacity vector. Consider the function

$$\begin{aligned} g(W) &= f(W) + C \cdot W \\ &= (C - B(A_W)) \cdot W + Q(A_W). \end{aligned}$$

Its gradient $\bigtriangledown g(W)$ is $C - B(A_W)$, hence our requirement $B(A_{W^*}) = C$ just means $\bigtriangledown g(W^*) = 0$. This corresponds to a global maximum of the smooth concave function g. So the problem we want to solve is: Find W^* such that $g(W^*)$ is maximized.

Finding the maximum of a concave and smooth n-variate function is a well-studied problem. In our case, we can exploit the fact that, for any given weight vector W, we can compute $g(W)$ and $\bigtriangledown g(W)$ from the power diagram with weights W. So a *gradient method* (see, e.g., [129]) for iteratively *approximating* W^* can be used. Starting, for example, with the weight vector $W_0 = 0$ (corresponding to the Voronoi diagram of S), we use the iteration scheme

$$W_{k+1} = W_k + t_k \bigtriangledown g(W_k).$$

If the step sizes t_k are chosen properly then W_k converges to the solution W^* at a superlinear rate. Intuitively, what happens is that the weights of those sites whose region measures are too small (too large) are increased (respectively, decreased) at each step.

If S is a set of n sites in the plane, and ϱ is the uniform distribution in the square, each step can be carried out in $O(n \log n)$ time: For the current weight vector W_k, we need $O(n \log n)$ time to construct the power diagram of S and W_k, and time $O(n)$ is needed in addition to calculate the area and the integral of squared distances for each region within the unit square. The space requirement is optimal, $O(n)$.

6.4.3. *More applications*

We conclude this section with some applications of the discrete least-squares clustering problem. As before, let S and X be finite sets of n sites and m points in $\mathbf{R}^d$, respectively.

For $Y \subset X$ and $s \in S$, define the *variance* of the cluster Y with respect to the site s as $\sum_{x \in Y} \delta(x, s)^2$. Then a constrained least-squares clustering $L : X \to S$ is just a clustering for X such that the clusters have prescribed sizes and the sum of cluster variances is minimized. Besides being optimum in the above sense, these clusters have the important property that their convex hulls are pairwise disjoint: Distinct clusters are contained in different regions of a power diagram, and power regions are convex. *Hull-disjointness* is a natural and desirable property of clusters which, for instance, eases the classification of new points. Simple examples show that replacing variance by the sum of distances destroys hull-disjointness.

If we define the *profit* of cluster Y with respect to site s as $\sum_{x \in Y} x \cdot s$, then L maximizes the sum of cluster profits for given cluster sizes: We have $\delta(x, s)^2 = x^2 + s^2 - 2x \cdot s$, and the sum of the first two terms is independent of the assignment, provided capacity constraints are satisfied. This concept is motivated by the following *transportation problem*: Interpret a point $x = (x_1, \ldots, x_d)$ as a truck loaded with x_i units of the ith good, and a site $s = (s_1, \ldots, s_d)$ as a market that sells the ith good at price s_i per unit. Choose the site capacities according to the attractiveness of the markets, and L will tell you where each truck should go in order to achieve maximal profit for these capacities.

The next application makes use of the property that constrained least-squares clusterings are *invariant* under translation and scaling.

Observation 6.1. *Let $\sigma > 0$ and $\tau \in \mathbf{R}^d$, and consider a least-squares clustering $L : X \to S$ with capacities c. Then $\sigma L + \tau$ is a least-squares clustering of X to $\sigma S + \tau$ subject to c.*

Consider the special case that S and X are of equal cardinality n, and let $L : X \to S$ be a least-squares clustering subject to $c(s) = 1$ for all $s \in S$. L is called a *least-squares matching* in this case. Define a (*one-to-one*) *least-squares fitting* as the least-squares matching $L_* : X \to \sigma S + \tau$ where the value of L_* is minimal over all positive scaling factors σ and all translation vectors τ. Observation 6.1 tells us that $L_*^{-1}(\sigma s + \tau) = L^{-1}(s)$ for all $s \in S$. Thus, when computing the least-squares fitting L_*, we can first calculate and fix the matching L, as a least-squares matching of X to S, and then determine the optimizing values of σ and τ for this matching, which is an easy analytic task.

In this way, a least-squares fitting of size n in $\mathbf{R}^2$ can be computed in $O(n^3)$ time and $O(n)$ space, and a least-squares fitting of m points to n sites subject to given capacities can be computed in $O(n^2 m \log m + nm \log^2 m)$ time and $O(m)$ space; see [97] for details. This improves the best known least-squares matching algorithm (a variant of Vaidya's algorithm [689]), with $O(m^{2+\varepsilon})$ time and $O(m^{1+\varepsilon})$ space, for $m > n^2$.

Recently, *partial least-squares matchings* in $\mathbf{R}^2$ have been studied in Rote [608]. We are given some point set X (the 'pattern') of size m, where m is now smaller than the size n of the set S of sites (the 'ground set'). Consider the assignment $A : X \to S$ that minimizes

$$\sum_{x \in X} d(x, A(x))^2$$

for site capacities which are not fully prescribed but only constrained to $c(s) \in \{0, 1\}$. Unlike the fully constrained case (covered in Lemma 6.1), the assignment A now depends on the geometric position of X. Site capacities will be 1 or 0 as demanded by optimality. Note that A is an injective mapping. In view of an efficient *pattern matching* algorithm, the following question arises: How many different optimal assignments are there when the pattern X is *translated* in the plane? As a partial answer, $m(n - m) + 1$ is a tight bound when X is moved along a straight line. This implies a polynomial-time algorithm for this restricted case.

If X is moved but slightly, the assignment will most likely remain unchanged. Now a partition of the plane into cells can be defined, such that each cell reflects all translations of X with the same optimal assignment. This partition is just the classical Voronoi diagram of S if X is a singleton set (i.e., $m = 1$), and its cells are convex polygons for general m; see Figure 6.7 for the case $m = 6$.

Henze *et al.* [403] showed that the size of this *partial-matching partition* is $O(m!\ m^2 n^4)$ (and polynomial in n for any fixed dimension d). The conjectured complexity is $O(m^2 n^2)$, as this is also the complexity of the

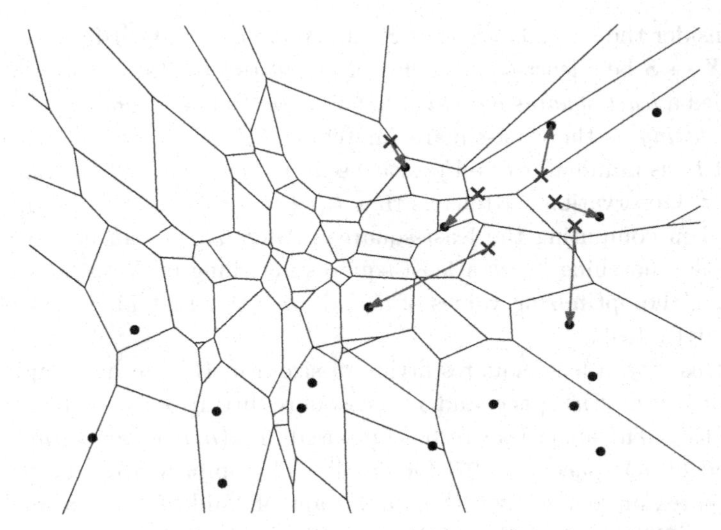

Figure 6.7. Partial-matching partition of the plane. A pattern (×) of size 6 is matched to a ground set of 18 sites (•); from [608].

overlay of m Voronoi diagrams, each for n sites. The *overlay* of two convex partitions C_1 and C_2 of the plane is the (convex) partition obtained by superimposing C_1 and C_2; new vertices are created by edges that cross, hence new edges and polygons arise as well.

We mention that every partial-matching partition is actually a *power diagram*, as a convex projection polyhedron in 3-space can be defined.

Another related assignment problem has been investigated in Brieden and Gritzmann [167]. In a *balanced fractional clustering*, each of the m points $x \in X$ is allotted an 'area' $a(x)$, and each of the n sites (cluster centers) $s \in S$ has a given capacity $c(s)$. Here $a(x)$ and $c(s)$ are real-valued, and $m \geq n$. (In a more relaxed variant, some small interval $[c^-(s), c^+(s)]$ of valid capacities per site may be specified instead.) Areas and capacities obey the condition

$$\sum_{x \in X} a(x) = \sum_{s \in S} c(s).$$

Now, fractions $\xi_1(x), \ldots, \xi_n(x)$ of each point x, with $\sum \xi_i(x) = 1$, are to be assigned to the sites $s_1, \ldots, s_n$ such that

$$\sum_{x \in X} \xi_i(x)a(x) = c(s_i) \quad (\text{or} \quad \in [c^-(s_i), c^+(s_i)]) \quad \text{for all } i.$$

This clustering problem is motivated by an application in farmland consolidation. The n farmers residing at sites $s_1, \ldots, s_n$ cultivate a total of m lots $x \in X$ with individual areas $a(x)$, and they want to exchange lots so as to move them closer together for efficiency purposes. Of course, the original farm sizes $c(s_i)$ should (almost) stay the same by reassignment.

In its integer version, deciding the existence of a balanced fractional clustering is *NP-hard*; see Borgwardt *et al.* [154].

As is shown in [167], certain types of balanced fractional clusterings are induced by power diagrams of the sites in S, and in particular, by so-called *centroidal power diagrams*, where the sites coincide with weighted centroids of their clusters; cf. Subsection 9.3.4.

6.5. Higher-order Voronoi diagrams

Higher-order Voronoi diagrams are natural and useful generalizations of the classical Voronoi diagram. Given a set S of n point sites in d-space $\mathbf{R}^d$ and an integer k between 1 and $n-1$, the *order-k Voronoi diagram* of S, for short $V_k(S)$, partitions the space into regions such that each point within a fixed region has the same k closest sites. $V_1(S)$ just is the (closest-site) Voronoi diagram $V(S)$ of S.

Many of the applications of the classical Voronoi diagram (in Chapter 8 and other chapters) are meaningful in their 'order-k' versions as well, like *k-nearest neighbor search* and *largest* 'almost' *empty circle* placement; or higher-order Voronoi diagrams apply directly, like for *clustering* and *classification* problems (Section 8.4).

The regions of $V_k(S)$ are convex polyhedra, as they arise as the intersection of halfspaces of $\mathbf{R}^d$ bounded by symmetry hyperplanes of the sites. A subset M of k sites in S has a non-empty region in $V_k(S)$ iff there is a sphere that encloses M but no site in the complement $S \setminus M$. In fact, the region of M in $V_k(S)$ just is the set of centers of all such spheres.

Figure 6.8 illustrates a planar order-2 Voronoi diagram. Two differences between this and the classical Voronoi diagram are apparent. A region need not contain its defining sites, and the bisector of two sites may contribute more than one edge (or facet, in higher dimensions) to the diagram.

6.5.1. *Farthest-site diagram*

For the largest admissible value $k = n-1$, a subset M of S of size $n-1$ has to be closest to a point $x \in \mathbf{R}^d$ in order to claim x for its region. In other words, the singleton site $\{p\} = S \setminus M$ has to be farthest from x. For this reason, the diagram $V_{n-1}(S)$ is commonly called the *farthest-site Voronoi diagram* of S. It contains, for each site $p \in S$, the region of all points x for which p is the farthest site in S.

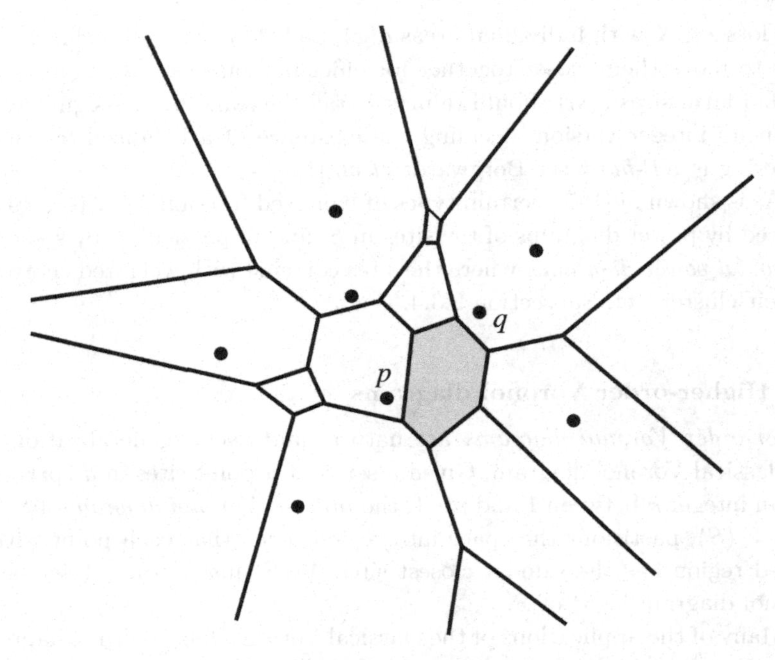

Figure 6.8. Region of $\{p, q\}$ in the order-2 Voronoi diagram $V_2(S)$.

This diagram has special structural properties. For example, only sites lying on the boundary of the *convex hull* of S, and exactly those, have non-empty regions in $V_{n-1}(S)$. This is because a site interior to the convex hull can never be the farthest from any point x in $\mathbf{R}^d$. Moreover, all regions are *unbounded*, as each of them contains an unbounded part of some ray emanating from the defining site and pointing 'into' S. In the plane, this implies that the edge graph of $V_{n-1}(S)$ is a tree. It consists of exactly $2h-3$ edges and $h-2$ vertices, if h denotes the number of *extreme points* in S (and general position is assumed; the size of $V_{n-1}(S)$ is less, otherwise). For an illustration, see Figure 6.10 in Subsection 6.5.4, where several farthest-site Voronoi diagrams are drawn in dashed style.

Exact upper bounds on the size of farthest-site Voronoi diagrams in $\mathbf{R}^d$ for $d \geq 3$ are derived in Seidel [628].

Several methods of construction apply to $V_{n-1}(S)$. Let us first consider the lifting approach taken in Subsection 6.2.2, where distances to the sites $p_i \in S$ are described by hyperplanes $\pi(p_i)$ in $\mathbf{R}^{d+1}$. Clearly, the weights $w(p_i)$ have all to be put to 0 now, as we deal with Euclidean distances rather than with power functions.

As farthest distances are to be used, the *lower envelope* of the hyperplanes $\pi(p_1), \ldots, \pi(p_n)$ will vertically project to $V_{n-1}(S)$. Consequently, by duality this implies that the *upper part* of the *convex hull* of their set of poles in $\mathbf{R}^{d+1}$, $\{\lambda(p_i) \mid p_i \in S\}$, corresponds to the *farthest-site Delaunay complex*, $\mathrm{DT}_{n-1}(S)$, of S. This simplicial complex includes a d-simplex spanned by $d+1$ sites in S if and only if its circumsphere encloses the entire set S. Observe that such simplices are exclusively spanned by extreme points of S, such that no vertices of the complex $\mathrm{DT}_{n-1}(S)$ are interior to the convex hull $\mathrm{conv}(S)$ of S in $\mathbf{R}^d$.

For example, in the planar case, $\mathrm{DT}_{n-1}(S)$ is just a triangulation of the vertices of the convex polygon $\mathrm{conv}(S)$. We called it the *farthest-site Delaunay triangulation*. This structure has an optimality property, as was shown by Eppstein [321]: It minimizes *angularity*, and thus, the smallest occurring angle, among all possible triangulations for the convex hull points. By contrast, the Delaunay triangulation $\mathrm{DT}(S)$ *maximizes* angularity; see Section 4.2.

Algorithmically, the discussion above implies that both structures $V_{n-1}(S)$ and $\mathrm{DT}_{n-1}(S)$ *come for free*, when we compute a convex hull in $\mathbf{R}^{d+1}$ in order to obtain the classical (closest-site) Voronoi diagram $V(S) = V_1(S)$ or the Delaunay complex $\mathrm{DT}(S) = \mathrm{DT}_1(S)$ in $\mathbf{R}^d$, as done in Sections 3.5 and 6.2.

When we are *not* interested in closest-site structures, the farthest-site Delaunay triangulation in $\mathbf{R}^2$ is preferably computed directly, with a simple *ear-clipping algorithm* which also achieves an $O(n \log n)$ runtime, and which we shall sketch below.

Let C be a convex polygon. An *ear* of C is a triangle which shares two edges with C. Each triangulation T of C (and of any simple polygon) has at least two ears, as these correspond to leaves in the dual tree of T. The algorithm now constructs the $h-2$ triangles building $\mathrm{DT}_{n-1}(S)$ ear by ear. Recall that $h \leq n$ denotes the number of extreme points of S, that is, the number of vertices of the polygon $\mathrm{conv}(S)$.

Initially, by scanning the boundary of $\mathrm{conv}(S)$, one of its ears, Δ, is identified whose circumcircle encloses S. The largest of the h encountered circles will have this property. The ear Δ is clipped from $\mathrm{conv}(S)$, and the two newly created ears adjacent to Δ are examined, and their circumradii put into a *priority queue* Q, which initially holds the h circumradii checked before. In this way, at any time the currently largest radius stored in Q gives us the next ear to be clipped off, and with it, one more triangle of $\mathrm{DT}_{n-1}(S)$. It is clear that only $O(h)$ insertions and maxima deletions take place in Q. Either operation can be performed in time $O(\log h)$. So, the $O(n \log n)$ time spent for computing $\mathrm{conv}(S)$ dominates the runtime. The diagram $V_{n-1}(S)$, if desired, can be derived from $\mathrm{DT}_{n-1}(S)$ in additional $O(h)$ time.

A similar ear-clipping algorithm is used in Devillers [262], for re-triangulating the 'hole' that results from *removing* a site p and its incident faces from the (classical) Delaunay triangulation $\mathrm{DT}(S)$. (Rather than being convex, this hole is a *star-shaped* object; all its parts are visible from p.) He also bases his clipping sequence on some *shelling order* of the triangles, in $\mathrm{DT}(S \setminus \{p\})$. This order can be found by considering planes through the projections of the ear triangles onto a vertical paraboloid in 3-space (see Section 3.5), because these planes intersect the rotation axis of the paraboloid in some shelling order of the Delaunay triangles. The method works in general dimensions d, and constitutes a corrected version of Heller's planar algorithm [401]. Shelling orders of convex polytopes play a role in the computation of higher-dimensional convex hulls (and thus, of Delaunay, Voronoi, and power complexes); see e.g. Seidel [633].

Farthest-site Voronoi diagrams are meaningful for distance functions other than the Euclidean, and for sites more general than points. In fact, the planar shelling algorithm described above generalizes within the same runtime bound, $O(n \log n)$, to the *farthest-line segment Voronoi diagram*; see Aurenhammer *et al.* [95] and Section 5.1.

We remark that the maximal size of generalized farthest-site Voronoi diagrams may be drastically smaller than that of their closest-site counterparts. This is briefly discussed at the beginning of Subsection 7.4.3.

6.5.2. *Hyperplane arrangements and k-sets*

The family of *all* occurring higher-order Voronoi diagrams for a given set S of point sites in d-space $\mathbf{R}^d$, $V_1(S), V_2(S), \ldots, V_{n-1}(S)$, is closely related to an *arrangement of hyperplanes* in $\mathbf{R}^{d+1}$. This fact has been observed in the seminal paper by Edelsbrunner, O'Rourke, and Seidel [309]. We describe this relationship in the more general setting of power diagrams, by defining an *order-k power diagram*, $\mathrm{PD}_k(S)$, for a set S of weighted point sites in an analogous way; see Aurenhammer [87, 93].

Recall from Section 6.2 that the power function, $\mathrm{pow}(x, p)$, with respect to a site p in $\mathbf{R}^d$ can be expressed by a hyperplane $\pi(p)$ in $\mathbf{R}^{d+1}$.

The set of hyperplanes $\{\pi(p) \mid p \in S\}$ dissects $(d+1)$-space into a polyhedral cell complex called a *hyperplane arrangement*. Arrangement cells are convex, and can be classified according to their relative position with respect to the hyperplanes in $\{\pi(p) \mid p \in S\}$. A cell C is said to be of *level k* if exactly k hyperplanes are vertically above C. For example, the *upper envelope* of $\{\pi(p) \mid p \in S\}$ (i.e., the pointwise maximum of these linear

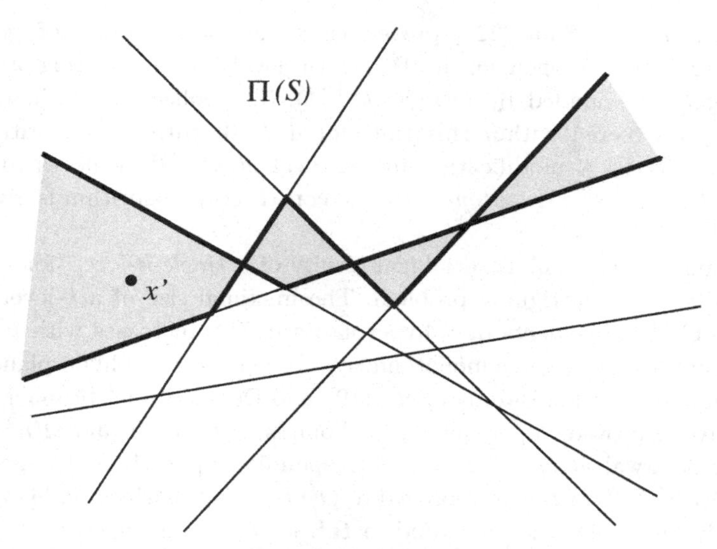

Figure 6.9. Arrangement of six lines in $\mathbf{R}^2$ with level-2 cells shaded. Each point x' in a fixed level-2 cell projects vertically to a particular region of the respective order-2 power diagram.

functions) bounds the only cell, $\Pi(S)$, of level 0. All cells of level 1 share some facet with $\Pi(S)$, so that their vertical projection gives the (order-1) power diagram $\text{PD}(S)$.

More generally, the cells of level k project to the regions of $\text{PD}_k(S)$, for each k between 1 and $n-1$. To see this, let $x' \in \mathbf{R}^{d+1}$ be a point in some k-level cell C. Then k hyperplanes $\pi(p_1), \ldots, \pi(p_k)$ are above x', and $n-k$ hyperplanes $\pi(p_{k+1}), \ldots, \pi(p_n)$ are below x'. That means that, for the vertical projection x of x' onto $\mathbf{R}^d$, we have $\text{pow}(x, p_i) < \text{pow}(x, p_j)$ for $1 \leq i \leq k$ and $k+1 \leq j \leq n$. Hence x is a point in the region of $\{p_1, \ldots, p_k\}$ in $\text{PD}_k(S)$. Figure 6.9 gives an illustration for $d+1=2$.

Hyperplane arrangements are well-investigated geometric objects, concerning their combinatorial as well as their algorithmic complexity; see e.g. Edelsbrunner *et al.* [309, 311], and especially the monograph by Edelsbrunner [300]. Their size does not vary with the particular placement of the hyperplanes, and an optimal insertion-based construction algorithm in general dimensions exists. From the discussion above, we obtain the following theorem.

Theorem 6.7. *Let S be a set of n (weighted or unweighted) point sites in $\mathbf{R}^d$. The family of all higher-order power diagrams (or Voronoi diagrams) for S realizes a total of $\Theta(n^{d+1})$ faces, and it can be computed in optimal $\Theta(n^{d+1})$ time.*

Clarkson and Shor [229] proved that the total number of cells in an hyperplane arrangement in $\mathbf{R}^d$, with levels not exceeding a fixed value of k, is bounded by $O(n^{\lfloor d/2 \rfloor} k^{\lceil d/2 \rceil})$. The collection of these cells can be constructed within this time for $d \geq 4$, with the algorithm in Mulmuley [554]. A modification by Agarwal *et al.* [10] achieves roughly $O(nk^2)$ time in $\mathbf{R}^3$. An *output-sensitive* construction algorithm is given in Mulmuley [555].

Bounding the combinatorial complexity of a *single k-level* (for a fixed value of k) is an intriguing problem. The maximal size of a k-level, over all possible arrangements of n hyperplanes in $\mathbf{R}^d$, increases with $k \leq \frac{n}{2}$, though not necessarily in a monotonic way for a *given* set of hyperplanes. To mention a few results, Barany *et al.* [112], and Dey and Edelsbrunner [270], respectively, proved upper and lower bounds of $O(n^{8/3})$ and $\Omega(n^2 \log n)$ for $\mathbf{R}^3$. Agarwal *et al.* [9] gave a corresponding upper bound dependent on k, $O(nk^{5/3})$, which was improved to $O(nk^{3/2})$ in Sharir *et al.* [643]. The currently best known lower bound in $\mathbf{R}^d$, for $d \geq 3$, is $\Omega(n^{d-1} \cdot 2^{c\sqrt{\log n}})$; see Sharir *et al.* [643] and Nivasch [566].

All the results above apply to order-k power diagrams in one dimension lower. (Note that, interestingly, a high-order power diagram has a larger worst-case size than a high-order Voronoi diagram; cf. Subsection 6.5.3.)

Moreover, these results also apply to so-called *k-sets* of a set S' of n points in the same dimension, by the polarity transform between points and hyperplanes described in Section 6.2. A k-set is a subset of k points of S' that can be separated from S' by a hyperplane. In fact, k-sets have been objects of long-lasting investigation in combinatorial geometry, as is documented in Goodman and O'Rourke's handbook [371].

Polarity preserves the relative position between points and hyperplanes. With this, we observe that each point x' in a fixed k-level cell will dualize into some hyperplane $h(x')$ that cuts off a unique k-set from the corresponding set S' of poles (the dualized arrangement hyperplanes), because $h(x')$ has the same k points of S' below it, as long as x' stays inside the same cell.

As each dividing hyperplane splits S' into two subsets of cardinalities k and $n - k$ respectively, for some value of k, the number of k-sets of S' trivially equals the number of its $(n - k)$-sets. Now observe that each k-set M of S' bijectively corresponds to an *unbounded* region in the power diagram $\mathrm{PD}_k(S')$ (in the *same* dimension), because points x far from S' but still closest to M have to exist. We conclude that the number of unbounded regions in an order-k power (or Voronoi) diagram is exactly the same as in its order-$(n - k)$ counterpart — a fact not true for non-point sites like e.g. line segments [584]. Moreover, this number is, in the worst case, of the same order of magnitude as the number of *all* regions in an order-k power diagram in one dimension less.

Similar concepts are *k-facets*, which are oriented $(d-1)$-simplices spanned by d points in an n-point set S in $\mathbf{R}^d$, such that there are exactly k points of S in the positive open halfspace determined by such a simplex. The numbers of k-sets and k-facets lie within constant factors from each other. In dimension 3, the exact total number of *k-triangles*, for $1 \leq k \leq m$, has been determined in Aichholzer *et al.* [42] for the range $m < \frac{n}{4}$.

6.5.3. *Computing a single diagram*

Many practical applications ask for the computation of a *single order-k Voronoi diagram* $V_k(S)$ in the plane, for a fixed value of k. (Typically, k does not depend on $|S| = n$ but is a small constant.) Like in the classical Voronoi diagram $V(S)$, edges are pieces of perpendicular bisectors of sites. Vertices are centers of circles that pass through three sites. However, these circles are no longer empty; they enclose either $k-1$ or $k-2$ sites. Lee [495] showed that this diagram has $O(k(n-k))$ regions, edges, and vertices. It is easy to see that the regions of the order-2 Voronoi diagram $V_2(S)$ are in one-to-one correspondence with the edges of $V(S)$. Hence $V_2(S)$ realizes at most $3n-6$ regions in the plane.

Considerable efforts have been made to compute the single planar order-k Voronoi diagram efficiently. Different approaches have been taken in Lee [495], Chazelle and Edelsbrunner [197], Aurenhammer [92], Clarkson [224], Agarwal *et al.* [10], and Ramos [600]. In the last three papers, randomized runtimes of $O(kn^{1+\varepsilon})$, respectively $O(kn\log n + n\log^3 n)$ and $O(kn2^{c\log^* k} + n\log n)$ are achieved, which is close to optimal. For the L_1- and L_∞-*metrics*, an output-sensitive algorithm has been given in Liu *et al.* [509].

Below we describe a (roughly) $O(k^2(n-k)\log n)$ time *randomized incremental algorithm* by Aurenhammer and Schwarzkopf [102], that can be modified to handle arbitrary *on-line sequences* of site insertions, site deletions, and k-nearest neighbor queries. Though not being most time efficient, the algorithm profits from its simplicity and flexibility.

The heart of the algorithm is a *duality transform* that relates the diagram $V_k(S)$ to a certain *convex hull* in 3-space. This transform allows us to insert and also delete sites in a simple fashion by computing convex hulls. Let $M \subset S$ be any subset of k sites. M is transformed into a point $q(M)$ in 3-space, by taking the *centroid* (center of mass) of M and lifting it up vertically. More precisely,

$$q(M) = \frac{1}{k}\left(\sum_{p\in M} p, \sum_{p\in M} p^T p\right).$$

Now consider the set $Q_k(S)$ of all points that can be obtained from S in this way. That is, $Q_k(S) = \{q(M) \mid M \subset S, |M| = k\}$.

Lemma 6.2. *The part of the convex hull of $Q_k(S)$ that is visible from the plane (i.e., its lower part) is dual to $V_k(S)$.*

The lemma can be proved by first mapping each subset M of k sites into a non-vertical plane $\pi(M)$ in 3-space, given by

$$\pi(M) : x_3 = \frac{2}{k}\sum_{p\in M} x^T p - \frac{1}{k}\sum_{p\in M} p^T p$$

and then considering the *upper envelope*, $\Pi_k(S)$, of all these planes. It is not difficult to show that the facets of $\Pi_k(S)$ project vertically to the regions of $V_k(S)$. The lemma then follows from observing the *polarity* (cf. Section 6.2) between the planes $\pi(M)$ and the points $q(M)$.

Note the difference of embedding $V_k(S)$ for $k \geq 2$ in this way, compared to its embedding into arrangement levels described earlier. The convex polyhedral surface $\Pi_k(S)$ is not (the upper) part of the boundary of the respective k-level — which rather is nonconvex — except for $k = 1$ where both surfaces are the upper envelope of the same set of planes.

To construct $V_k(S)$, we could just compute the point set $Q_k(S)$, determine its convex hull, and then dualize its triangles, edges, and vertices that are visible from the plane. However, $Q_k(S)$ contains a point for *each* subset of k sites of S, and thus has cardinality $\binom{n}{k} = \Theta(n^k)$. On the other hand, only $O(k(n-k))$ points lie on the convex hull, as $V_k(S)$ has this many regions.

We use randomized incremental insertion of sites in order to compute this convex hull more efficiently. Let $S = \{p_1, \ldots, p_n\}$, and let C_i denote the visible part of the convex hull of $Q_k(\{p_1, \ldots, p_i\})$, for $k+1 \leq i \leq n$. Points of $Q_k(S)$ lying on the *convex triangular surface* C_i are called *corners* of C_i.

We start by determining C_{k+1}. $Q_k(\{p_1, \ldots, p_{k+1}\})$ contains $k+1$ points which can be calculated in time $O(k)$, so $O(k \log k)$ time suffices. The generic step of the algorithm is the insertion of site p_i into C_{i-1}, for $i \geq k+2$.

(1) Identify all triangles of C_{i-1} which are destroyed by p_i and cut them out. Let B be the set of corners on the boundary of the hole.
(2) Calculate the set P of all new corners created by p_i.
(3) Compute the convex hull of $P \cup B$, and fill the hole with the visible part Γ_i of this convex hull. This gives C_i.

Each triangle Δ of C_{i-1} is dual to a vertex of $V_k(S)$. This vertex is the center of a circle that passes through three sites in S. Triangle Δ will be destroyed if this circle encloses p_i. The destroyed triangles of C_{i-1} form a connected surface, say Σ_{i-1}. Hence, if we know one such triangle in advance, Σ_{i-1} can be identified in time proportional to the number n_i of its triangles. Moreover, the set P of new corners can be calculated easily from the edges of Σ_{i-1}, as each such edge gives rise to a unique corner.

Lemma 6.3. *Given C_{i-1} we can construct C_i in time $O(n_i \log n_i)$, provided we know a triangle of C_{i-1} that is destroyed by p_i.*

When looking for a *starting triangle* of Σ_{i-1}, we profit from another nice property of the duality transform: If the vertical projection Δ' of a triangle Δ of C_{i-1} contains p_i then Δ is destroyed by the insertion of p_i. This leaves us with the problem of locating p_i in the triangulation given by the planar projection of C_{i-1}.

In fact, we get the desired *point-location structure* nearly for free. Adapting a technique used in Guibas *et al.* [390] for constructing Delaunay triangulations, we do not remove the triangles of C_{i-1} that get destroyed by the insertion of p_i, but mark them as old. When marked old, each triangle gets a pointer to the newly constructed part Γ_i of C_i. The next site then is located by scanning through the '*construction history*' of C_i. The structure for point location *within* each surface Γ_j, $j \leq i$, which is needed in addition, is a byproduct of the randomized incremental convex hull algorithm in [390], which we used for computing Γ_j. (The latter structure is similar to the *Delaunay tree* structure in Section 3.2, designed for the same purpose.)

In summary, the order-k Voronoi diagram for n sites in the plane can be computed in expected time $O(k^2(n-k)) \log n + nk \log^3 n)$, and optimal (deterministic) $O(k(n-k))$ space, by an online randomized incremental algorithm.

Full details and the following extensions are given in [102]. Deletion of a site can be done in the reverse order of insertion, again by computing a convex hull. The history-based point location structure used by the algorithm can be adapted to support k-nearest neighbor queries (see Subsection 8.1.1). A *dynamic data structure* is obtained, allowing for insertions and deletions of sites in expected time $O(k^2 \log n + k \log^2 n)$, and k-nearest neighbor queries in expected time $O(k \log^2 n)$. This promises a satisfactory performance for small values of k.

6.5.4. *Cluster Voronoi diagrams*

The order-k Voronoi diagram allots regions to subsets of *fixed size* k in the underlying point set S. Let us mention, at the end of this section, a different type of Voronoi diagram, whose regions are also defined by subsets of S. Let $C_1, \dots, C_t$ be a *partition* of S, that is, $S = \bigcup_{i=1}^{t} C_i$ and $C_i \cap C_j = \emptyset$ for $i \neq j$. (Note that, by contrast, the subsets defining an order-k Voronoi diagram do highly overlap.) We define the *cluster Voronoi diagram* for $C_1, \dots, C_t$ as the Voronoi diagram for these subsets (also called clusters) under the *Hausdorff distance*

$$h(x, C_i) = \max\{d(x,p) \mid p \in C_i\}.$$

Cluster Voronoi diagrams have been mainly considered in the plane, under different names, like the *closest covered set diagram* in Abellanas *et al.* [5], or the *min–max Voronoi diagram* in Papadopoulou and Lee [581]. Sometimes also the name *Hausdorff Voronoi diagram* is used. Figure 6.10 gives an illustration.

By expressing distances to sites $p \in S$ as planes $\pi(p)$ as done earlier in this section (see also Section 6.2), the diagram can be obtained by vertically projecting the *upper envelope* of concave surfaces $\Sigma(C_i)$, $i = 1, \dots, t$, where each surface $\Sigma(C_i)$ is the *lower envelope* of the planes in $\{\pi(p) \mid p \in C_i\}$. Using this interpretation, Edelsbrunner *et al.* [304] bound the size of the diagram by $O(n^2 \alpha(n))$, and give a construction algorithm running in the same time bounds. (Here $\alpha(n)$ denotes the inverse of *Ackermann's function*, which is extremely slowly growing with n.) The size drops to $O(n)$ if the clusters are pairwise *convex-hull disjoint*. A lower bound is $\Omega(n^2)$, which is attained, for example, for 2-point clusters whose corresponding line segments intersect heavily. The running time is $O(n^2)$ if only 2-point or 1-point clusters are present.

Applying the same '*Voronoi surface*' interpretation, Huttenlocher *et al.* [415] show that the size of the cluster Voronoi diagram is only $O(nt \cdot \alpha(nt))$, and that the diagram can be computed in time $O(nt \cdot \log(nt))$, both size and runtime now being sensitive to the number t of clusters. Alternatively, Papadopoulou [579] analyzes the size as $O(n + \gamma)$, where γ counts the number of piercing cluster pairs, of a certain type. She also proposes a *plane sweep* construction algorithm, whose complexity depends on several parameters. This algorithm was adapted later, in an output-sensitive manner and also exploiting the 3D interpretation above,

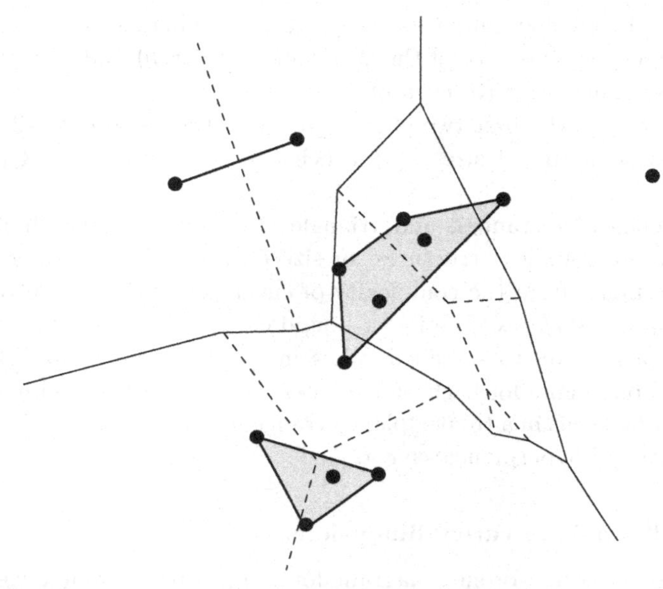

Figure 6.10. Voronoi diagram for 4 clusters under the Hausdorff distance. The region of each cluster C_i is partitioned by the *farthest-site Voronoi diagram* of its defining sites (i.e., by projected parts of the surfaces $\Sigma(C_i)$). The clusters are convex-hull disjoint in this example. Note that sites interior to cluster hulls do not contribute to the diagram and thus can be ignored.

in Xu *et al.* [709] to the case where clusters are (potentially intersecting) *isothetic* rectangles, and the Hausdorff distance is taken in the L_∞ metric. In fact, one obtains an *additively weighted* closest-site diagram (see Section 7.4) for rectangles. This diagram is motivated by the problem of critical area computation in VLSI design; see Papadopoulou and Lee [583].

Dehne *et al.* [250] gave an algorithm for the Voronoi diagram of hull-disjoint clusters of total size n, which runs in $O(n \log^4 n)$ time and $O(n \log^2 n)$ space. It is based on *divide & conquer* and therefore can be *parallelized* efficiently. The main difficulty lies in dealing with the merge chain in the conquer step, which can break into several, possibly cyclic, components. This cannot happen for the classical Voronoi diagram, see Section 3.3, which represents the special instance of singleton clusters $C_i = \{p_i\}$ for $i = 1, \ldots, n$. A recent *randomized incremental* approach, in Cheilaris *et al.* [186], achieves an (expected) runtime more than a $\log n$ factor faster, at optimal storage cost $O(n)$.

Another concept related to order-k Voronoi diagrams are so-called *2-site Voronoi diagrams*, considered in Barequet *et al.* [114]. The plane is now divided into regions minimizing some *measure* of a point x with respect to all possible pairs of sites $\{p, q\} \subset S$. This measure can be the sum, product, or difference of the two distances $d(x, p)$ and $d(x, q)$, or the area of the triangle xpq (or even others).

Interestingly, the first two choices just give the usual order-2 Voronoi diagram, a structure of size $O(n)$ which can be computed in $O(n \log n)$ time.

Difference of distances, and triangle area, on the other hand, lead to much more complex structures, of size $O(n^{4+\varepsilon})$. These upper bounds stem, once more, from the complexity of envelopes of the respective (non-linear) *Voronoi surfaces.* The latter bound can be sharpened to $\Theta(n^4)$, by exploiting a relation to so-called *zones* in an *arrangement of* $\binom{n}{2}$ *planes* in 3-space, one plane for each pair of sites. A *zone*, in this context, is the collection of all cells in a (hyper)plane arrangement which are cut by placing some additional hyperplane; see e.g. [300].

6.6. Medial axis in three dimensions

While the classical Voronoi diagram for point sites is well investigated and understood in higher dimensions, knowledge becomes quite sparse for sites of more general shape, concerning structural as well as algorithmic properties. This is already true for the seemingly harmless generalization from point sites to line segments or spheres.

The complexity of the *Voronoi diagram for n line segments* (and in particular, the *Voronoi diagram for straight lines*) in $\mathbf{R}^d$ may be as large as $\Omega(n^{d-1})$, as was observed by Aronov [67]. By a relation of Voronoi diagrams to *lower envelopes* of hypersurfaces (see Subsection 7.5.1), the results in Sharir [639] imply an upper bound of $O(n^{d+\varepsilon})$, for any $\varepsilon > 0$, on the size of this diagram. Based on that relation, a numerically robust algorithm for constructing the Voronoi diagram of straight lines in $\mathbf{R}^3$ has been designed recently in Hemmer *et al.* [402].

Surprisingly, such types of Voronoi diagrams are still small 'on the average'. Dwyer [297] showed that the *expected* combinatorial complexity of the Voronoi diagram of n uniformly distributed *k-flats* (affine k-dimensional subspaces of $\mathbf{R}^d$) is $O(n^{d/(d-k)})$ for $d \geq 3$. For example, the expected size of the Voronoi diagram for n lines (1-flats) in $\mathbf{R}^3$ is only $O(n\sqrt{n})$, and even less in higher dimensions.

6.6.1. *Approximate construction*

No better bounds than $\Omega(n^2)$ and $O(n^{3+\varepsilon})$ are known even in $\mathbf{R}^3$, which includes a case of particular interest in several applications, namely, the *medial axis* of a (generally nonconvex) polyhedron P in three dimensions. The arising *bisectors* contain, among other components, patches of parabolic and hyperbolic surfaces and have a fairly complicated structure. Regions are still *simply connected* (as in the plane, see Section 5.1), but even for three straight lines as sites, the induced structure gets so intricate that a separate paper has been devoted to its exploration; see Everett *et al.* [332]. The upper bound for the complexity of the diagram can be reduced to $O(n^{2+\varepsilon})$ if the straight line sites are confined to have only *constantly many orientations*; see Koltun and Sharir [470].

If the Euclidean distance function is replaced by some *polyhedral* (*convex*) *distance function* (Section 7.2), then the upper bound can be tightened to $O(n^2\alpha(n)\log n)$, even if convex polyhedra (of size $O(1)$) are allowed as sites; see Chew *et al.* [217] and Koltun and Sharir [469], respectively. A robust and practically efficient algorithm for computing the medial axis of a boundary-triangulated object in 3-space under a convex polyhedral distance function is described and implemented in Aichholzer *et al.* [26]. Figures 6.11 and 6.12 give an illustration of the output, when a tetrahedron is used as the unit ball for the convex distance function. Note that the resulting medial axis is a *piecewise linear* object now, and that, depending on the shape of the unit ball, the Euclidean medial axis will be resembled to a certain extent.

We will elaborate on the *approximate construction* of the Euclidean medial axis of a 3D object in this section — a notoriously difficult task with various algorithms and techniques proposed over the years. As the medial axis has evolved as a compact geometrical representation of shapes

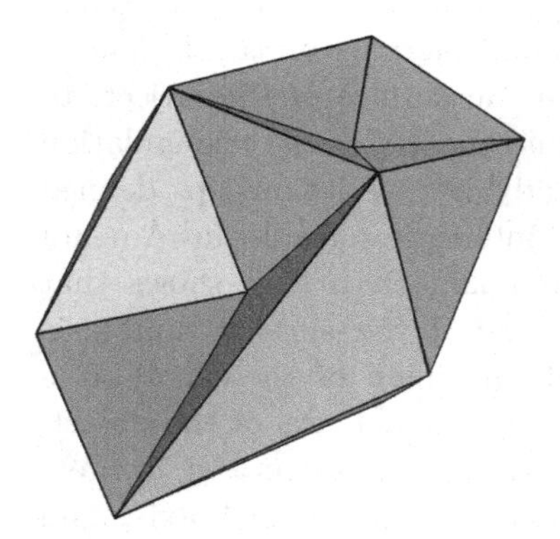

Figure 6.11. A nonconvex polyhedron with triangulated (i.e. simplicial) boundary.

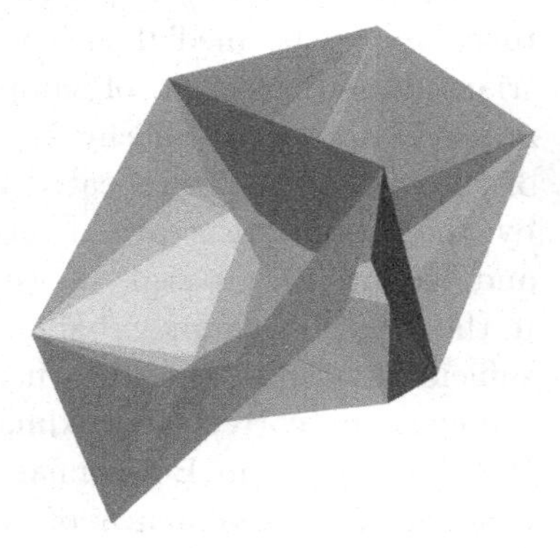

Figure 6.12. Piecewise linear medial axis under a tetrahedral distance function of Figure 6.11.

in diverse applications, there is need for an efficient and stable computation of this structure, or of suitable approximations thereof. Careful treatments of medial axes are given, from the geometric and stability point of view, in the survey article by Attali, Boissonnat, and Edelsbrunner [83], and from the applied point of view, in the recent book by Siddiqi and Pizer [654]. The latter source also covers the case of the *discrete medial axis*, i.e., its representation and computation as a *binary (pixel* or *voxel) image*; cf. Section 11.1.

Computing the *exact medial axis* of a polyhedron in 3-space is elaborate. A popular though time consuming approach are *tracing methods*, see e.g. Milenkovic [542], Sherbrooke *et al.* [646], and Culver *et al.* [237]. The edge skeleton of the medial axis is constructed piece by piece, basically at the cost of looking at the whole polyhedron at each step. These *continuous* methods do not require any sampling of the surface or volume of the input object. However, they are prone to numerical errors, due to the inherent algebraic complexity of the bisectors. Also, exactness turns out to be a disadvantage in certain applications, as small and unwanted features of the object (for example, small irregularities on the boundary) might get fully reflected in its medial axis.

Sampling methods, on the other hand, do not suffer from these drawbacks. In *spatial sampling*, the space of interest is subdivided into suitable cells, and an approximation of the medial axis is computed at the resolution of these cells; see Etzion and Rappoport [331] and Sud *et al.* [665]. This method is generally faster than continuous methods, though problems arise with appropriately handling (almost) degenerate situations, unless subdivision proceeds to a sufficiently fine level of detail.

Surface sampling approaches have received more attention. They try to compute the medial axis via the Voronoi diagram (or the Delaunay triangulation) of a set of sample points on the surface of the object. In an early approach, Sheehy *et al.* [645] refine the Delaunay triangulation of the sample points repeatedly, until the adjacency relationships defined by the medial axis are determined. An interesting result in Amenta and Kolluri [60], based on work in Amenta and Bern [58], shows that if the 'right' Delaunay balls are chosen, and if the surface sample is sufficiently dense, then the medial axis of the union of these balls is a topologically correct approximation of the true medial axis of the object. Related results, in Boissonnat and Cazals [140] and Amenta *et al.* [59], concern the convergence of Voronoi vertices to the medial axis, and the pruning of the Voronoi diagram with angle criteria (and others) in order to obtain an approximation of the medial axis; see also Dey and Zhao [271].

6.6.2. *Union of balls and weighted α-shapes*

Representing an object as a *union of balls* for approximate medial axis construction is a flexible and natural method, as it resembles the geometric definition of the medial axis as an infinite set of centers of *maximal inscribed balls*; see Subsection 5.5.2. This infinite set, or the corresponding collection of balls, is also called the *medial axis transform* (MAT) of an object, a term common in computer graphics and object modeling; see e.g. Siddiqi and Pizer [654].

The issues of how to choose an appropriate surface point sample whose Delaunay balls serve as a candidate set, and how to keep small the constructed set of balls, are treated among others in Amenta and Kolluri [60] and in Aichholzer *et al.* [35], respectively. The resulting algorithm simplifies the (topologically correct) medial axis automatically, via the choice of the approximating balls, and delivers a piecewise linear description. This method is also efficient, because the medial axis of a finite union of balls can be computed comparatively easily, applying an elegant structural result in Attali and Montanvert [86] and Amenta and Kolluri [61], which we describe next.

Let B be a finite set of balls in $\mathbf{R}^3$, and let U denote their union. By interpreting the center of each ball $b \in B$ as a point site p, and the squared radius of b as a weight for p, we can consider the *power diagram* (Section 6.2) of the resulting set of weighted sites.

We are interested in the faces of this power diagram that intersect the union of balls U. Each such j-face f, for $0 \leq j \leq 3$, corresponds to a dual simplex $\sigma(f)$ in the respective *regular tetrahedral complex*, namely, to the convex hull of the centers of the balls that define f. If *general position* is assumed, then $\sigma(f)$ is a $(3-j)$-simplex. Tetrahedra, triangles, line segments, and single points may occur as dual simplices. The collection of these simplices is called the *weighted α-shape*, $A(B)$, of the set B of balls, a concept introduced in Edelsbrunner [302] as a means for *molecular modeling* and *shape reconstruction.* See Figure 6.13 for a picture in one dimension lower.

$A(B)$ consists of various components, the full-dimensional ones (of dimension three in our case), and components of dimensions two, one, and zero. A face of $A(B)$ is termed *singular* if it is not part of any tetrahedron that contributes in forming the α-shape. Note that singular faces thus cannot be full-dimensional.

Theorem 6.8. *The medial axis of* $U = \bigcup_{b \in B} b$ *is composed of the singular faces of* $A(B)$, *plus the portion of the Voronoi diagram of* U*'s vertices that resides inside the non-singular parts of* $A(B)$.

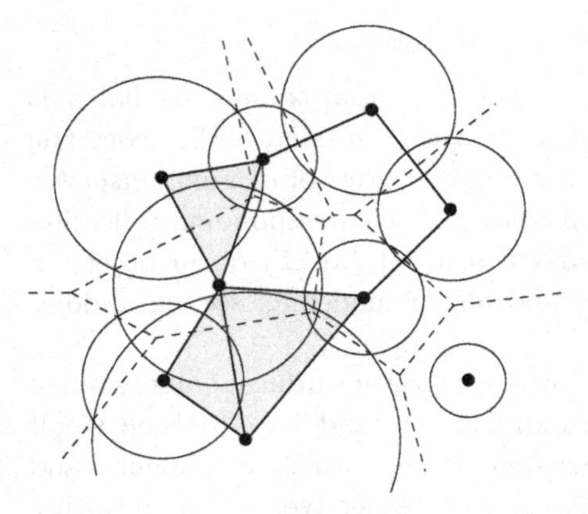

Figure 6.13. Power diagram for nine disks and their weighted α-shape. In this two-dimensional example, three line segments and three points (1 being isolated) occur as singular faces.

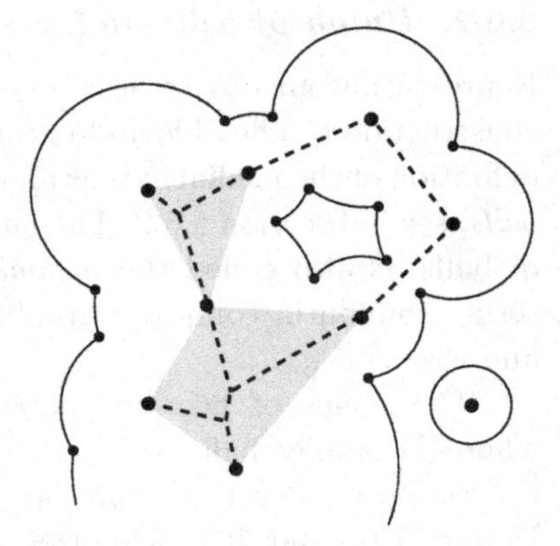

Figure 6.14. The medial axis of the union of these disks. Singular faces are copied, and non-singular parts are partitioned by the Voronoi diagram of the union's vertices.

Figure 6.14 gives an illustration. Theorem 6.8 does not only lead to construction methods for the medial axis of the union of balls in general dimensions (via power diagrams and Voronoi diagrams; Section 6.2), but it also makes explicit that this medial axis has a *piecewise linear* structure. Surprisingly though, its combinatorial complexity is unsettled; a trivial upper bound is $O(n^4)$, stemming from the $O(n^2)$ vertices of the balls' union. We refer the interested reader to the survey paper on the union of geometric objects by Agarwal *et al.* [15].

The level of approximation to the real medial axis (of the original 3D object) can be tuned by the number of Delaunay balls used; see Aichholzer *et al.* [35] for details.

In such a *medial ball representation* of an object, the Delaunay balls will usually intersect heavily. In the 2D case, Giesen *et al.* [364] could take advantage of this fact. They proved that, if the boundary of a planar shape P is sampled sufficiently densely, then the vertices of the resulting union U of Delaunay disks all need to be sample points. Consequently, the medial axis of U just consists of the edges and vertices of the sample points' Voronoi diagram that lie entirely inside P. It suffices to compute one Voronoi diagram, rather than the regular triangulation of the disks and the parts of the Voronoi diagram within its (non-singular) triangles.

In Subsection 8.2.2, an *unweighted* variant of α-shapes will be considered. In fact, if all ball radii are equal to a fixed value α, their power diagram is just the Voronoi diagram, $V(S)$, of the set S of their centers.

Now, if a face f of $V(S)$ intersects the boundary ∂U of their union U, any point $x \in f \cap \partial U$ is the center of a sphere of radius α that passes through f but does not enclose any other element of S. That is, the dual face $\sigma(f)$ (which is a face of $A(B)$) belongs to the Delaunay tetrahedrization DT(S), and $\sigma(f)$ is also part of the (unweighted) *α-shape of the point set S* according to its definition in Subsection 8.2.2.

This implies that the α-shape of a point set constitutes the boundary faces of the weighted α-shape for congruent radius-α balls. In $\mathbf{R}^2$, for example, it consists of all singular edges, plus all edges bounding the non-singular components.

Recall from Subsection 6.2.1 that the union of n balls is decomposed nicely by their power diagram. In $\mathbf{R}^2$, this leads to an $O(n \log n)$ algorithm for computing the union of n disks [107, 90] — even though their bounding circles may intersect in $\Theta(n^2)$ points. The *area of their union* can be calculated in additional $O(n)$ time. By contrast, the fastest known algorithm for computing the area of the union of n *triangles* runs in quadratic worst-case time. The latter task is one of the so-called *n^2-hard problems* listed in Gajentaan and Overmars [347]. Another important n^2-hard problem is testing a given set of n points in the plane for *general position*, with respect to collinearities. These problems are unlikely to have a subquadratic solution, and do so only if the following basic arithmetic problem (the 3SUM problem) does: Given a set M of n integers, do there exist numbers $a, b, c \in M$ with $a + b + c = 0$?

An intuitive reason why disks are easier to treat is that they are *homothets* of each other, while triangles in general are not. Another explanation is that disks and balls are *fat objects*, i.e., their ratio of diameter and *width* (size of smallest enclosing slab) is bounded by a constant (1 in that case). For fat convex objects, more efficient algorithms can be designed in many cases; see e.g. de Berg *et al.* [243] and Agarwal *et al.* [15].

Let us mention at this place that another flexible and versatile tool for making explicit the interaction between the balls in a finite set B can be based on power diagrams.

Generalizing to $\mathbf{R}^3$ a concept introduced in Cheng *et al.* [203], Edelsbrunner [303] defines the *ϱ-skin surface for B* as the envelope of an infinite set of balls derived from B by convex combination and shrinking (by a factor of ϱ). The family of all ϱ-skin surfaces, for $\varrho \in [0, 1]$, can be represented by a cell complex $\mathcal{C}$ in a horizontal strip in $\mathbf{R}^4$, defined by 'linearly interpolating' between the regular simplicial complex of B situated in the hyperplane $x_4 = 0$, and the power diagram of B situated in the hyperplane $x_4 = 1$. Now, intersecting $\mathcal{C}$ with $x_4 = \varrho$ gives a complex of convex three-dimensional cells — either shrunk copies of regular simplices or power cells, or polyhedra filling the gaps between. These cells decompose the ϱ-skin surface into spherical and hyperbolic patches.

Skin surfaces are smooth manifolds that vary continuously with the centers and radii of the balls. Application in *molecular modeling* and *shape morphing* are described in [303]. In particular, they allow to define a 'morphing distance' between shapes which, in the 2D case [203], can be used for automatic *pattern recognition* (e.g., for the classification of Chinese characters).

Constructing the family of power diagrams that arise during the morphing process is discussed in Chen and Cheng [199].

6.6.3. *Voronoi diagram for spheres*

Let us observe that the medial axis of a finite union U of balls is *not* a subset of the seemingly similar Voronoi diagram for their n bounding spheres. The former structure is defined by distances to the boundary ∂U of U, whereas the latter structure is based on distances to the individual spheres, and is of interest and importance it its own right.

The *Voronoi diagram for spheres* is a tessellation of the entire space, and is non-linear also in the exterior of its defining spherical sites. The bisector of two spheres is a sheet of a hyperboloid of rotation, bending towards the smaller sphere. (If both spheres have the same radius, their bisector is a hyperplane.) The diagram is composed of hyperbolic and possibly linear patches and has quite a complicated combinatorial structure. Though regions are *star-shaped* — as each of them is visible from the respective sphere center — they typically are multiply adjacent, and boundary facets may arise that are not simply connected.

In many applications, the focus is on the part of the diagram exterior to U. Given its importance in practice, for example in molecular biology (Will [706] and Angelov *et al.* [63]) and in material sciences (Goede *et al.* [365] and Naberukhin *et al.* [560]), the Voronoi diagram for n spheres in $\mathbf{R}^d$ has received only moderate attention.

A size of $O(n^{\lfloor d/2 \rfloor+1})$ can be shown, by its relationship to power diagrams in $\mathbf{R}^{d+1}$ proved in Aurenhammer and Imai [98], which can also be used to construct the diagram efficiently. More recently, algorithms similar in spirit but more convenient, have been devised in Will [706], Boissonnat and Karavelas [146], and Boissonnat and Delage [141]. They construct the diagram region by region, exploiting that a single region corresponds to a certain *convex hull of spheres*, which in turn can be computed by intersecting a power diagram with the unit sphere in one dimension higher. Interestingly, and unlike the classical Voronoi diagram or the power diagram, a single region can be as complex as the entire diagram, hence of size $\Theta(n^{\lfloor d/2 \rfloor+1})$ in $d \geq 3$ dimensions, giving $\Theta(n^2)$ in $\mathbf{R}^3$.

Different methods of construction exist for $d = 3$, being more friendly to implementation though theoretically slower; see Kim and Kim [453] and references therein. They first construct the classical Voronoi diagram for the centers of the spheres, and then let the centers gradually expand one by one into spheres of the required sizes, thereby updating their regions in the diagram.

Note that the Voronoi diagram for spheres is just the *additively weighted* (Euclidean closest-site) *Voronoi diagram* with the sphere centers as its sites, the weights being the sphere radii. Its two-dimensional variant, the Voronoi diagram for n circles, is of linear size because it consists of n star-shaped planar regions. We discuss this useful structure, along with several of its applications, in Section 7.4.

Chapter 7

GENERAL SPACES & DISTANCES

So far we have mainly discussed Voronoi diagrams of sites in d-space that are defined with respect to the *Euclidean distance function*. Now we want to generalize the space in which the sites are situated and/or the distance measure used; but we shall only discuss the case of point sites, apart from a few important exceptions.

The main questions are which of the structural properties the standard Voronoi diagram enjoys will be preserved, and if the remaining properties are strong enough to apply one of the algorithmic approaches introduced in Chapter 3 for computing the Voronoi diagram.

7.1. Generalized spaces

7.1.1. *Voronoi diagrams on surfaces*

Since the surface of earth is not flat, it seems very natural to ask about Voronoi diagrams of point sites on curved surfaces in 3-space. The distance between two points on the surface is the minimum Euclidean length of a curve that connects the points and runs entirely on the surface. Such a curve will be called a *shortest path* or a *geodesic path*.

Brown [170] has addressed the Voronoi diagram for point sites on the surface of a *sphere* in 3-space. Here great circles play the role of lines in the Euclidean plane. In fact, the bisector of two points is a great circle, and the shortest paths are segments of great circles, too. (One can show that the only other metric space in which all bisector segments are shortest paths is the hyperbolic space; see Busemann [174].)

For each pair of antipodal points on the sphere there is a continuum of shortest paths connecting them. But this does not affect this (*geodesic*) *Voronoi diagram* of n points; it can be computed in optimal $O(n \log n)$ time and linear space, by adaption of the algorithms mentioned in Chapter 3; see, in particular, the paper by Na *et al.* [559].

Given a line ℓ and a point p not on ℓ, in Euclidean geometry one can draw exactly one line through p that does not intersect ℓ, the line parallel to ℓ. In *hyperbolic geometry* many such lines exist. One model of hyperbolic geometry named after *Felix Klein* is given by the interior of the unit sphere; lines correspond to chords, and the distance between two points p and q is given by

$$d(p,q) = \operatorname{arccosh} \frac{1 - p \cdot q}{\sqrt{(1-|p|^2)(1-|q|^2)}}.$$

Nielsen and Nock [563] have observed that in this model bisectors are hyperplanes. Consequently, by the results in Subsection 6.2.3, *hyperbolic Voronoi diagrams* in Klein's model are power diagrams. Nilforoushan and Mohades [564] and, very recently, Bogdanov *et al.* [138] have considered the *Poincaré model* of hyperbolic geometry, where lines correspond to diameters and to circular arcs meeting the boundary of the unit sphere at a right angle. The latter paper discusses how to extract from the Euclidean Delaunay complex those simplices forming the *hyperbolic Delaunay complex*, in general dimensions d. Remarkably, these simplices can be computed with rational arithmetic for rational input points, a property well suited for implementation. That the hyperbolic Voronoi diagram can be derived from its Euclidean counterpart has first been observed in Devillers *et al.* [265]. It can be constructed in $O(n \log n)$ time and $O(n)$ space in the two-dimensional case.

Applications arise, for example, in meshing of high-genus surfaces, in information theory, and in routing on wireless ad hoc networks. See Tanuma *et al.* [679], who also consider several generalizations, e.g., for spheres and segments as sites.

Quite different is the situation on the surface of a *cone*, which we will discuss next. In order to determine the bisector of two points p and q, we can cut the cone along a halfline emanating from the apex, and unfold it; in Figure 7.1 the halfline diametrically opposed to p has been chosen. Since curve length does not change in this process, each shortest path on the cone that does not cross the cut is transformed into a shortest path in the plane, i.e., into a line segment. In order to represent those shortest paths that cross the cut, we add to the unfolded cone two more copies, as shown in Figure 7.1(ii). Now the shortest path on the cone from some point x to site q corresponds to the shortest one of the line segments $\overline{qx}$, $\overline{q'x}$, and $\overline{q''x}$. This explains why the unfolded bisector $B(p,q)$ consists of segments of the planar bisectors of p, q and p, q'.

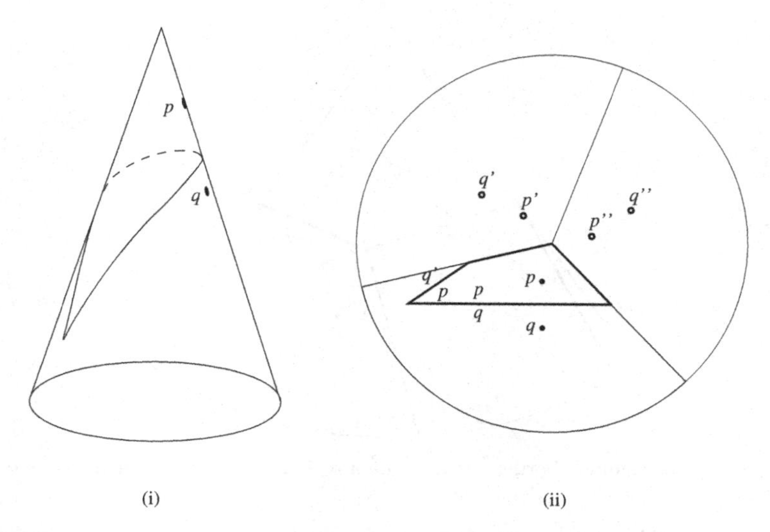

Figure 7.1. A cone sliced and unfolded, showing the bisector of p and q.

In spite of this strong connection to the plane, the Voronoi diagram of points on a cone has structural properties surprisingly different from the planar Voronoi diagram. If the unfolded cone forms a wedge of angle less than 180° then the bisector of two points can be a closed curve. If three points p, q, r are placed, in this order, on a halfline emanating from the apex of such a cone, the bisector $B(q, r)$ fully encircles $B(p, q)$ which in turn encircles the apex. This causes the Voronoi region of q in $V(\{p, q, r\})$ to be *not simply connected.*

Also, two Voronoi regions can be adjacent in more than one Voronoi edge. Such a situation is shown in Figure 7.2, on the unfolded cone. The bisectors of the three points cross twice, at the Voronoi vertices v and w; the latter happens to lie on the cut. (It is interesting to observe that none of these phenomena occurs on the sphere, although there, too, bisectors are closed curves and cross twice.)

Despite its more complex structure, the Voronoi diagram of n points on the surface of a cone can be constructed in optimal time $O(n \log n)$ and space $O(n)$, using a *sweep circle* that expands from the apex; see Dehne and Klein [247]. This approach works without unfolding the cone.

Mazón and Recio [526] have independently pointed out the algebraic background of the unfolding procedure illustrated by Figure 7.1(i), and obtained the following generalization. Let P denote the Euclidean plane or the sphere, and let G be a *discrete group of motions* on P, i.e., a group of bijections that leave the distance between any pair of points of

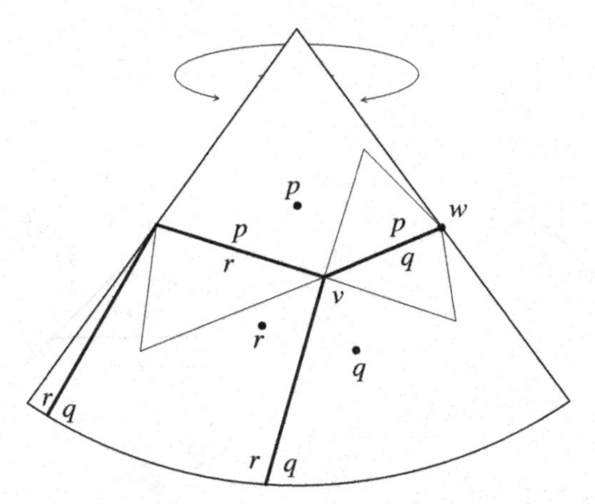

Figure 7.2. The common border of the Voronoi regions of q and r consists of two edges.

P invariant. We assume for G that, for each point $p \in P$, there exists a constant c satisfying

$$p \neq g(p) \Longrightarrow d(p, g(p)) \geq c$$

for all motions $g \in G$.

Examples in the plane are the group generated by a rotation of rational angle about some given point, or the group generated by two translations that move each point a fixed distance to the right and a fixed distance upwards, respectively. In the 19th century, mathematicians have completely classified all discrete groups.

Two points $p, p' \in P$ are *equivalent* if there exists a motion in G that takes p to p'; the equivalence class, $[p]$, of p is called the *orbit* of p. The *quotient space*, P/G, consists of all orbits. In order to geometrically represent P/G one starts with a connected subset of P that contains a representative out of every orbit; equivalent points must be on the boundary. Such a set is called a *fundamental domain*, if it is convex. The following lemma provides a nice way of obtaining a fundamental domain; a proof can be found in Ehrlich and Im Hof [318].

Lemma 7.1. *Let p be a point of P that is left fixed only by the unit element of G. Then its Voronoi region* $\mathrm{VR}(p, [p])$ *is a fundamental domain.*

In Figure 7.1(ii), for example, the point set $\{p, p', p''\}$ is the orbit of p under a clockwise rotation by 120°. The Voronoi region $\mathrm{VR}(p, \{p, p', p''\})$ equals the master copy of the unfolded cone, as drawn by solid lines. Each interior point x is the only point of $[x]$ contained in this region, only the points on the boundary (i.e., the cut) of the unfolded cone are mapped into each other by rotation. If we identify these two halflines we obtain the cone depicted in Figure 7.1(i), a model of the quotient space of the Euclidean plane over the cyclic group of order 3.

In a similar way we would obtain a rectangle as the fundamental domain of the group of two translations mentioned above, and identifying opposite edges would result in a *torus*.

If we want to compute the Voronoi diagram of a set S of point sites on a surface associated with such a quotient space P/G, we can proceed as follows. Let S_0 denote a set of representatives of S in a fundamental domain $D \subset P$. First, we compute the Voronoi diagram $V([S_0])$, where $[S_0]$ denotes the union of the orbits of the elements of S_0, an infinite but periodic set. Due to [526], $V([S_0])$ can be obtained by applying the motions in G to the Voronoi diagram of a finite set of points of S_0 and translated copies of S_0.

Theorem 7.1. *There exists a finite subset U of $[S_0]$ such that the identity $V([S_0]) = [V(U) \cap D]$ holds.*

If one removes from $V([S_0])$ all Voronoi edges that separate points of the same orbit and intersects the resulting structure with the fundamental domain D, the desired diagram $V(S)$ results, after identifying equivalent points. Although the set S_0 can be constructed effectively, it seems hard to establish an upper bound for the efficiency of this step.

Ehrlich and Im Hof [318] have studied, from a differential geometrist's point of view, structural properties of the Voronoi diagram in such *Riemannian manifolds* where any two points are connected by a unique shortest path.

Of particular interest, though of different flavor, is the discrete case of a *polyhedral* (i.e., *polygonal*) *surface* in 3-space, of combinatorial size m. Here, constructing Voronoi diagrams is closely related to computing single-source shortest paths by the *continuous Dijkstra technique*. (For Dijkstra's method, and graph search methods in general, see e.g. the book by Gibbons [363].) If we place those points of a surface patch in the same region whose shortest paths to the source have the same combinatorial structure (i.e., they visit the same edges and vertices in the same order) then Voronoi-like decompositions result. Conversely, the continuous Dijkstra algorithm can often be implemented to run efficiently for n sources simultaneously, so that for each point on the surface the nearest source is known; see, e.g., Schreiber and Sharir [622] for a recent result.

In general, the continuous Dijkstra technique allows the Voronoi diagram of n sites to be computed in $O(N^2 \log N)$ time and $O(N^2)$

space, where $N = \max\{m, n\}$; see Mitchell *et al.* [547]. Moet *et al.* [549] and Aronov *et al.* [68] have studied the complexity of *bisectors* and Voronoi diagrams on triangulated terrains. It turns out that, under realistic assumptions on the size and slope of the m terrain triangles, and on the density of their projections, bisectors consist of only $O(m)$ segments, and Voronoi diagrams are of complexity $O(n + m\sqrt{n})$. Driemel *et al.* [287] recently even improved this to linear, $O(n + m)$, if uniform distribution of the n sites is assumed.

An algorithm for computing the *Delaunay triangulation* on the surface of a compact polyhedron in 3-space is developed in Indermitte *et al.* [423], along with a possible modeling application in biology. They propose a *flipping* procedure and show that it halts after a finite number of flip operations.

7.1.2. *Specially placed sites*

The above setting (geodesic Voronoi diagrams on surfaces) should not be confused with the scenario where only the *sites* are confined to lie on a specific surface, and the classical Euclidean Voronoi diagram (or Delaunay triangulation) in 3-space is sought. The latter scenario arises in the study of realistic input models, and is motivated by *surface reconstruction algorithms* that compute the Delaunay triangulation of a set of sample points.

For *polyhedral surfaces*, the complexity of the 3D Delaunay triangulation is $O(n \text{ polylog } n)$ for random samples, when sites are uniformly distributed on the surface, see Golin and Na [368], in contrast to the $O(n)$ complexity for uniform distribution in 3-space, see Dwyer [296]. (The function polylog n stands for $(\log n)^c$, for some constant $c > 1$.) For random samples on certain surfaces of bounded curvature, including *cylinders* of constant radius and height, a complexity of $O(n\sqrt{n \log n})$ is shown in Erickson [328]. The case of cylinders is settled in Devillers and Goaoc [263], with a tight bound of $\Theta(n \log n)$.

As a recent result in $\mathbf{R}^d$, Amenta *et al.* [57] prove that for n point sites 'well distributed' on the boundary of a k-dimensional nonconvex polyhedron ($2 \le k \le d-1$), the size of the *Delaunay simplicial complex* is $O(n^{(d-1)/k})$. They even improve this bound slightly, to asymptotic tightness.

A *flipping* algorithm for the 2D Delaunay triangulation on the *flat torus* (i.e., periodic copies of the square, cf. Subsection 7.1.1) is described in Telley [681]. The 3D Delaunay triangulation on the flat torus (periodic cube) is treated algorithmically in Caroli and Teillaud [183]. They adapt the *incremental site insertion* method, and mention various application areas, including astronomy, material engineering, biomedical computing, and fluid dynamics.

In fact, Voronoi diagrams for *periodically placed sites* have been among the very first instances considered. Carl Friedrich Gauss, Gustav Lejeune Dirichlet, and later Georgi Feodosjewitsch Voronoi [693, 694] observed that quadratic forms have an interpretation in terms of Voronoi diagrams for parallelohedrally ordered sites in space. Moreover, certain tilings of space by convex polyhedra, generated by motions that form a *crystallographic group*, can be interpreted as Voronoi diagrams. Their regions are all congruent copies of each other, and have been called *special fundamental domains* by Arthur Schönflies [620] and *plesiohedra* by Boris Delone (Delaunay) [252]. In other words, such congruent '*space-filler*' polytopes can be found by means of computing Voronoi diagrams for carefully chosen sites. There has been a hunt for space-fillers with many facets over the years. We refer to the survey articles on tilings and geometric crystallography by Schulte [623] and Engel [320] for more material on this topic.

An important special case are Voronoi diagrams whose sites form a lattice. A *lattice* in d-space $\mathbf{R}^d$ is specified by its basis, i.e., a set of vectors $\{v_1, \ldots, v_d\}$ that span $\mathbf{R}^d$. The sites are obtained by integer combinations of the vectors from the lattice basis. The regions of such a Voronoi diagram are all translates of each other; they are *centrally symmetric* (convex) *polytopes* in $\mathbf{R}^d$. Thus, it is meaningful to talk of *the* Voronoi region of a given lattice — also called its *prototile* — or of its dual *lattice Delaunay polytope.*

Constructing the prototile $\mathrm{VR}(L)$ of a lattice L basically amounts to singling out its neighboring sites. This is a complicated task in higher dimensions; computing the number of vertices of $\mathrm{VR}(L)$ is an *#P-hard problem.* See Dutour *et al.* [294], who also give an algorithm that works efficiently in low dimensions. Earlier construction methods can be found in Viterbo and Biglieri [692]. Given its prototile, many parameters of a lattice can be determined. For example, its packing radius equals the *inradius* of $\mathrm{VR}(L)$, and its covering radius equals the *circumradius* of $\mathrm{VR}(L)$. Main applications of Voronoi diagrams for lattices arise in information theory and cryptography. Lattice Delaunay polytopes for positive *quadratic-form distances* (Subsection 7.4.5) are treated in Dutour and Rybnikov [293]; see also [692, 294] for further references and applications.

7.2. Convex distance functions

In numerous applications the Euclidean metric does not provide an appropriate way of measuring distance. In the following sections, we consider the Voronoi diagram of point sites under distance measures different from the Euclidean. We start with convex distance functions, a concept that generalizes the Euclidean distance but slightly. Whereas this generalization does not cause serious difficulties in the plane, surprising changes will occur as we move to 3-space.

7.2.1. *Convex distance Voronoi diagrams*

Let C denote a compact convex set in the plane that contains the origin in its interior. Then a *convex distance function* can be defined in the following way. In order to measure the distance from a point p to some point q, the set C is translated by the vector p. The halfline from p through q intersects the boundary of C at a unique point q'; see Figure 7.3. Now one puts

$$d_C(p,q) = \frac{d(p,q)}{d(p,q')}.$$

By definition, C equals the *unit disk* of d_C, that is, the set of all points q satisfying $d_C(0,q) \leq 1$. Sometimes, C is also referred to as the *shape* that defines d_C (cf. Subsection 7.2.2), or as the *calibration figure* or *Eichfigur* (a name used in Hermann Minkowski's work of 1927 on number theory [544]). The value of $d_C(p,q)$ does not change if both p and q are translated by the same vector.

One can show that the *triangle inequality* $d_C(p,r) \leq d_C(p,q) + d_C(q,r)$ is fulfilled, with equality holding for collinear points p, q, r. By definition, we have $d_C(p,q) = d_{C'}(q,p)$, where C' denotes the reflected image of C about the origin. We can define the Voronoi diagram based on an arbitrary convex distance function by associating, with each site p, all points x of the plane such that $d_C(p,x) \leq d_C(q,x)$ holds for all other sites q.

If the set C is (*point-*) *symmetric* about the origin then d_C is called a *symmetric convex distance function*. This is equivalent to saying that the function $q \mapsto d_C(0,q)$ is a *norm* in the plane.

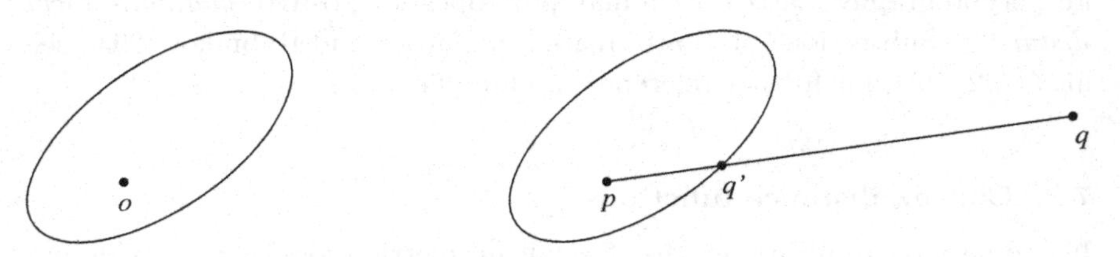

Figure 7.3. Defining a convex distance function d_C with unit disk C.

Well-known is the family of L_p (or *Minkowski*) *norms*, $1 \le p < \infty$, defined by the equation

$$L_p(q,r) = \sqrt[p]{|q_1 - r_1|^p + |q_2 - r_2|^p}, \quad \text{for } q = (q_1, q_2),\ r = (r_1, r_2).$$

Whereas L_2 is the *Euclidean distance*, the L_1-*norm* $L_1(q,r) = |q_1 - r_1| + |q_2 - r_2|$ is called the *Manhattan distance* of q and r, because it equals the minimum length of a path from q to r that follows a rectangular grid. Its unit circle is shown in Figure 7.4(i). If index p tends to ∞, the value of $L_p(q,r)$ converges to the *L-infinity norm*, $L_\infty(q,r) = \max\{|q_1 - r_1|, |q_2 - r_2|\}$, also called the *maximum norm*. The unit circle of L_∞ equals the aligned square $[-1,1]^2$; therefore, L_1 and L_∞ are related by a 45° rotation of the plane.

In Figure 7.4(ii) the bisector of two points under the Manhattan distance L_1 is shown. Another possible situation can be obtained by rotating picture (ii) by 90°. But if p and q are the diagonal vertices of an aligned square then their bisector $B(p,q)$ is no longer a curve: it consists of two quarterplanes connected by a line segment; see Figure 7.4(iii).

If the unit disk C is *strictly convex*, that is, if its boundary contains no line segment, this phenomenon cannot occur. In fact, we can construct the *bisector* of p and q in the way depicted in Figure 7.5. Let T and B denote the outer tangents to the two copies C_p, C_q of C, and let t, t' and b, b' be the respective tangent points. Consider a parallel line L between T and B. It intersects the boundaries of C_p, C_q in four points. Two of them, l and l', are facing each other. The halflines from p through l and from q through l' intersect in a point of the bisector $B(p,q)$, and each point of $B(p,q)$ can be obtained this way. Thus, line L moving from T to B establishes a

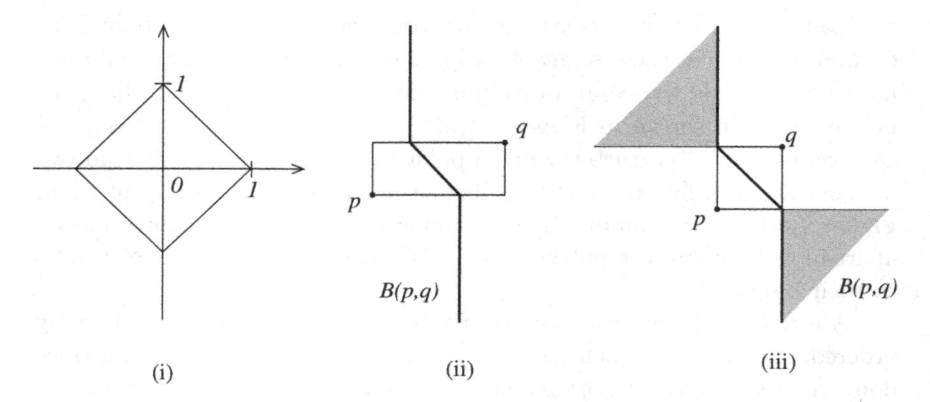

Figure 7.4. Unit circle and bisectors of the Manhattan distance L_1.

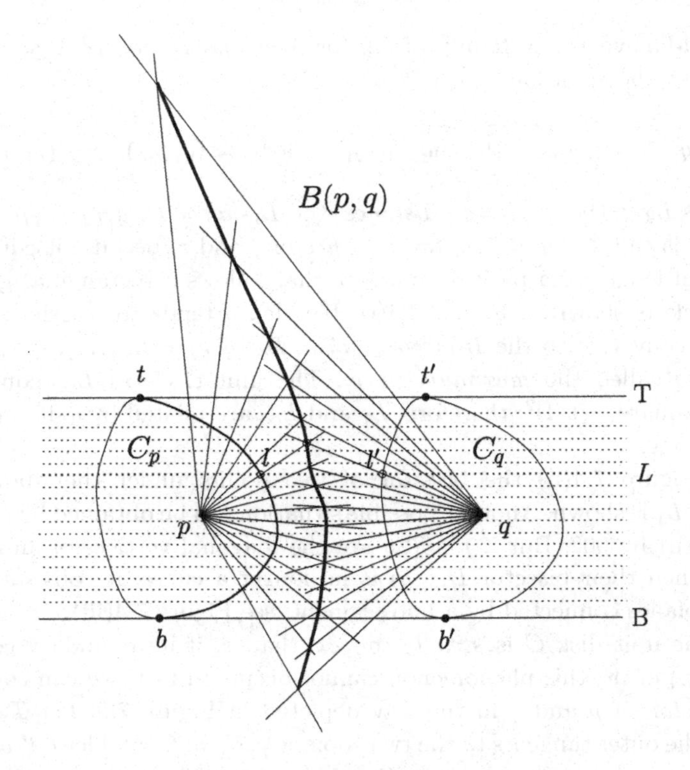

Figure 7.5. Constructing the bisector of p and q.

homeomorphism of $B(p,q)$ and the boundary part of C_p between t and b, which is homeomorphic to a line. Therefore, $B(p,q)$ is homeomorphic to a line, too.

We observe that the above argument would fail if C were flat at the top, so that t and t' were contained in horizontal line segments. Indeed, as C_p and C_q expand, these segments would, at some time, overlap, and their intersection would write out a two-dimensional bisector part. This happens in the situation shown in Figure 7.4(iii). To avoid this phenomenon, one can (symbolically) perturb the given point sites in such a way that no two of them define a line parallel to a line segment in the boundary of C. In Figure 7.4(iii), for example, if we imagine p and q to be slightly moved apart in x-direction, the polygonal curve drawn by three bold lines results as their bisector.

Alternatively, one can assume that the point sites in S are linearly ordered. If $p \prec q$ then the common boundary of the dominance domain $D(q,p)$ and $B(p,q)$ is chosen as a bisecting curve for p and q, i.e., the set $B(p,q)$ is added to the region of p with respect to q. This choice leads to a consistent definition of Voronoi diagrams. We shall employ it in Definition 7.5 in Section 7.3.

Although the bisector of two points can be described quite easily, it may have a complicated shape. Corbalan *et al.* [233] gave an example of a *strictly convex* set C, with a *smooth* boundary, and of two points p, q whose bisector under d_C 'wiggles' infinitely often between two parallel lines. (An object is called *smooth* if it has a unique tangent everywhere.) By slightly translating p and q, one obtains two bisector curves, $B(p, q)$ and $B(p', q')$, that cross infinitely often. When constructing Voronoi diagrams, some care is needed in dealing with this complication.

But even though the bisector $B(p, q)$ may wiggle, it is still confined to the bent strip defined by the halflines from p through t and b, and from q through t' and b'; see Figure 7.5 for notation.

For each strictly convex distance function, the *strict triangle inequality* holds: We have $d_C(p, r) < d_C(p, q) + d_C(q, r)$ unless p, q, r are collinear. Moreover, two circles with respect to d_C intersect in at most two points (thus being so-called *pseudo-circles*), and two *related* bisectors $B(p, q), B(p, r)$ intersect in at most one point; proofs can be found in Icking *et al.* [418].

Voronoi regions based on convex distance functions are in general not convex, as the example of the Manhattan distance shows. But they are still *star-shaped*, as seen from their sites: For each point $x \in D(p, q)$, the line segment $\overline{px}$ is also contained in $D(p, q)$. This follows from a more general fact shown in Lemma 7.4 later.

The star-shapedness of the Voronoi regions, together with the convexity of the circles, is a property strong enough for applying the *divide & conquer* algorithm; cf. Section 3.3. Hwang [416], Lee [493], and Lee and Wong [499] have studied the Voronoi diagram based on L_p norms (sometimes called L_p-*Voronoi diagram* for short); they provided algorithms that run within $O(n \log n)$ many steps. Here and in the sequel, a *step* not only denotes a single operation of a Real Random Access Machine (cf. [596]) but also an elementary operation like computing the intersection of two bisector curves.

Widmayer *et al.* [704] have described an optimal algorithm for computing the Voronoi diagram of a distance function based on a convex m-gon, C. This generalizes the Manhattan distance to an environment where motions are restricted to a *finite set of orientations*, given by the rays from the origin through the vertices of C. The diagram has a size of $O(mn)$ in this case, if individual line segments on bisectors are counted as edges.

Eventually, Chew and Drysdale [215] studied two-dimensional Voronoi diagrams based on general convex distance functions d_C. Aiming for an

$O(n \log n)$ divide & conquer algorithm, one crucial point is in the merge step. If we use a split line to subdivide the site set S into subsets L and R, the *merge chain* $B(L, R)$, consisting of all edges of the Voronoi diagram $V_{d_C}(S)$ that separate L-regions from R-regions, need not be connected. Fortunately, each of its connected components turns out to be an *unbounded* chain, not a loop. Therefore, a starting edge of each component can be found at infinity, as in the Euclidean case; see Section 3.3.

It is well known that each symmetric convex distance function d_C is *equivalent* to the Euclidean distance d, in the sense that for suitable constants a and A, the inequalities

$$a \cdot d(p, q) \leq d_C(p, q) \leq A \cdot d(p, q)$$

hold for all points p, q. In particular, a sequence of points p_i converges to some limit point p under d_C iff it does so under the Euclidean distance. One might wonder if these similarities cause the Voronoi diagrams of d and d_C to have similar combinatorial structures. A counterexample is shown in Figure 7.6; the regions of p and q are adjacent under L_2 but not under L_1.

Corbalán *et al.* [232] have provided systematic answers to the above question. Let d_C, d_D denote two symmetric and strictly convex distance functions whose unit circles are smooth. If for each set S of at most four points in the plane, $V_{d_C}(S)$ and $V_{d_D}(S)$ have the same structure as embedded *planar graphs*, then D must be a scaled version of C. More generally, if for each set S of at most four points the Voronoi diagram

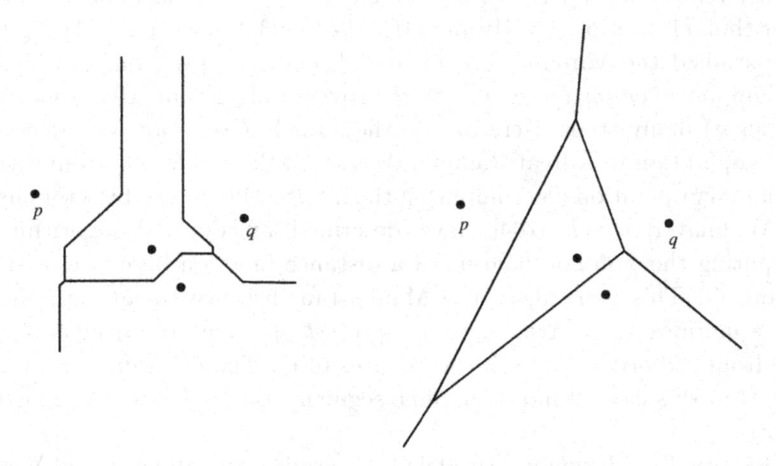

Figure 7.6. Four point sites whose Voronoi diagrams based on the Manhattan and the Euclidean distance have a different combinatorial structure.

$V_{d_C}(S)$ has the same combinatorial structure as $V_{d_D}(f(S))$, for some bijection f of the plane, then f is linear and $f(C) = D$ holds, up to scaling. Conversely, if f is a linear bijection of the plane, then the Voronoi diagram of $f(S)$ with respect to the convex set $f(C)$ can be obtained by applying f to $V_{d_C}(S)$, for each site set S.

The image of the Euclidean circle under a linear bijection is an ellipse. The above result implies that convex distance functions based on ellipses are the only ones whose Voronoi diagrams can be obtained from the Euclidean Voronoi diagram of a transformed set of sites. The following theorem gives an even stronger reason why such a reduction is not possible for unit circles other than ellipses.

Theorem 7.2. *Let C be a strictly convex, compact set, symmetric about the origin, which is not an ellipse. Then there exists a set of nine points in the plane whose Voronoi diagram with respect to d_C has a structure no Euclidean Voronoi diagram can achieve.*

In the proof given in [232], a Voronoi diagram based on d_C is constructed whose *dual tessellation* (see Subsection 7.2.2) is either of the topological shape shown in Figure 7.7(i), or of similar type. Let us assume that it can be realized by a Euclidean Delaunay tessellation; cf. Chapter 2. By construction, r lies outside the circumcircle of q, s, z, so that we have $\beta + \beta' < \pi$; see Figure 7.7(ii). Similarly, $\delta + \delta' < \pi$ holds. Since p lies on the circumcircle of w, q, z we have $\alpha + \alpha' = \pi$, and $\gamma + \gamma' = \pi$ holds for the same reason. The primed angles at z add up to 2π. Therefore, we obtain

$$\alpha + \beta + \gamma + \delta < 2\pi.$$

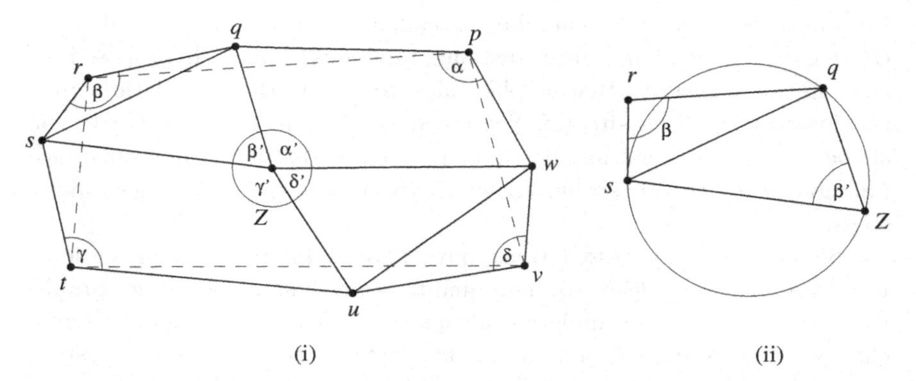

Figure 7.7. The graph in (i) is not a Euclidean Delaunay tessellation. In (ii) we have $\beta + \beta' < \pi$ iff r lies outside the circle. Equality holds iff r lies on the circle.

But this is impossible because the points different from z must be in *convex position* (because the convex hull of a point set equals the boundary of the unbounded face of its Delaunay triangulation), so that each of the angles at p, r, t, v includes an angle of the quadrilateral drawn by dashed lines. These angles add up to 2π.

7.2.2. *Shape Delaunay tessellations*

The Delaunay tessellation of a set S of sites can be defined in two ways: As the *geometric dual* of the Voronoi diagram of S, if we connect two sites in S by a dual edge iff their Voronoi regions are adjacent; or by using the *empty circle property*: An edge $\overline{pq}$ between two sites $p, q \in S$ is a Delaunay edge iff there exists some circle that passes through p and q but encloses no other site in S.

For the Euclidean distance, these two definitions are equivalent. However, for general convex distance functions d_C, this is no longer true, because the symmetry property is not guaranteed. More specifically, in the definition of the bisectors $B(p, q)$ of the diagram $V_{d_C}(S)$ (Subsection 7.2.1), distances *from* the sites are measured, whereas for the empty-circle property, distances are taken from the center of the convex shape C, i.e., *towards* the sites. Since we have $d_C(p, q) = d_{C'}(q, p)$, where C' denotes the reflected image of C about its center, the dual of $V_{d_C}(S)$ is just the (edge-based) 'empty-circle' Delaunay tessellation for the shape C'. This fact has a surprising consequence: The dual of $V_{d_C}(S)$, and thus, the combinatorial structure of $V_{d_C}(S)$, is invariant under movements of the center of C. The geometric structure changes, of course.

Drysdale [288] considered such so-called *shape Delaunay tessellations*, and showed how the *divide & conquer* approach in Guibas and Stolfi [394] (see also Section 3.3) can be extended to run, in optimal time $O(n \log n)$, for general convex distance functions. Ma [514] proposed an *incremental algorithm* that applies flips to repair the tessellation after the insertion of each site (cf. Section 3.2). Skyum [656] constructs the shape Delaunay tessellation, based on a *smooth* and *strictly convex* distance function, using the *sweep line approach* (Section 3.4), in $O(n \log n)$ many steps.

Shape Delaunay tessellations have several interesting applications. For example, they lead to high-quality and dynamic spanner graphs (Subsection 8.2.4). They undergo only a nearly quadratic number of discrete changes when the sites are *moved* on algebraic trajectories of constant description; see Agarwal *et al.* [12], and Section 9.1 on Voronoi diagrams for moving sites.

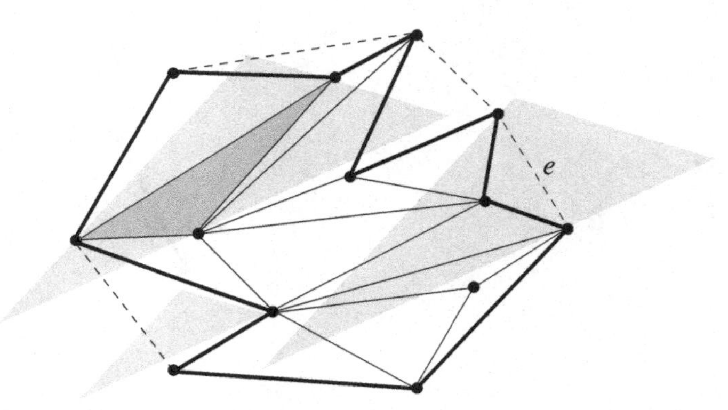

Figure 7.8. A shape Delaunay tessellation where C is an obtuse triangle (whose copies are shown lightly shaded). The support hull is emphasized. Note the direction-sensitivity of the tessellation's triangles according to C.

If the shape C is not strictly convex and smooth, then the shape Delaunay tessellation, for short $\mathrm{DT}_C(S)$, is *not* a full triangulation of the convex hull of S. Consult Figure 7.8: An edge e between two sites in S that can serve as a chord of any non-empty homothet of C will not be part of $\mathrm{DT}_C(S)$. (A *homothet* of C is an object that can be obtained from C by translation and scaling; rotations are not allowed.) The subset of the plane covered by $\mathrm{DT}_C(S)$ is called the *support hull* of S (with respect to d_C) in [288], generalizing the hull concept in Lee [493] for L_p-metrics. The support hull is connected, as it has to contain a certain spanning tree of $\mathrm{DT}_C(S)$, see later, and $\mathrm{DT}_C(S)$ is a full triangulation therein.

We remark that if the definition of $\mathrm{DT}_C(S)$ were based on its *triangles* rather than on its *edges* (as above), then more edges were possibly included, dual to unbounded parallel edges of $V_{d_{C'}}(S)$. In particular, the corresponding support hull could enlarge.

Flipping edges in shape Delaunay tessellations is considered in detail in Paulini [588]. In principle, the empty shape property can be used to decide 'Delaunayhood' of an edge (cf. Section 4.2), but some care has to be taken. Let e be any edge spanned by the point set S. For the given convex shape C, consider its homothetic copy of infinite size, and denote with C_L and C_R its two translates having the endpoints of e on the boundary, from the left and the right side, respectively. (Without loss of generality, we assume that edge e is not horizontal.) We distinguish three disjoint (open) domains in the plane $\mathbf{R}^2$ as shown in Figure 7.9, defined as

$$I = C_L \cap C_R, \quad II = \mathbf{R}^2 \setminus (C_L \cup C_R), \quad \text{and}$$
$$III = (C_L \cup C_R) \setminus (C_L \cap C_R).$$

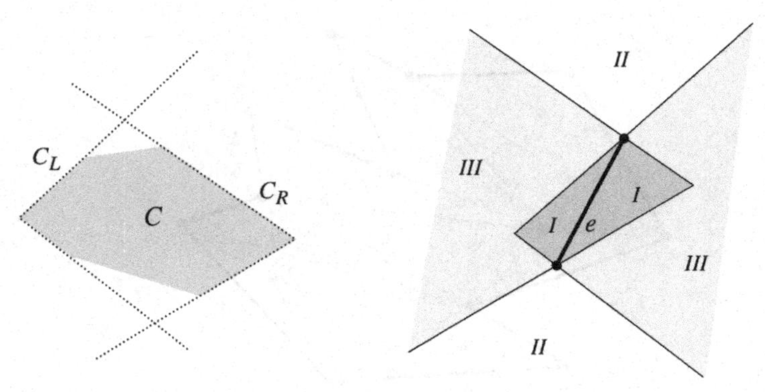

Figure 7.9. Infinite copies of a polygonal shape C (left), and the three domains they induce for a particular edge e (right).

Clearly, if the shape C is smooth (for example, a circular disk, as for the classical Delaunay triangulation), then C_L and C_R are just the two open halfplanes bounded by the straight line ℓ supporting e, and the distinction above is not meaningful, because $I = II = \emptyset$, and $III = \mathbf{R}^2 \setminus \ell$ is the only domain.

For the non-smooth case, if domain I contains any point from S, then edge e cannot be an edge of $\mathrm{DT}_C(S)$. Also, no point in domain II can form a valid triangle with edge e, whereas a point p in domain III can do so, because an (otherwise empty) homothet covering e and p might exist. In the case where S, apart from e's endpoints, lies entirely within the domain II, edge e is part of the support hull of S, from both sides. Indeed, this situation might happen for all valid edges, and $\mathrm{DT}_C(S)$ then degenerates to a tree, containing no valid triangles.

It is well known that any given triangulation T of S can be transformed into the (classical) Delaunay triangulation $\mathrm{DT}(S)$ of S, by applying a sequence of edge flips based on the 'local Delaunayhood' of an edge (see Section 4.2 and Subsection 6.3.3). Local Delaunayhood is defined via the *empty-circle property*, but can be equivalently characterized by an angle criterion. Let Q be any convex quadrangle in T, given as the union of two triangles Δ_1 and Δ_2 in T. These triangles share an edge e. We flip e, and consider the two triangles Δ_3 and Δ_4 that the quadrangle Q is split into now. Then, if e was not *locally Delaunay*, the smallest angle arising in Δ_1 and Δ_2 is smaller than the smallest angle arising in Δ_3 and Δ_4. That is, the *angularity* of T increases lexicographically. This property is, unfortunately but not surprisingly, lost for general convex shapes C, and a related criterion that maintains the analogy between $\mathrm{DT}(S)$ and $\mathrm{DT}_C(S)$

is of interest. Indeed, for DT(S) we also have the inequalities

$$\max\{r_1, r_2\} < \max\{r_3, r_4\} \quad \text{and} \quad \min\{r_1, r_2\} < \min\{r_3, r_4\},$$

where r_i denotes the *circumradius* (radius of the *circumcircle*) of triangle Δ_i for $i = 1, 2, 3, 4$. While there are counterexamples for the min-inequality for general convex shapes C, the max-inequality still holds [588]. Thus, each flip in the triangulation T lexicographically reduces the sorted vector of triangle circumradii (i.e., scaling factors of C). It also can be shown, by adapting the proof of Theorem 4.3 in Section 4.2, that T can always be flipped into $\mathrm{DT}_C(S)$: Arbitrarily creating edges which are locally Delaunay with respect to C will terminate in a unique structure. As a consequence, $\mathrm{DT}_C(S)$ minimizes the *largest circum-shape*, for all possible triangulations of the point set S.

Note that the *smallest enclosing shape* of a triangle need not be unique. Alonso *et al.* [48] discuss this and related issues in detail.

Lambert [482] elaborated on the question about which flip operations, based on a given set of convex shapes, guarantee a unique locally optimal triangulation (i.e., one where no improving flips do exist). He proved that the underlying set of shapes has to be a homothetic family, as it is the case for $\mathrm{DT}_C(S)$.

The algorithmic question of how many improving flips are required to transform a given triangulation T of S into $\mathrm{DT}_C(S)$ remains open. (Of course, the lower bound of $\Omega(n^2)$ for the classical Delaunay triangulation in Subsection 6.3.3 carries over, if C can be any convex shape.) Of interest is also the corresponding *recognition problem* for shape Delaunay tessellations: Given T, does there exist a convex shape C such that $T = \mathrm{DT}_C(S)$? This problem was recently shown to be *NP-hard* [607].

The Delaunay triangulation is not only useful because its triangles optimize angles and circumradii (and other quantities), but also because it contains various subgraphs which arise in different applications. The minimum spanning tree, the nearest neighborhood graph, and the Gabriel graph are among these subgraphs; see Section 8.2. When considering whether such structures extend to general convex distance functions d_C, we encounter two problems: First of all, the function d_C is not a metric unless the shape C is (*point-*)*symmetric*. Also, d_C depends on the chosen center v of C, whereas the triangulation $\mathrm{DT}_C(S)$ does not. Still, a convex distance function can be constructed that is center-independent and symmetric, and which yields the desired subgraph property for *minimum spanning trees* and *Gabriel graphs* [588], as we describe next.

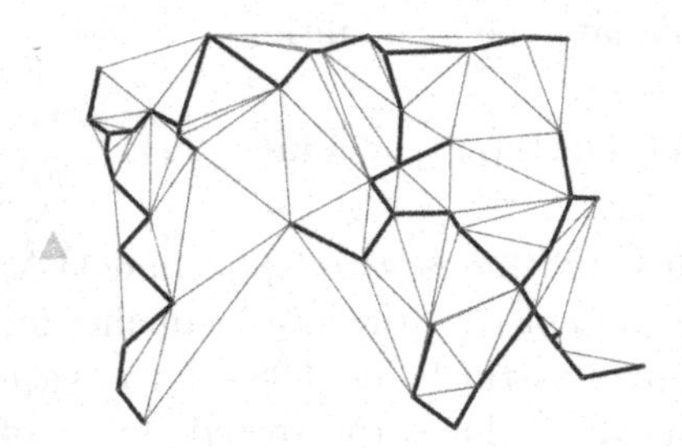

Figure 7.10. $\mathrm{DT}_C(S)$ for triangle shape C contains the minimum spanning tree for the metric m_C.

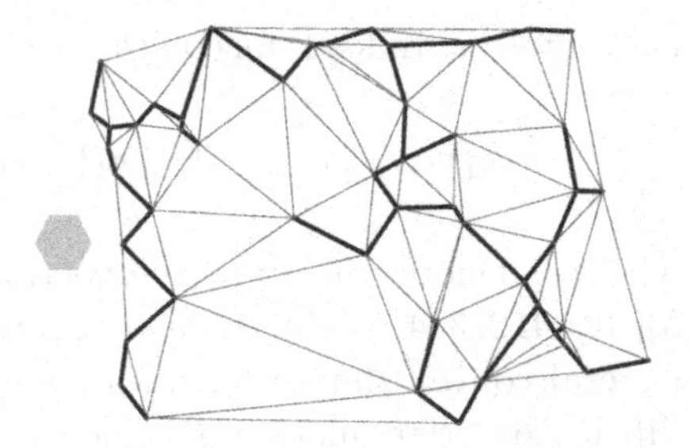

Figure 7.11. $\mathrm{DT}_{\hat{C}}(S)$ for Minkowski sum shape $\hat{C}$ contains the same minimum spanning tree (from [588]).

Assume that C is centered at the origin, and denote with C_v, for $v \in C$, the translation of C by v. Given two points p and q, we define the distance function

$$m_C(p,q) = \min_{v \in C} d_{C_v}(p,q).$$

Observe that $m_C(p,q)$ equals the Euclidean distance $d(p,q)$, divided by the length of the longest line segment contained in C and parallel to the edge $\overline{pq}$. Thus the concept of the, in general, non-symmetric *radius* (see the beginning of Subsection 7.2.1) is replaced by the concept of symmetric *diameter*. That is, $m_C(p,q)$ is just the scaling factor of the smallest homothet of C that covers $\overline{pq}$. To see that we really deal with a convex distance function, observe further that

$$m_C(p,q) = d_{\hat{C}}(p,q), \quad \text{with } \hat{C} = \bigcup_{v \in C} C_v.$$

The symmetric convex shape $\hat{C}$ is just the *Minkowski sum* of C and its reflected shape. (The Minkowski sum of two sets A and B in the plane is defined as $A \oplus B = \{a + b \mid a \in A,\ b \in B\}$. Loosely speaking, $A \oplus B$ is the union of all translates of B that overlap with A.)

Coming back to the question of subgraphs, we can now use the following facts: For any symmetric shape, and in particular for $\hat{C}$, the equality $m_{\hat{C}}(p,q) = \frac{1}{2}d_{\hat{C}}(p,q)$ holds, such that the two metrics $m_{\hat{C}}$ and $d_{\hat{C}}$ yield the same minimum spanning tree. Further, we have $m_{\hat{C}}(p,q) = \frac{1}{2}m_C(p,q)$, for general convex shapes C. In conclusion, the minimum spanning tree with respect to the metric $m_{\hat{C}}$, which is contained in $\mathrm{DT}_{\hat{C}}(S)$ because $m_{\hat{C}}$ is symmetric, is the minimum spanning tree of S with respect to m_C. This tree is indeed a subgraph of the original shape Delaunay triangulation, $\mathrm{DT}_C(S)$, as can be easily proven by using the empty-shape property of $\mathrm{DT}_C(S)$. Figures 7.10 and 7.11 illustrate these facts. Notice that $\mathrm{DT}_C(S)$

and $\mathrm{DT}_{\hat{C}}(S)$ can be quite different triangulations, even concerning their support hulls.

Similar subgraph properties also hold for the *Gabriel graph* of S (a supergraph of the minimum spanning tree), because this graph is defined by emptiness of the smallest homothet, $C_{\overline{pq}}$, of C that covers an edge $\overline{pq}$. Interestingly, the Gabriel graph coincides with $\mathrm{DT}_C(S)$ for any *triangular* shape C, because then the 'forbidden' domain I for an edge $\overline{pq}$ (compare Figure 7.9) just is $C_{\overline{pq}}$.

7.2.3. *Situation in 3-space*

The definition of a convex distance function, and in particular of the L_p-norms, can easily be extended to dimensions higher than 2. In this subsection we study *convex distance functions in* 3*-space.* Such a distance function is based on a convex, compact set C in 3-space which contains the origin in its interior. We call such a unit sphere C *good* if it is, in addition, *strictly convex, symmetric*, and *smooth.*

Some of the pleasant properties of two-dimensional convex distance functions still have their counterparts in 3-space. For example, the *bisector* $B(p,q)$ of two points is a surface homeomorphic to the plane, and two such bisector surfaces $B(p,q), B(p,r)$ have an intersection homeomorphic to the line, or empty. In these aspects, convex distance functions in 3-space do not differ from the Euclidean distance. But whereas a Euclidean sphere is uniquely determined by four points in space, this is no longer true for convex distance functions. In Icking *et al.* [418] the following result has been shown; consult Figure 7.12.

Theorem 7.3. *For each* $n > 0$ *there exist a good convex set* C *and four points in* 3*-space, such that there are* $2n + 1$ *homothetic copies of* C *containing these points in their boundaries. This number does not decrease when the four points are independently moved within small neighborhoods.*

The center v of a scaled and translated copy of C containing p, q, r, s in its boundary is of the same d_C-distance to each of these points. Hence, v is a Voronoi vertex in the diagram $V_{d_C}(\{p,q,r,s\})$. Therefore, Theorem 7.3 implies that there exists *no global upper bound* to the number of vertices of the Voronoi diagram of four points in 3-space, that holds for arbitrary convex distance functions.

If it is known that for a particular convex distance function d_C no more than k homothetic copies of C can pass through any four points in *general position*, then, obviously, the complexity of the Voronoi diagram of n points

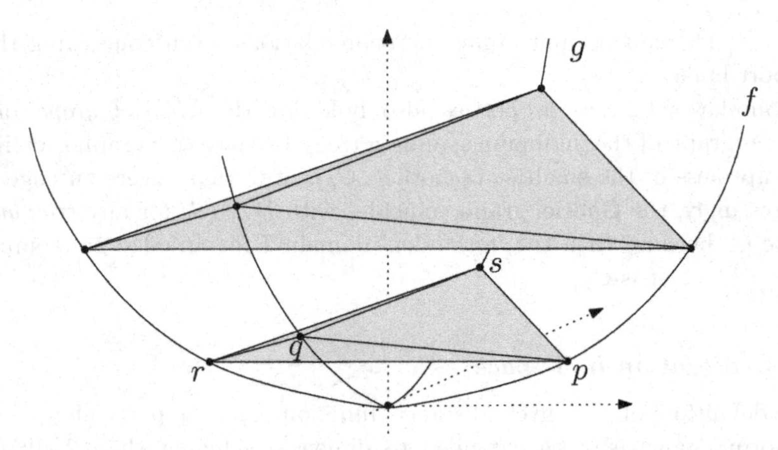

Figure 7.12. As vertices r and p move upward along the convex curve f, while the tetrahedron T defined by $\{r, p, q, s\}$ stays homothetic, vertices q and s write out a convex curve, g. The convex hull of f and g forms the lower half of a convex body C. Since C touches the vertices of infinitely many homothetic copies of T, an infinite number of homothetic copies of C are passing through $\{r, p, q, s\}$. This construction can be tuned to yield a proof of Theorem 7.3.

based on d_C is in $O(kn^4)$, because at least four Voronoi regions meet at a Voronoi vertex.

Lê [487] has obtained the following result on the number k for the family of *L_p-norms*, using the fact that they are defined by algebraic equations of degree p. The proof uses results from the theory of additive complexity, see Benedetti and Risler [120]. This and the subsequent results require that the sites be in general position.

Theorem 7.4. *There exists an upper bound to the number of L_p-spheres in d-space that can pass through $d+1$ given points, which depends only on d but not on p.*

The fourth power of n in the general upper bound of $O(kn^4)$ for Voronoi diagrams in dimension 3 is believed to be far off. For semi-algebraic distance measures of constant degree, a general result by Sharir [639] on lower envelopes implies an upper complexity bound of only $O(n^{d+\varepsilon})$. This result applies to the L_p family, but the constant in O tends to ∞ as p grows.

For the three-dimensional Euclidean Voronoi diagram, the true worst-case complexity is only $\Theta(n^2)$; see Section 6.1. Improving on previous work by Boissonnat *et al.* [147] and Tagansky [673], Icking and Ma [417] were able to generalize this tight bound to *polyhedral convex distance functions.*

Theorem 7.5. *Let C be a fixed convex polyhedron in 3-space. Then the Voronoi diagram with respect to d_C of n points is of complexity $O(n^2)$.*

The constant in O is proportional to $k^3\alpha(k)$, where k denotes the complexity of C, and α denotes the (extremely slowly growing) inverse of *Ackermann's function.* The proof makes use of the fact that such edges of bisector surfaces, that are generated by an edge of C, can be of at most k different slopes.

Boissonnat *et al.* [147] have also shown that the Voronoi diagrams of n points in d-space based on the L_∞*-norm*, or on a *simplex* as the unit sphere, are both of complexity $\Theta(n^{\lceil d/2 \rceil})$.

The unit spheres for the L_∞-norm and the L_1-norm are *hypercubes* and *cross-polytopes* (i.e., duals of hypercubes), respectively. (A hypercube in d-space is the Cartesian product of d mutually orthogonal line segments of equal length; for example, a square in the plane, or a cube in 3-space where the dual cross-polytope is the regular octahedron.) In contrast to hypercubes, cross-polytopes have a large number of facets in high dimensions, which partially explains why tight bounds for L_1-diagrams are only available for dimensions up to three.

For the complexity of the *Voronoi diagram for lines* in 3-space under the *Euclidean* distance only an $O(n^{3+\varepsilon})$ upper bound is known [639] (unless the *number of orientations* is bounded [470]). Confirming the impression that polyhedral distance functions are somewhat simpler than the Euclidean distance, Chew *et al.* [217] were able to provide a smaller bound for lines under a polyhedral convex distance function.

Theorem 7.6. *Let C denote a convex polyhedron of constant complexity in 3-space. Then the Voronoi diagram of n lines based on d_C is of complexity $O(n^2\alpha(n)\log n)$. A lower bound is given by $\Omega(n^2\alpha(n))$.*

Koltun and Sharir [469] have generalized Theorem 7.6 to *line segments.* For more general sites, they obtained the following.

Theorem 7.7. *Let C denote a convex polyhedron of constant complexity in 3-space. Then the Voronoi diagram of polyhedral sites of total complexity n has a size of $O(n^{2+\epsilon})$.*

This includes the practically relevant case of the *medial axis* of a (nonconvex) polyhedron P in 3-space, under a convex polyhedral distance function. If some point in the 'unit ball' C is fixed as its center, then the medial axis is the union of the centers of all *homothets* of C that can be inscribed in P. Defined this way, the medial axis is a piecewise linear set, which can be constructed efficiently; see Aichholzer *et al.* [26].

The reader interested in purely mathematical properties of convex distance functions is referred to the survey by Martini *et al.* [517, 518] and to the monograph by Thompson [684].

7.3. Nice metrics

Convex distance functions do not apply to environments where the distance between two points can change under translations. This happens in the presence of obstacles, or in cities whose streets do not form a regular grid. In order to model such environments in a more realistic way, we can use the concept of a metric.

In this section we consider point sites and Voronoi diagrams that are based on a *metric*, m, in the plane. (However, a metric can be defined in arbitrary spaces). It associates with any two points, p and q, a non-negative real number $m(p, q)$ which equals 0 if and only if $p = q$ holds. Moreover, we have *symmetry* $m(p, q) = m(q, p)$, and the *triangle inequality* $m(p, r) \leq m(p, q) + m(q, r)$. The set

$$\{z \in \mathbf{R}^2 \mid m(a, z) < r\}$$

will be called an (open) m-*disk* of radius r centered at point a.

General metrics are quite a powerful modeling tool. Suppose there is an air-lift between two points a and b in the plane. Then

$$m(p, q) = \min \begin{cases} d(p, q), \\ d(p, a) + f + d(b, q), \\ d(p, b) + f + d(a, q), \end{cases}$$

describes how fast one can travel from p to q; we assume that going by car takes time equal to the Euclidean distance, d, whereas the flight between a and b takes time $f < d(a, b)$. It is easy to verify that the function m is a metric in the plane: We could warp the plane so that a and b have Euclidean distance f in 3-space, and glue on a straight handle connecting them; the function m describes the lengths of *shortest paths* in the resulting space.

7.3.1. *The concept*

Voronoi regions with respect to the *air-lift metric* m need not be connected. Suppose site q lies on the line segment $\overline{ab}$, as shown in Figure 7.13. In order to construct the Voronoi diagram of the two sites a and q we use the *expanding waves* approach mentioned in Chapter 2, and let two Euclidean

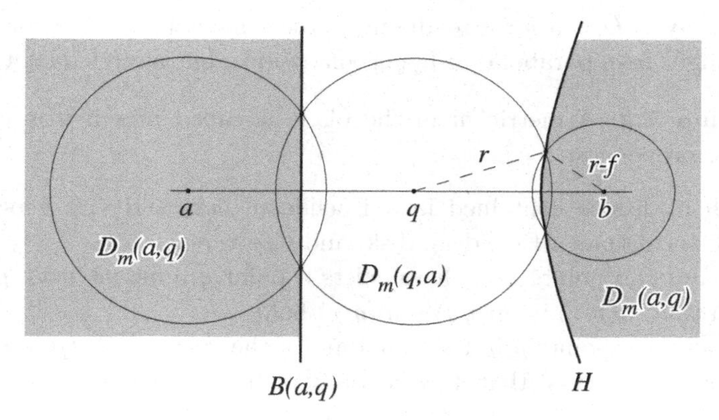

Figure 7.13. If there is an air line connecting a and b, the Voronoi region of a with respect to q is disconnected.

circles expand from a and q at the same speed. Their intersections form the Euclidean bisector $B(a,q)$. As soon as the radius of these circles has reached f, the duration of the flight from a to b, a new circle starts growing from b. Subsequently, its radius is always by f smaller than the radius r of the other two circles. The intersections of the circles expanding from q and from b form a hyperbola, H; it is the locus of all points x satisfying $d(q,x) - f = d(b,x)$.

Such Voronoi diagrams of n points with *additive weights* are also known as the *Voronoi diagram for circles.* They often appear quite unexpectedly. One can construct them in time $O(n \log n)$ by different algorithms; see Section 5.5, and especially Subsection 7.4.1.

Clearly, the region of a with respect to air-lift metric m, $D_m(a,q)$, consists of the two parts shaded in Figure 7.13; they are not connected. In order to exclude such phenomena we restrict ourselves to a subclass of metrics introduced in Klein and Wood [467]. First, we define the type of bisector curve we would like to work with.

Definition 7.1. A curve J in the plane is called *nice* if its stereographic projection to the sphere can be completed to a Jordan curve through the north pole. (A *Jordan curve* on the sphere, or in the plane, is a closed curve without self-intersections.)

A nice curve J in the plane is *simple* (i.e., no self-intersections), unbounded, and homeomorphic to a straight line. It splits the plane into two unbounded, open domains, D_1 and D_2. Both of its ends tend to infinity. Just as each point p on J can be accessed from D_i, $i = 1, 2$, by an arc that runs entirely in D_i but for its endpoint p, the same holds for the north pole. Any straight line, parabola, or hyperbola is nice, but a circle is not.

Definition 7.2. A metric m in the plane is called *nice* if it enjoys the following properties:

(1) Each m-disk is contained in a Euclidean disk, and vice versa; each m-disk contains a Euclidean disk, and vice versa.
(2) For any two points p, r, there exists a point q different from p and r such that $m(p,r) = m(p,q) + m(q,r)$ holds.
(3) For any two points p, q, the boundary of the bisector $B_m(p,q)$ consists of two nice curves (that may partially or fully coincide).

In the sequel we shall discuss some implications of these properties that will be relevant for Voronoi diagrams later on.

Property (1) of Definition 7.2 ensures that 'close' and 'far' mean the same under m as in the Euclidean metric d, only the quantification of *how* close or far may differ. The open sets are the same for either metric.

Moreover, the plane endowed with the metric m is *complete*, in the following sense. If a sequence $(a_i)_{i>0}$ fulfills the *Cauchy condition* under metric m (that is, if for each $\varepsilon > 0$ there is an index threshold N such that $m(a_i, a_j) < \varepsilon$ holds for all $i, j \geq N$), then a limit point a^* exists in the plane to which the sequence converges.

Property (2) implies that one can always find a point q different from p, r that makes the triangle inequality

$$m(p,r) \leq m(p,q) + m(q,r)$$

an equality. Such a point q is said to lie 'between' p and r. Once q is found, one can find points in between p and q, and between q and r, and so on. Thanks to the completeness guaranteed by (1), these points form, in the limit, a path π from p to r, which has a remarkable property.

Definition 7.3. A path π is called *m-straight* if for any three consecutive points a, b, c on π the equality $m(a,c) = m(a,b) + m(b,c)$ holds.

Theorem 7.8. (Menger [541]) *Let m be a complete metric space that fulfills Property* (2) *of Definition* 7.2. *Then for any two points p, r there exists an m-straight path connecting them.*

The following observation shows why m-straight paths are useful. As in the standard definition of arc length, one can put many consecutive points on a path π and add up their m-distances. If a limit exists, for arbitrarily fine resolution, it is called the *m-length* of π. For an m-straight path from p

to r, its m-length equals $m(p,r)$, as a direct consequence of Definition 7.3. No other path from p to r can have a smaller m-length. Thus, m-straight paths are *shortest paths* under the metric m, the true analogue of straight segments in the Euclidean metric.

Some authors call m-straight paths *geodesics*, others use this name for the wider class of paths that cannot be shortened locally.

Property (3) of Definition 7.2 refers to the metric bisector

$$B_m(p,q) = \{z \in \mathbf{R}^2 \mid m(p,z) = m(z,q)\}.$$

It may not be a nice curve itself, as the example depicted in Figure 7.4(iii) shows. In this case, one of the two boundaries of $B_m(p,q)$ will be chosen as a bisecting curve, and we want this boundary to be a nice curve; see Definition 7.5 below.

Lemma 7.2. *Each* convex distance function *is nice.*

Proof. Let d_C be a convex distance function based on a convex set C. Then Property (1) of Definition 7.2 is obviously fulfilled, as the d_C-disks are scaled copies of C. Property (3) follows from the discussion of Figure 7.5. Property (2) is also guaranteed; each interior point q of the line segment connecting p with r lies in between p and r. □

More generally, d_C-straight paths of convex distance functions can be described as follows. Suppose that the halfline from p to r cuts the boundary of C_p, the translated copy of C centered at p, at some point c. If c is interior point of a straight segment $\overline{vw}$ in the boundary, then exactly the v- and w-*monotone paths* from p to r are d_C-straight. If c is not an interior edge point, the line segment $\overline{pr}$ is the only d_C-straight path from p to r; see Figure 7.14.

A statement converse to Lemma 7.2 also holds. If a nice metric m is invariant under translations, and if $m(p,r) = m(p,q) + m(q,r)$ holds for

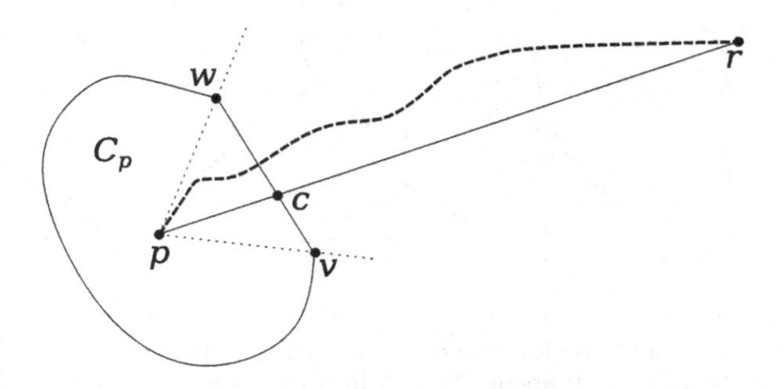

Figure 7.14. A d_C-straight path in a convex distance function.

any three consecutive points p, q, r on a line, then m is a convex distance function.

More complex examples of nice metrics are *composite metrics* that result from assigning different convex distance functions to the regions of a planar subdivision, or the *Karlsruhe* (or *Moscow*) *metric*; see Klein [461]. Here a center point, 0, is fixed and only such paths are allowed that consist of pieces of circles around 0 and of pieces of radii from 0. The Karlsruhe metric $m(p, q)$ denotes the minimum Euclidean length of an allowed path from p to q.

It depends on the angle between p and q if the shortest connecting path runs through the center, or around it; see Figure 7.15(i). The bisector of two points consists of up to eight segments that are pieces of straight lines or hyperbolas if written in polar coordinates. In (ii), the Voronoi diagram of eight point sites under the Karlsruhe metric is shown.

Clearly, in the example of the *air-lift metric* considered earlier, Property (2) of Definition 7.2 is violated: There is no point q satisfying $m(a, b) = m(a, q) + m(q, b)$ different from one of the two airports, a and b. (The points in mid-air do not count, as they are not in $\mathbf{R}^2$.)

7.3.2. *Very nice metrics*

Lemma 7.2 in the preceding subsection can be sharpened in the following way.

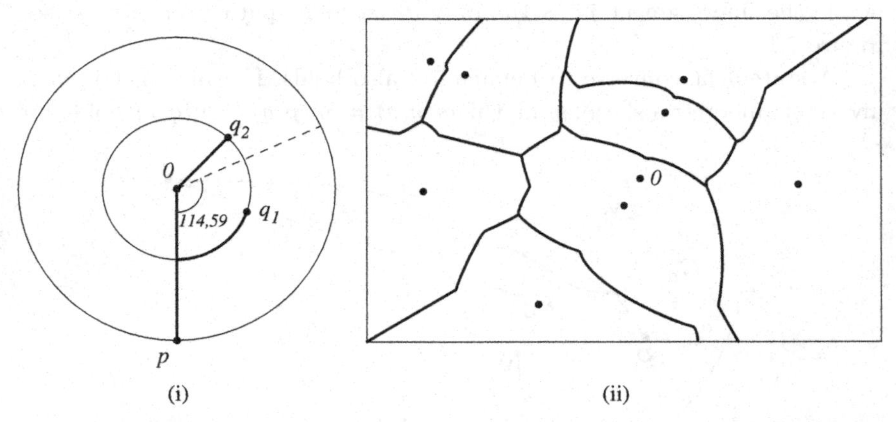

Figure 7.15. The shortest path in the Karlsruhe metric runs through the center iff the angle between p and q exceeds about 114, 59°. In (ii) a Voronoi diagram in the Karlsruhe metric is shown.

Definition 7.4. A nice metric m is called *very nice* if, for any three points a, b, c, one of the following properties is fulfilled.

(1) Each m-straight path from a to b contains c.
(2) Each m-straight path from a to c contains b.
(3) There exist m-straight paths π_b from a to b, and π_c from a to c, that have only point a in common.

Definition 7.4 clearly applies to the Euclidean metric; a, b, and c are collinear, or the line segments from a to b and c are disjoint, except for point a. From the above characterization of d_C-straight paths, the following generalization is immediate.

Lemma 7.3. *Each convex distance function is very nice.*

The Karlsruhe metric, however, is nice, but not very nice. All straight paths from a point p to points in a wedge 'behind' the center, O, are passing through O, but they do not contain each other.

Now we define the Voronoi diagram $V_m(S)$ based on a nice metric m. Let '$\prec$' be some order relation on S, for example, the lexicographical order.

Definition 7.5. (i) For $p, q \in S$, let

$$R_m(p,q) = \begin{cases} C_m(p,q) = \{z \in \mathbf{R}^2 \mid m(p,z) \le m(z,q)\} & \text{if } p \prec q \\ D_m(p,q) = \{z \in \mathbf{R}^2 \mid m(p,z) < m(z,q)\} & \text{if } q \prec p \end{cases}$$

denote the halfplane associated with p. Let $R_m(q,p)$ be its complement, the halfplane associated with q.

(ii) Define

$$\mathrm{VR}_m(p,S) = \bigcap_{q \in S \setminus \{p\}} R_m(p,q)$$

to be the Voronoi region of p with respect to S under m, and let

$$V_m(S) = \bigcup_{p,q \in S,\ p \neq q} \overline{\mathrm{VR}_m(p,S)} \cap \overline{\mathrm{VR}_m(q,S)}$$

denote the Voronoi diagram of S under m. (Here $\overline{A}$ denotes the *closure* of a set A, that is, the union of A with its boundary.)

By this definition, the border $J(p,q)$ between the halfplanes of p and q equals the common boundary of $B_m(p,q)$ with $D_m(q,p)$, if $p \prec q$, and the common boundary of $B_m(p,q)$ and $D_m(p,q)$, if $q \prec p$. By Definition 7.2(3), both of these boundaries are nice curves.

Definition 7.5 puts each Voronoi edge in the region of the smaller of the two sites it separates. An advantage of this technical modification becomes apparent in Lemma 7.4.

Let us generalize the standard notion of star-shapedness to general metrics.

Definition 7.6. A set V is called *m-star-shaped* as seen from point p in V if, for each point $x \in V$, each m-straight path from p to x is fully contained in V.

Lemma 7.4. *Let m be a nice metric. Then each Voronoi region $VR_m(p, S)$ is m-star-shaped as seen from its site p, hence pathwise connected.*

Proof. Since m-star-shapedness is stable in intersections, we need only prove it for each set $R_m(p, q)$, according to Definition 7.5. Suppose that $R_m(p, q) = C_m(p, q)$ holds. Let π be an m-straight path from p to some point $x \in C_m(p, q)$, and suppose that some point y on π does not belong to $C_m(p, q)$. That is, we have $m(p, x) \leq m(x, q)$ but $m(p, y) > m(q, y)$. By the straightness of π we obtain

$$\begin{aligned} m(p, x) &= m(p, y) + m(y, x) \\ &> m(q, y) + m(y, x) \geq m(q, x), \end{aligned}$$

which is a contradiction. Hence, $\pi \subset C_m(p, q)$. The case where $R_m(p, q) = D_m(p, q)$ is analogous. Theorem 7.8 ensures that for each $x \in \mathrm{VR}_m(p, S)$ an m-straight path from p to x exists. Together, these facts imply that $\mathrm{VR}_m(p, S)$ is a pathwise connected set. □

Regions that are *path-connected* are often just called *connected*. We are using either term, as appropriate from the context. Observe that path-connectedness is weaker than *simple connectedness*: all connecting paths might run around some 'holes' in the region. (To give a formal definition, a set M in the plane is called *simply connected* if every *Jordan curve* $J \subset M$ can be contracted to a point in M.)

A Voronoi diagram all of whose regions are connected is sometimes referred to as an *orphan-free Voronoi diagram.*

Sometimes it is just a single point that makes a Voronoi region connected. An example for such a cut-point in the Karlsruhe metric is shown in Figure 7.16. Since p, q, r have the same distance from the center, bisector $B(p, q)$ consists of the halfline that emanates from O and separates p, q, and of a 2-dimensional wedge 'on the other side' of center O, and the same holds for bisector $B(q, r)$. If tie-break order $q \prec p \prec r$ is chosen,

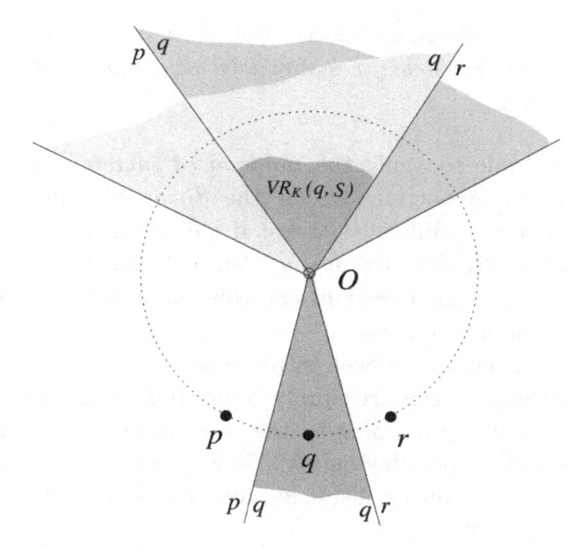

Figure 7.16. The region of q has a cut-point in the Karlsruhe metric if $q \prec p \prec r$.

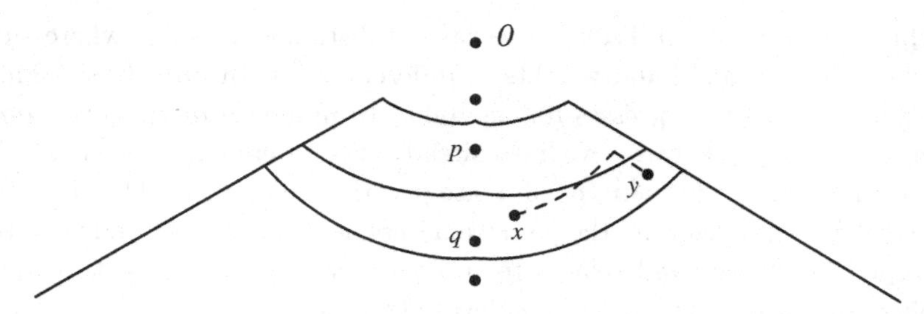

Figure 7.17. The unique shortest path between x and y in the Karlsruhe metric does not stay within the Voronoi region of q.

the darkly shaded Voronoi region $\mathrm{VR}(q, S)$, where $S = \{p, q, r\}$ results, according to Definition 7.5.

Whereas the star-shapedness of the Voronoi regions is preserved in nice metrics, the example of the Karlsruhe metric m shows that the convexity of the Euclidean Voronoi regions has no counterpart. In Figure 7.17, the Voronoi diagram of four point sites on the same radius is shown. There exists exactly one m-straight path connecting the points x, y of the Voronoi region of q, but this path is not contained in the region of q.

That the Voronoi regions under a nice metric m are m-star-shaped is a property strong enough for computing the Voronoi diagram efficiently. The following has been shown by Dehne and Klein [248].

Theorem 7.9. *The Voronoi diagram of n point sites under a nice metric in the plane can be constructed within $O(n \log n)$ many steps, using the* sweep line approach.

It is also possible to apply the *randomized incremental* construction method introduced in Section 3.2, or the *divide & conquer* technique. With the latter a new difficulty arises: If we use a line to subdivide S into subsets L and R, then the *merge chain* $B_m(L, R)$ of Voronoi edges separating L-regions from R-regions can contain cycles that would not be detected during the merge phase.

In [461] two criteria have been introduced for $B_m(L, R)$ to be *acyclic*. For example, if the m-circles are simply connected we can use, as a divider of S, any curve ℓ homeomorphic to the line whose intersection with all m-circles is connected (possibly empty). This generalizes a result by Chew and Drysdale [215] on convex distance functions, where each straight line possesses this property.

7.4. Weighted distance functions

In this section we shall discuss a variant of distance measures where sites can be assigned individual weights in different ways. Intuitively speaking, the weight of a site expresses its capability to influence its neighborhood. A prominent representative we have already encountered in Section 6.2: the power function $d(x, p)^2 - w(p)$ of a site p with weight $w(p)$. The resulting *weighted* Voronoi diagram, the *power diagram*, is an important and versatile concept that unifies and relates to many other geometric structures and applications; most of Chapter 6 is devoted to them.

7.4.1. *Additive weights*

Also other types of weighted Voronoi diagrams appear, sometimes quite unexpectedly, in many applications. As an introductory example, consider a polygonal chain $T = (p_1, p_2, \ldots, p_n)$ on n vertices in the plane, modeling a railroad track. Its *dilation*,

$$\delta(T) = \max_{1 \le i < j \le n} \frac{d(T_i^j)}{d(p_i, p_j)},$$

measures the maximum detour encountered when using the train instead of travelling as the crow flies. Here, $d(T_i^j)$ denotes the Euclidean length of the subchain $(p_i, p_{i+1}, \ldots, p_j)$ of T.

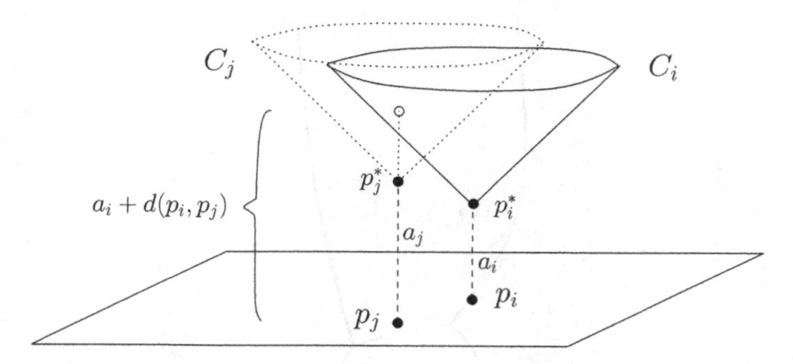

Figure 7.18. An arrangement of two cones.

Suppose we want to test if $\delta(T) \leq \kappa$ holds, for some threshold value $\kappa > 1$. This means, for each i we want to know if

$$\kappa \geq \frac{d(T_i^j)}{d(p_i, p_j)} = \frac{d(T_1^j) - d(T_1^i)}{d(p_i, p_j)}$$

holds, or equivalently,

$$d(p_i, p_j) + a_i \geq a_j \quad \text{with } a_m = \frac{d(T_1^m)}{\kappa} \tag{7.1}$$

for each $j > i$. This formula has an interesting geometric interpretation. Let C_m denote the upright cone of angle 90° in 3-space whose apex p_m^* is positioned at height a_m above point p_m in the xy-plane; see Figure 7.18. Clearly, $d(p_i, p_j) + a_i \geq a_j$ holds if, and only if, point p_j^* lies below cone C_i. Since $a_i \geq a_j$ whenever $i \geq j$, we have $\delta(T) \leq \kappa$ if and only if each apex p_j^* appears on the *lower envelope* of the cones C_j.

In Section 3.5 we have seen how Voronoi diagrams can be obtained from lower envelopes of so-called *distance cones*. One can imagine that from each site p a circle expands at unit speed, forming a 90° cone C_p in 3-space. The lower envelope of the arrangement of these cones, when projected vertically to the xy-plane, equals the Voronoi diagram of the sites $p \in S$.

The above *dilation problem* leads us to study more general arrangements, where each cone C_p has been lifted up by a certain distance, a_p, above the xy-plane, modeling that expansion of circles start at individual times. We may assign to each site p the weighted distance function $d(p, z) + a_p$, which describes the vertical distance from $z \in \mathbf{R}^2$ to cone C_p. The projection of the lower envelope of the arrangement of all cones C_p, where $p \in S$, equals the Voronoi diagram of S under these distance functions.

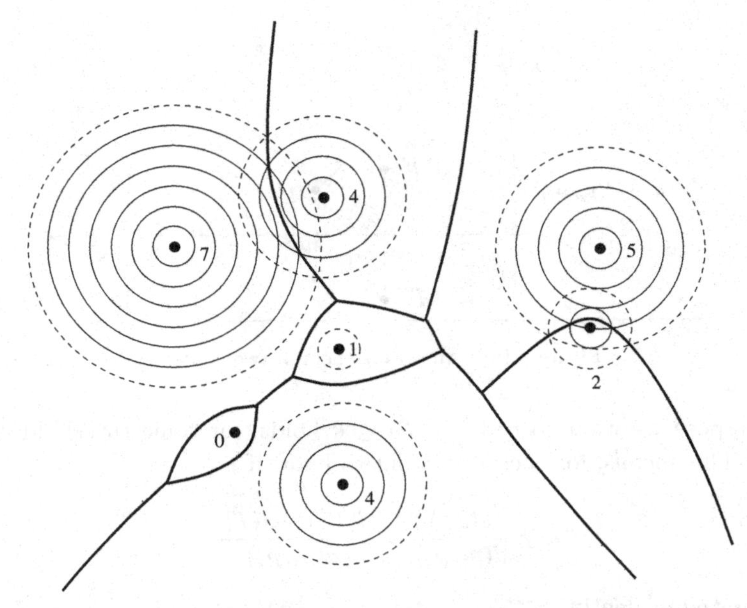

Figure 7.19. Additively weighted Voronoi diagram for seven sites. Numbers express site weights (radii) $w_p = -a_p$. (Dashed circles σ_p are intersections of the cones C_p with the xy-plane.) A 'hot' real-world interpretation is that *grass fires* starting at locations p at times a_p meet in the hyperbolic region borders (Calabi and Hartnett [176]). In a 'cool' interpretation the edges of the diagram are viewed as the *interference pattern* of waves emanating from pebbles p tossed at times a_p into a quiet pond.

The same structure also appears as the *Voronoi diagram for circles* in the Euclidean metric. A point $z \in \mathbf{R}^2$ outside a circle σ_p with radius w_p around site p has a distance of

$$d_p(z) = d(p, z) - w_p$$

to σ_p. If z lies inside σ_p we consider its distance to σ_p as negative. Then, the Voronoi diagram of the circles σ_p is the *additively weighted Voronoi diagram* of S, with site weights w_p.

Figure 7.19 depicts an instance of this diagram. Time-shifted 'wave' expansions are indicated by arrays of co-centric circles, with outermost circles σ_p.

The bisector of two weighted sites p and q is a hyperbola bent towards p if $w_p < w_q$, or a straight line if $w_p = w_q$. The Voronoi region of p is empty if σ_p fully lies inside some other circle σ_q. All Voronoi regions are *simply connected*, in fact, *star-shaped* as seen from the respective sites, by the same argument as in Lemma 7.4. Hence, the additively weighted Voronoi diagram is of linear complexity; cf. Lemma 2.3. Note, however, that two regions in this diagram may be adjacent in as many as $O(n)$ edges — for example if many sites of small weight are placed near the bisector of two heavily weighted sites.

The additively weighted Voronoi diagram is also known as *Johnson–Mehl model* in mineralogy [433]. In computational geometry, it has been studied, e.g., by Lee and Drysdale [496] and Sharir [638]. Fortune [344] has provided a construction algorithm that runs in time $O(n \log n)$, using the *sweep-line* approach (Section 3.4). Also, the divide & conquer algorithm in [25] (Section 5.5) applies, which can handle arbitrary circular arcs.

Thus, one can solve the aforementioned dilation problem in time $O(n \log n)$. We just construct the additively weighted Voronoi diagram of the chain's vertices p_i, equipped with the weights $w_i = -a_i$ of Formula 7.1, and check if each p_i has a non-empty Voronoi region; see Agarwal *et al.* [14] and Grüne *et al.* [385] for further results on the dilation problem.

We have already encountered the additively weighted Voronoi diagram in the context of *air-lift metrics* in Section 7.3. Another surprising application is in the *Monge–Kantorovich transportation problem*; see Gangbo and McCann [350] and Caffarelli *et al.* [175]. Imagine an area A with a set of n construction sites, p_i. Area A is covered with sand, according to some Lebesgue-continuous probability distribution (the uniform distribution, for example). Each site has a demand for a share of $\sigma_i > 0$ of sand, where $\sum_i \sigma_i = 1$. How should sand be transported to sites, in order to minimize transportation cost?

If one assumes that the cost of moving a unit of sand from z to p is given by a strictly convex function $h(p - z)$, then the optimum solution equals the projection of the lower envelope of an arrangement of h-cones (*distance cones*)

$$C_p = \{(z, h(p - z)) \mid z \in \mathbf{R}^2\} + a_p.$$

In general, the additive values a_p must be determined by numerical methods.

A simple proof for rather general distance measures can be found in Geiß *et al.* [358]. If $h(z) = |z|^2$, then a *power diagram* represents the optimum solution; see Aurenhammer *et al.* [97] and Section 6.4. Related is the *minimum-cost load balancing* problem studied in Aronov *et al.* [69], who show that the respective optimal partition of a convex planar domain is induced by an additively weighted Voronoi diagram.

To see one more occurrence of this diagram, imagine some *tilted plane* T in $\mathbf{R}^3$, passing through the origin O. As T is not horizontal, distances on T may be defined as the Euclidean distance plus a multiple $k > 0$ of the signed difference in height. The cost of 'moving' on T thus depends not only on the Euclidean distance but also on how much upwards or downwards the movement has to travel, simulating the situation when driving a vehicle on T. This so-called *skew distance* is direction-sensitive and, in particular, non-symmetric, and is investigated in Aichholzer *et al.* [31]. Its 'unit circle' C is a conic with focus O, directrix the horizontal line $y = 1/k$, and eccentricity k. Thus C is an ellipse for $0 < k < 1$, and unbounded otherwise, namely a parabola for $k = 1$, and a branch of a hyperbola for $k > 1$.

Now consider a set S of n point sites on T, without loss of generality in its y-positive halfplane. To each site $p \in S$, assign the value $k \cdot p_y$ as its weight ($p_y > 0$ denotes the y-coordinate of p). Then the additively weighted Voronoi diagram for these sites coincides with the Voronoi diagram for the skew distance on T. As a consequence, this so-called *skew Voronoi diagram* can be constructed in $O(n \log n)$ time and $O(n)$ space.

A similar concept is the so-called *boat-sail distance* in Sugihara [667], whose definition underlies a given flow field (of a river, for example). Its Voronoi diagram can be computed by a plane sweep (Section 3.4) in $O(n \log n)$ time for homogeneous flow fields of any speed, but tends to get quite complex for anisotropic flows, where approximate construction methods exist; see Nishida and Sugihara [565].

7.4.2. *Multiplicative weights*

In the circle expansion model explained before, it is natural to ask what happens if circles expand from their sites at different speeds. If we assign to each site p a *multiplicatively weighted* distance function

$$d_p(z) = \frac{d(p, z)}{w_p}$$

where $w_p > 0$, then sites of larger weights emit circles that expand more quickly, resulting in *distance cones*

$$C_p = \{(z, d_p(z)) \mid z \in \mathbf{R}^2\}$$

of larger opening angles. This has quite drastical consequences for the *multiplicatively weighted Voronoi diagram*, which results from projecting the lower envelope of the cones to the plane. We will discuss this in the following.

Since flat cones are always penetrated by steeper cones, the bisector of two sites must be a closed curve, circumscribing the site of lesser weight. Apollonius [64] discovered the following.

Lemma 7.5. *The locus*

$$B(p,q,r) = \left\{ z \in \mathbf{R}^2 \middle| \frac{d(p,z)}{d(q,z)} = r \right\}$$

of all points z in the plane, whose Euclidean distances to p and q have a fixed ratio $r \neq 1$, is a circle, later named Apollonius circle *after him.*

Proof. Apollonius knew that an inner angular bisector divides the opposite edge of a triangle to the proportion of its adjacent edges. Thus, in Figure 7.20

$$\frac{d(p,e)}{d(e,q)} = \frac{d(p,z)}{d(z,q)} = r$$

holds, and analogously

$$\frac{d(p,e')}{d(q,e')} = \frac{d(p,z)}{d(z,q)} = r$$

is fulfilled for the outer angular bisector $\overline{ze'}$ of the triangle (p,q,z). By construction, $\overline{ze}$ and $\overline{ze'}$ form a right angle. Thus, by *Thales' theorem* (which asserts that every angle in the half-circle is a right angle), there

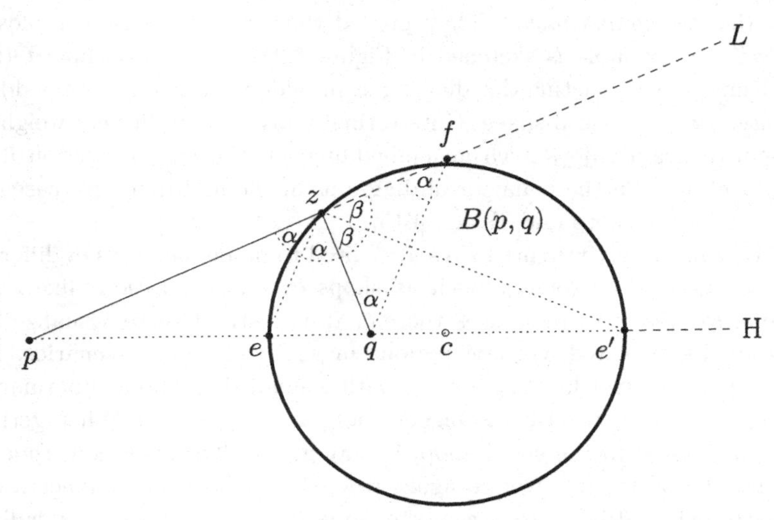

Figure 7.20. An Apollonius circle bisecting the multiplicatively weighted sites p and q in the plane.

exists a circle passing through e, e' and z. From

$$d(p,e) = \frac{r}{r+1} d(p,q) \quad \text{and} \quad d(e,q) = \frac{1}{r+1} d(p,q)$$

together with

$$r = \frac{d(p,e')}{d(q,e')} = \frac{d(p,q) + d(q,e')}{d(q,e')} = \frac{d(p,q)}{d(q,e')} + 1,$$

the diameter and center of this circle can be computed to

$$d(e,e') = d(e,q) + d(q,e') = \frac{2r}{r^2-1} d(p,q),$$

$$d(q,c) = d(e,c) - d(e,q) = \frac{1}{r^2-1} d(p,q).$$

It follows that this circle, $C(r)$, depends only on p, q, and ratio r, but not on the particular point $z \in B(p,q,r)$. Therefore, $B(p,q,r) \subseteq C(r)$. Since circles $C(r)$ and $C(r')$ are disjoint if $r \neq r'$, $B(p,q,r) = C(r)$ follows, for each $r \neq 1$. □

With circular bisectors, Voronoi regions are no longer simply connected. Even worse, regions can become *disconnected*, causing the complexity of the Voronoi diagram to grow. (One should observe that the proof of Lemma 7.4 does not extend to multiplicative weights.)

In [96], Aurenhammer and Edelsbrunner have investigated the multiplicatively weighted Voronoi diagram of n points, which is also referred to as the *Apollonius model.* They proved that its worst-case complexity is $\Theta(n^2)$; an example is sketched in Figure 7.21. Also, an optimal $O(n^2)$ algorithm for constructing this diagram is provided, based on its embedding in *power diagrams* in 3-space. (Interestingly, also the additively weighted Voronoi diagram enjoys such an embedding; cf. the end of Section 6.2.) Statistical data on the behavior of diagrams in the multiplicative case can be found in Sakamoto and Takagi [615].

Multiplicatively weighted sites are suited to model facilities of different attractiveness to customers, such as shops or service stations like radio senders; see [96] and references therein. It is instructive to visualize the effect of disconnected Voronoi regions in such real-world scenarios. For example, a customer located close to both a small shop and a supermarket will usually choose to visit the bigger, more attractive shop. When getting sufficiently close to the small shop, it may get his/her attention, though. At last, the situation reverses again when both shops are further away. Similarly, when driving in a car, the same (strong) radio sender will be dominantly received several times, after having passed by weaker senders.

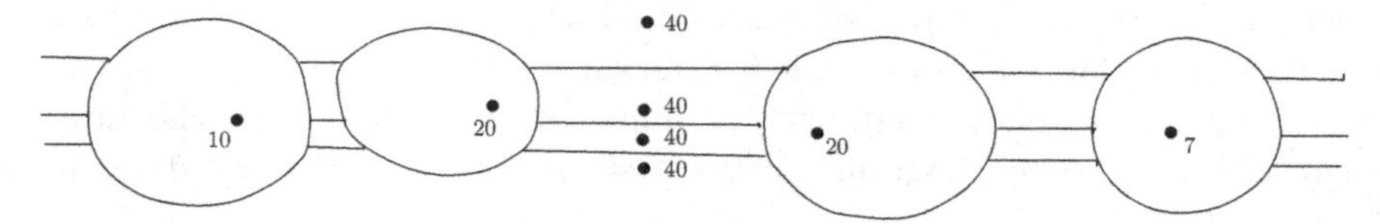

Figure 7.21. A Voronoi diagram of n point sites with multiplicative weights can be of complexity $\Theta(n^2)$. Numbers indicate site weights.

Diverse applications exist, in economy, geographical information systems, and physics; see Gambini *et al.* [349], Boots [152], and Reitsma *et al.* [604] who also mention different areas of use.

In biology, Bock *et al.* [135] used multiplicatively weighted Voronoi diagrams in order to describe the dynamic behavior of cell tissue. They model inner cell bodies as disks of radius w_i, and consider the weighted Voronoi diagram of the disks' centers, p_i, with multiplicative weights w_i. In order to avoid disconnected cells, the Voronoi region of p_i is restricted to a Euclidean disk of radius $w_i \cdot R$, for some constant $R > 1$. While this modification causes all regions to be bounded, there may now be 'holes' in the cell structure, that is, areas that do not belong to any region. Interestingly, the following property holds.

Lemma 7.6. *Let $w_{\min}$ and $w_{\max}$ denote the smallest and largest multiplicative weight, respectively. If*

$$R \leq \frac{w_{\max} + w_{\min}}{w_{\max} - w_{\min}}$$

holds for the restriction radius R, then all restricted Voronoi regions are star-shaped, *as seen from their sites p_i, hence connected. That is, the diagram becomes* orphan-free.

In a more realistic model, cells are represented by ellipses rather than circles; see Bock [136]. This approach leads to *anisotropic* Voronoi diagrams that will be introduced in Subsection 7.4.4.

Multiplicatively weighted Voronoi diagrams also arise as special cases of a structure called *angular Voronoi diagram*. Given a set of n sites of general shape (one-dimensional or two-dimensional objects), the plane is divided into regions from where a particular site is seen under the largest angle, compared to the remaining sites. If the sites are disks of various radii, a multiplicatively weighted Voronoi diagram results. In fact, the locus of all points x that 'see' two given disks at the same angle is a circle. The case of line segment sites is treated in Asano *et al.* [78]. The best known upper bound is now only $O(n^{2+\varepsilon})$ instead of $O(n^2)$. Applications to *mesh improvement* are plausible, based on finding vertex placements that minimize the largest visual angle for the surrounding edges.

Angular Voronoi diagrams are quite different from the visibility-constrained Voronoi diagram (Section 5.4) or the polar Voronoi diagram (Subsection 8.5.1).

7.4.3. *Modifications*

Adapting to practical needs — not least for more realistic modeling of natural phenomena — weighting schemes have been modified, sometimes (but not always) with drastical changes of the resulting weighted Voronoi diagram.

As a 'harmless' modification, other than the Euclidean metric can be used. For instance, Lee and Torpelund-Bruin [501] apply multiplicatively weighted Voronoi diagrams for the L_p-metrics in the design of decision support systems.

Of interest are also *farthest-site* weighted Voronoi diagrams, where each site is associated with the region at farthest (rather than at closest) weighted distance; cf. Subsection 6.5.1. In fact, the *geometric* properties of the diagrams remain unchanged, because they are still projections of envelopes of cones — now the *upper envelope* has to be taken. *Combinatorially*, a big difference may occur in the multiplicative case, as has been shown in Lee and Wu [500]: the worst-case size reduces to $O(n)$. A similar reduction in size can be observed for planar Voronoi diagrams in the L_1-metric (Section 7.2; from $O(n)$ to constant size), and for the Voronoi diagram for crossing line segments (Section 5.1; from $O(n^2)$ to $O(n)$ size). On the other hand, power diagrams and additively weighted Voronoi diagrams retain a linear worst-case size in their farthest-site variants.

The weighting schemes discussed in the preceding subsections are obviously combinable. For example, a mixture of the power function and the multiplicatively weighted distance, sometimes called a *mixed weighted distance*,

$$\Delta(x,p) = \frac{d(x,p)^2 - w_1(p)}{w_2(p)}$$

arises in the context of *clustering* as a certain intra-cluster measure; see Inaba *et al.* [421], and Section 8.4 for geometric clustering methods related to Voronoi diagrams.

A mixed weighted distance of the same type also applies to the following *graph drawing* problem. Eppstein [324] recently proved that every planar graph with maximum degree 3 has a planar drawing in which the edges are circular arcs that meet at equal angles around every vertex. Such drawings are called *Lombardi drawings* and have perfect angular resolution. His construction is based on a Voronoi diagram for circle packings that is invariant under *Möbius transformations* (which are known to be angle-preserving; see e.g. Bern and Eppstein [127]). The weighting scheme used is

$$\Delta'(x,p) = \frac{d(x,p)^2 - r(p)^2}{2r(p)},$$

where p and $r(p)$ are the center and radius, respectively, of a circle in the packing. Figure 7.22 gives a small example of a Lombardi drawing.

Interestingly, the same construction can be used to investigate the graph-theoretic properties of *soap bubbles.* In the arising spherical cell complexes, angles around edges and vertices are equally spaced as well, by *Plateau's laws* which have been proved by Taylor [680] in 1976. From their piecewise spherical shape, soap bubbles actually look like the regions of a 3D multiplicatively weighted Voronoi diagram, the weights reflecting pressure conditions; but clearly, no disconnected regions can occur. Whether this naive conjecture is justified is still unsettled, though. The graphs formed by two-dimensional 'soap bubbles' have been recently characterized in Eppstein [325], based on his previously mentioned results in [324].

Here may be the right place to observe that the *distance cone* (lower envelope) *model* for Voronoi diagrams is not identical to the *growth model* (i.e., the *wavefront model*), where the expansion of circles, or of any other appropriate shapes, ceases when getting in contact with each other. The latter model cannot treat Voronoi diagrams with disconnected regions.

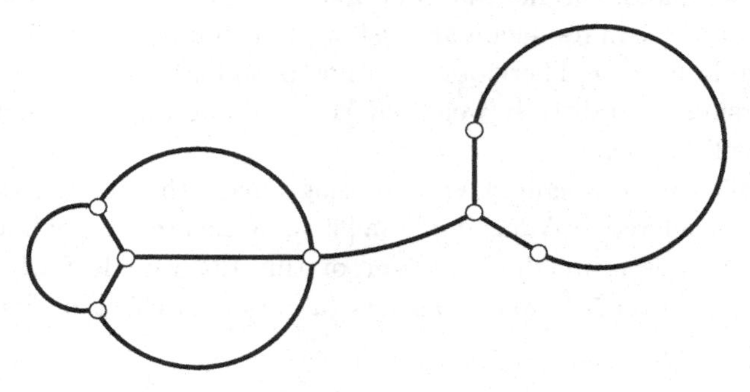

Figure 7.22. Lombardi drawing of a graph with seven vertices. It contains vertices of degree 2, 3, and 4, with circular edges spaced at angles π, $\frac{2\pi}{3}$, and $\frac{\pi}{2}$, respectively.

A discussion of this issue, in the context of *straight skeletons*, is given in [38].

In fact, when modeling certain phenomena in the natural sciences, letting cease the expansion of circles once they start overlapping is more appropriate. For example, crystals growing from sites at different speeds will still produce connected regions. The same applies to colonial growth patterns of plants.

Schaudt and Drysdale [617] first studied this interesting variant and called it the *crystal growth Voronoi diagram*. Regions in this diagram may wrap around other sites' regions — hence they may not be simply connected — and *shortest path lengths* from sites to points are taken to measure distance then, considering completed regions as obstacles — hence regions will always be connected. This 'dynamic mixture' of multiplicatively weighted Voronoi diagram and *geodesic Voronoi diagram* (Section 5.4) leads to a quite complicated structure. Its worst-case size is $\Theta(n^2)$, even though there are only n regions, because the complexity of the site separators can be high, combinatorially as well as geometrically. (For instance, a quadratic size occurs if sites of increasing weights are situated on a straight line.) An exact $O(n^3)$ time construction algorithm is given in the paper above; however, separators have to be determined by numerical methods, as no closed form is available.

Kobayashi and Sugihara [471, 472] present a maybe more practical *digital algorithm*, which approximates the crystal growth Voronoi diagram in pixel representation (cf. Section 11.1). They also mention another application, concerning collision-free path planning for a robot avoiding enemy attacks.

We note that for *additive weights*, i.e., same growth rate for all circles but possibly different starting times, there is no difference between the growth model and the distance cone model: Regions are star-shaped in the additively weighted Voronoi diagram, and geodesic paths from a site p to any of the points x in its region are just straight-line segments, because x is always visible from p. Therefore, it is not meaningful to define a 'crystal growth version' of additively weighted Voronoi diagrams; no change in the structure will happen.

Multiplicatively weighted Voronoi diagrams with *visibility constraints* have been considered in Wang and Tsin [698], in the presence of line segment obstacles. As the unweighted variant of this diagram is fairly complex already, see Section 5.4, weighting has (almost) no effect on its size and construction time.

7.4.4. *Anisotropic Voronoi diagrams*

Here we consider an interesting and recently introduced diagram, where each site p is weighted by an individually assigned distance function, instead of a real number $w(p)$. The resulting *anisotropic Voronoi diagram* is investigated in Labelle and Shewchuk [479]. Each site p is allotted with a positive definite quadratic form, given by a square matrix $M_p = F_p^T F_p$, where the positive definite matrix F_p maps the physical plane into a rectified plane with the usual length, area, and angle measures, *as seen by site* p. In particular, the distance between two points x and y in the plane (from p's view) is

$$d_p(x, y) = \sqrt{(x-y)^T M_p (x-y)}.$$

The Voronoi region of p is defined to contain all points x in the plane where $d_p(x,p) \leq d_q(x,q)$, for all sites q different from p. Anisotropic Voronoi diagrams are a generalization of multiplicatively weighted Voronoi diagrams, where the matrices M_p are just multiples of the unit matrix I,

$$M_p = w(p) \cdot I$$

for multiplicative site weights $w(p)$. They exhibit an according behavior, with their regions being possibly *disconnected* and *multiply adjacent.* Bisectors are general quadratic curves, i.e., all types of conics can occur, as opposed to (Apollonius) circles. See Figure 7.23 for an illustration.

The diagram's combinatorial complexity is bounded by $O(n^{2+\varepsilon})$, by an interpretation in the lower envelope model in $\mathbf{R}^3$. Surprisingly, a totally different embedding in 5 dimensions exists; see Boissonnat *et al.* [150]. The anisotropic Voronoi diagram can be obtained by intersecting a certain manifold and a *power diagram* in $\mathbf{R}^5$.

In the main fields of application, namely, interpolation of functions and numerical modeling [479], an appropriate *dual* of the diagram is sought, in order to allow to minimize the number of triangles in the anisotropic mesh while retaining a good accuracy in computations. However, the geometric dual of an anisotropic Voronoi diagram is *not* a triangulation of the given sites, in general, due to the presence of disconnected and/or multiply adjacent regions. Conditions under which the dual is a correct triangulation (the so-called *anisotropic Delaunay triangulation*) are given in [479], along with respective construction algorithms that insert extraneous sites at well-chosen locations.

An alternative site insertion algorithm for computing anisotropic Delaunay triangulations, based on power diagrams, is proposed in [150].

Figure 7.23. Anisotropic Voronoi diagram for 10 sites. Regions are patterned by iso-contours of the distance function for the closest site. While every site is contained in its region, not every face of the diagram contains a site, due to region disconnectedness (from [479]).

See also Canas and Gortler [178], where further sampling and distribution conditions on the sites are given, in order to make the regions in an anisotropic Voronoi diagram connected (i.e., *orphan-free*). In another paper [179], they show that orphan-free anisotropic Voronoi diagrams have duals that are indeed triangulations.

Naturally, anisotropy in the plane can be specified in various ways, leading to different types of generalized Voronoi diagrams. Du and Wang [290] use an anisotropic metric field to define, for each point $x \in \mathbf{R}^2$, an individual anisotropic distance function. In contrast to [479], Voronoi regions are computed by using the complete metric. Kunze *et al.* [478] generate regions by means of *geodesic Voronoi diagrams* (Subsection 7.1.1) on parametric surfaces. A surface embedding of the plane in higher-dimensional space is also used in Kapl *et al.* [441], who define the distance of two points in $\mathbf{R}^2$ as the Euclidean distance in $\mathbf{R}^d$ of the two corresponding points on the respective surface. This has the advantage that regions can be obtained by intersecting the surface with a classical Voronoi diagram in $\mathbf{R}^d$. The embedding is generated, via a fitting procedure, from a given *distance graph* in $\mathbf{R}^2$ that specifies the anisotropy. One more way of introducing anisotropy is by weighting differently certain subsets of the plane; see Subsection 7.6.1.

7.4.5. *Quadratic-form distances*

Let us say at this place a few more words about general *quadratic-form distances* in $\mathbf{R}^d$, whose positive instances have been used to define the anisotropic Voronoi diagram in $\mathbf{R}^2$. Such distances are intimately related to the weighted concept of power diagrams. They are of the form

$$Q(x, y) = (x - y)^T \cdot M \cdot (x - y),$$

where M is a fixed non-singular $d \times d$ matrix. We may assume that, without loss of generality, M is symmetric, as we have $Q(x, y) = \frac{1}{2}(x - y) \cdot (M + M^T) \cdot (x - y)$. Interesting classes of Voronoi diagrams are induced, with surprising applications.

Separators are hyperplanes in $\mathbf{R}^d$. Therefore, such diagrams are polyhedral cell complexes, which are sometimes called *affine Voronoi diagrams* in the literature. We observe that the choices $M = I$ and $M = -I$ produce the classical Voronoi diagram and its farthest-site variant, respectively.

Unified treatments are given in Edelsbrunner and Seidel [310] via Voronoi surfaces in $\mathbf{R}^{d+1}$ (Subsection 7.5.1), and in Aurenhammer and Imai [98] via power diagrams in $\mathbf{R}^d$ (Subsection 6.2.3), asserting that the class of diagrams obtainable by Q *coincides* with the class of affine Voronoi diagrams and also with the class of power diagrams.

For $d = 2$ and $M = \left(\begin{smallmatrix} 0 & 1 \\ 1 & 0 \end{smallmatrix}\right)$, the distance $Q(x, y)$ describes the area of the isothetic rectangle with diagonal $\overline{xy}$. The obtained diagram proves useful in finding the largest empty axis-aligned rectangle among a finite set of points in the plane; see Chazelle *et al.* [196]. More recently, Aurenhammer *et al.* [104] gave attention to the *quasi-Euclidean distance* generated by the special *Jordan matrix*

$$M = \mathrm{diag}(1, \ldots, 1, -1).$$

This distance is capable of embedding the Voronoi diagram for *oriented spheres* in $\mathbf{R}^{d-1}$ into a power diagram in $\mathbf{R}^d$. Namely, if the dth coordinates of two points $p, q \in \mathbf{R}^d$ are interpreted as 'radii' (which are possibly negative), and the first $d - 1$ coordinates of p and q specify centers, then $Q(p, q)$ expresses the squared tangent length between the respective two oriented spheres in $\mathbf{R}^{d-1}$.

The aforementioned rectangle area related distance transforms to the two-dimensional instance of Q with $M = \mathrm{diag}\,(1, -1)$ by a rotation of $\frac{\pi}{4}$.

Motivation for studying quasi-Euclidean distances comes from special relativity theory, see e.g. Benz [123] and Woodhouse [707], where one describes events as points in the quasi-Euclidean space whose 'metric' is governed by the matrix M, and whose isometric mappings are the *Lorentz transformations.* The line spanned by two events p and q can be time-like, or space-like, or light-like — these cases correspond to $Q(p,q) < 0$, $Q(p,q) > 0$, and $Q(p,q) = 0$, respectively. The value of $Q(p,q)$ is a Lorentz invariant: If positive, it has an interpretation as square of distance, and if negative, then the square root of its absolute value is the proper time experienced by a particle whose life line is the line segment $\overline{pq}$.

Several physically meaningful modifications are considered in [104]. For example, the absolute-valued variant $\Delta_A(p,q) = |Q(p,q)|$ models a setting where lifetime is regarded as a *positive* distance. Or, putting

$$\Delta_\infty(p,q) = \begin{cases} Q(p,q) & \text{if } \geq 0, \\ \infty & \text{otherwise} \end{cases}$$

lets oriented spheres to which squared tangent lengths are negative not take part in the local formation of Voronoi regions — a 'space-driven' variant. When using

$$\Delta_0(p,q) = \begin{cases} Q(p,q) & \text{if } > 0, \\ 0 & \text{otherwise,} \end{cases}$$

negative squared tangent length means being as close as possible. That is, lifetime of particles is irrelevant again, apart from the fact that it restricts the diagram to domains where distances to all oriented spheres are positive.

Figure 7.24 depicts a partition of the plane induced by the distance variant Δ_A and three oriented spheres in $\mathbf{R}^1$, interpreted as point sites in $\mathbf{R}^2$. Two types of separators arise, namely, for the equation $Q(x,p) = Q(x,q)$, which gives straight lines, as well as for the equation $Q(x,p) = -Q(x,q)$, which gives hyperbolas. Colors gray, light gray, and white symbolize ownership of region parts by individual sites. The diagram has a size of $\Theta(n^2)$ and it can be constructed within this time bound, by a close structural connection to the underlying orthogonal straight line grid G, that reflects the sign pattern of Q for the given n sites (drawn in dotted style).

The same grid G also structures the diagram for the distance Δ_∞, illustrated in Figure 7.25. In fact, the diagram is a linear refinement of G; each of its grid cells Z is partitioned by the power diagram of exactly those sites whose distances to Z are positive. (The same color code for ownership applies; the topmost and the bottom-most cell, shown hatched, belong to

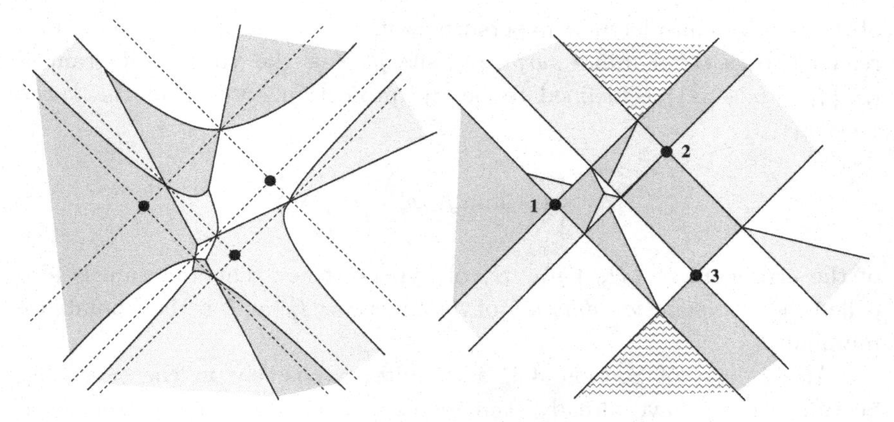

Figure 7.24. Distance Δ_A.

Figure 7.25. Distance Δ_∞.

no site.) A combinatorial upper bound of $O(n^3)$ results which, however, we do not consider as tight.

For applications in physics, diagrams in space-time $\mathbf{R}^4$ are needed. The sign grid for Q now is an arrangement of homothetic rotational double-cones, rather than of linear double-wedges, which complicates matters. Its size is $\Theta(n^4)$, and that of a power diagram is $\Theta(n^2)$ in the worst case (by Theorem 6.1), piling up to a — probably not tight — upper complexity bound of $O(n^6)$. Of somewhat simpler nature is the diagram induced by the distance Δ_0. It basically consists of two power diagrams in $\mathbf{R}^4$, each clipped by a cone. A combinatorial complexity and construction time of $\Theta(n^2)$ can be shown.

7.5. Abstract Voronoi diagrams

One might wonder if there exist some intrinsic properties most Voronoi diagrams have in common, and if these properties are a foundation strong enough for a unifying approach to the definition and to the construction of Voronoi diagrams.

7.5.1. *Voronoi surfaces*

Edelsbrunner and Seidel [310] have introduced a very general approach. It does not explicitly use the concepts of sites and distance functions. Rather, the site set, S, is a mere set of indices $\{p, q, \ldots\}$. A fixed domain, X, is given, and for each $p \in S$ a real-valued function $f_p : X \to \mathbf{R}$. The graph

$$H_p = \{(x, f_p(x)) \mid x \in X\}$$

of f_p can be considered a hypersurface in the space $X \times \mathbf{R}$, sometimes referred to as the *Voronoi surface* of site p. Now the Voronoi diagram of $(S, \{f_p \mid p \in S\})$ is defined to be the projection onto X of the *lower envelope*

$$L = \left\{ \left(x, \min_{p \in S} f_p(x) \right) \middle| x \in X \right\}$$

of the arrangement $\{H_p \mid p \in S\}$ of hypersurfaces. This diagram is also called the *minimization diagram* of the functions f_p, as L is their pointwise minimum.

At the end of Section 3.5, and quite extensively in the preceding Section 7.4, we have already studied several examples of this approach. Section 6.2 shows that there exist even *linear* functions f_p whose envelope projects onto the Voronoi (or power) diagram of S. Moreover, the order-k Voronoi (or power) regions of S are the projections of certain cells of the corresponding arrangement; see Section 6.5.

To establish this connection between Voronoi diagrams and *arrangements of hypersurfaces* in a space one dimension higher is one of the great advantages of this approach. As one implication, an upper bound of $O(n^{d+\varepsilon})$ on the combinatorial complexity of general classes of Voronoi diagrams follows from a result in Sharir [639] on envelopes of hypersurfaces; see Sections 6.6 and 7.2. In general, the computation of lower envelopes, and the analysis of their combinatorial complexity, is a central problem in computational geometry; see e.g. the book by Sharir and Agarwal [642].

7.5.2. *Admissible bisector systems*

A different approach to planar Voronoi diagrams has been suggested in Klein [461] and improved on in Klein *et al.* [462]. Again, there are no physical sites but a finite index set, S. Instead of distance functions or their graphs in 3-space, *bisecting curves* (*separators*) are used as primary objects.

Let us assume that for any two different indices p, q in S, a bisecting curve $J(p, q) = J(q, p)$ is given, which cuts the plane into two open domains, $D(p, q)$ and $D(q, p)$. We assume that these curves are *nice*, in the sense of Definition 7.1. This implies in particular that each curve $J(p, q)$ and the domains $D(p, q)$, $D(q, p)$ are unbounded. In the subsequent figures, the index p denotes the 'halfplane' $D(p, q)$.

The Voronoi region $\mathrm{VR}(p, S)$ is defined as the intersection of the open domains $D(p, q)$, where $q \in S \setminus \{p\}$. The *abstract Voronoi diagram* (*AVD* for short) $V(S)$ of $\{J(p,q) \mid p,q \in S,\ p \neq q\}$ is defined as the edge graph constituting the complement of all Voronoi regions.

Definition 7.7. The bisector system $\mathcal{J} = \{J(p,q) \mid p,q \in S,\ p \neq q\})$ is called *admissible* iff, for each subset $S' \subset S$ of size at least 3, the following conditions are fulfilled.

(1) The Voronoi regions $\mathrm{VR}(p', S')$, where $p' \in S'$, are *path-connected.*
(2) Each point of the plane lies in the closure of a Voronoi region $\mathrm{VR}(p', S')$.

Figure 7.26(i) shows an admissible curve system for $S = \{p,q,r\}$, and the resulting abstract Voronoi diagram (ii). We observe that the requirements of Definition 7.7 allow for some phenomena that cannot occur for Euclidean bisectors of points. For example, point a lies in the intersection of two bisecting curves $J(p,r)$ and $J(q,r)$, but not on the third curve, $J(p,q)$. Point w does lie on all three bisecting curves, but it is not a Voronoi vertex. The intersection of two bisecting curves, $J(p,r)$ and $J(q,r)$,

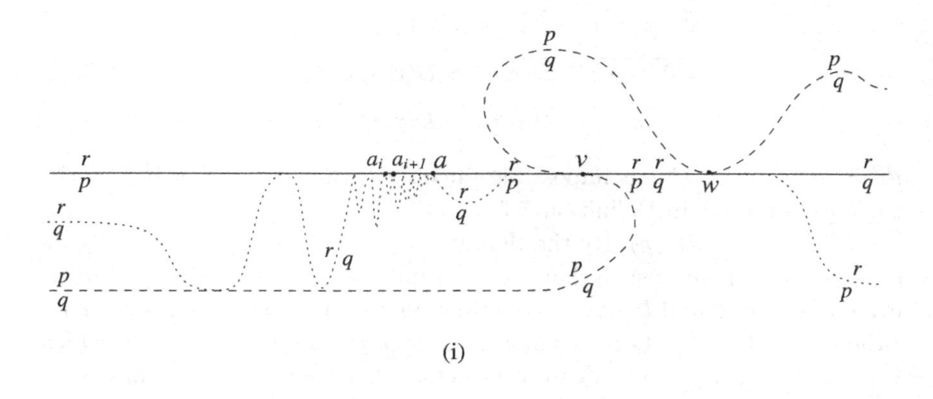

(i)

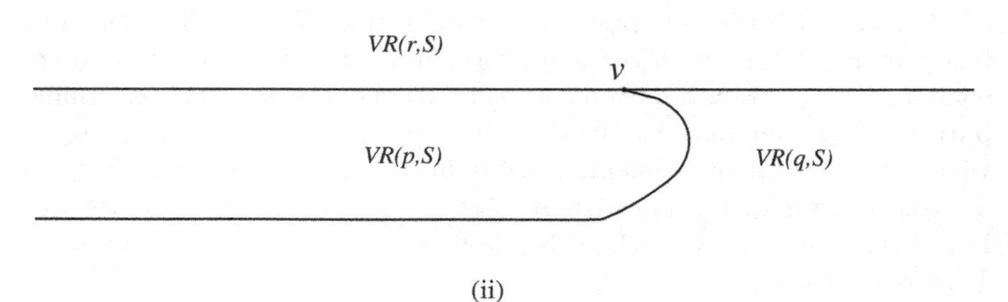

(ii)

Figure 7.26. An admissible curve system (i) and the resulting abstract Voronoi diagram (ii).

consists of an infinite number of isolated points that converge towards point a.

Still, the conditions stated in Definition 7.7 are strong enough to obtain a satisfying structure theory and efficient algorithms for AVDs. Some well-known structural properties of 'concrete' Voronoi diagrams can be derived directly. For example, the Voronoi diagram consists of all points in the plane that belong to the closures of two or more Voronoi regions; cf. Definition 2.1. With concrete distance functions, if a point z is closer to site p than to q, and closer to q than to r, then, trivially, z is closer to p than to r. The following lemma states an analogous fact for abstract Voronoi diagrams.

Lemma 7.7. *Let $\mathcal{J}$ be an admissible system, and let $p, q, r \in S$. Then, $D(p,q) \cap D(q,r) \subseteq D(p,r)$ holds.*

Proof. Let $z \in D(p,q) \cap D(q,r)$. Point z must be contained in one of the sets $D(r,p)$, $J(r,p)$, $D(p,r)$. If z were contained in $D(r,p)$, it could not lie in *any* of the closed Voronoi regions

$$\overline{\mathrm{VR}(q,S')} \subseteq \overline{D(q,p)} = D(q,p) \cup J(q,p),$$
$$\overline{\mathrm{VR}(r,S')} \subseteq \overline{D(r,q)} = D(r,q) \cup J(r,q),$$
$$\overline{\mathrm{VR}(p,S')} \subseteq \overline{D(p,r)} = D(p,r) \cup J(p,r)$$

for $S' = \{p,q,r\}$. This is impossible since these three sets cover $\mathbf{R}^2$, by the third requirement in Definition 7.7.

Suppose $z \in J(r,p)$. By the *Jordan property* of $J(r,p)$, there exists an arc α with endpoint z such that $\alpha \setminus \{z\}$ is fully contained in $D(r,p)$. With z, even a neighborhood U of z is contained in the open set $D(p,q) \cap D(q,r)$. Inside U, path α contains a point $z' \in D(p,q) \cap D(q,r) \cap D(r,p)$, which leads to the same contradiction as before. Consequently, the third case applies, that is, $z \in D(p,r)$. □

Lemma 7.7 is useful in proving that an abstract Voronoi diagram $V(S)$ is a *planar graph* with $O(|S|)$ many faces. Its vertices are just the points contained in the closures of three or more Voronoi regions. For algorithmic purposes, one can turn $V(S)$ into a bounded structure, by encircling it with a large closed curve, as proposed in Section 2. In this way it becomes a 2-connected planar graph with at most $n+1$ faces, all of whose vertices have degree at least 3. Each such graph can be obtained as an abstract Voronoi diagram.

Applying the concept of abstract Voronoi diagrams to a concrete situation is facilitated by the following result.

Theorem 7.10. *A curve system $\mathcal{J}$ is admissible if the properties stated in Definition 7.7 are fulfilled for each subset of 3 sites of S.*

The conditions in Definition 7.7 are met, for example, by the bisectors of *power diagrams* (Section 6.2), of sites with *additive weights* (Section 7.4), and of non-intersecting *line segments* or convex polygons (Section 5.1); see Meiser [539]. Another example has been provided by Edelsbrunner *et al.* [304] and Abellanas *et al.* [5]. They studied disjoint compact and convex sites in the plane, and their *Hausdorff Voronoi diagram* based on a *convex distance function* d_C, which puts each point x into the Voronoi region of the first site that is fully contained in a shape C expanding from x; cf. Section 7.2. It can be shown that the resulting regions are *simply connected.*

Also, the types of Voronoi diagrams studied by McAllister *et al.* [527], Ahn *et al.* [22], Karavelas and Yvinec [446], Abellanas *et al.* [6], Aichholzer *et al.* [38], and Bae and Chwa [109] are special cases of abstract Voronoi diagrams.

As a general result, we have the following lemma for the class of all *very nice metrics* (in the sense of Definition 7.4). We include the proof which has not yet appeared elsewhere.

Lemma 7.8. *If m is a very nice metric then its bisecting curves defined by Definition 7.5 form an admissible system.*

Proof. To prove that the resulting Voronoi regions are pathwise connected, let $z \in \mathrm{VR}(p, S)$. Since $\mathrm{VR}(p, S)$ is open, there is a closed disk D such that $z \in D \subset \mathrm{VR}(p, S)$ holds. Due to

$$|m(p, v) - m(p, w)| \leq m(v, w)$$

and because of Definition 7.2(1), the function $g(w) = m(p, w)$ is continuous. Thus, g attains its maximum value in the compact set D at some point $c \in D$. Let π_c denote an *m-straight path* from p to c. As homeomorphic image of a line segment, π_c cannot be space-filling: there must be a point $b \in D$ that does not lie on π_c. Conversely, c itself cannot be situated on an m-straight path from p to b, by maximality. Therefore, Definition 7.4(3) applies, that is, there are m-straight paths ρ_b and ρ_c from p to b and c, respectively, that have only point p in common. They are contained in the closure of $\mathrm{VR}(p, S)$, by Lemma 7.4.

If we clip ρ_b and ρ_c at their first points situated on the boundary ∂D of D, they form, together with a segment of ∂D, a *Jordan curve* whose interior, I, is contained in $\mathrm{VR}(p, S)$, and we can easily route a path from z to p through I.

The proof of Property (2) of Definition 7.7 is by straightforward case analysis, using Theorem 7.10 and Lemma 13 in [462]. □

Thus, the AVD machinery applies to all very nice metrics in the plane. If a metric is nice, but not very nice, Lemma 7.8 may not hold; for example, the curves $J(p, q)$ and $J(q, r)$ shown in Figure 7.16 cannot form an admissible system because $D(q, p) \cap D(q, r)$ would be disconnected. In this case, a more involved version [461] of AVDs could be applied, or one could resort to perturbation techniques.

There are other concrete Voronoi diagrams which are not covered by the AVD concept, because their bisecting curves fail to fulfill some of the properties stated in Definition 7.7 above. For example, in the *closest-polygon Voronoi diagram*, bisectors may be closed loops while its regions are still path-connected. In the *farthest-segment Voronoi diagram* investigated by Aurenhammer *et al.* [95], bisectors are unbounded simple curves, as required in Definition 7.7, but Voronoi regions may be disconnected. In a more general setting, namely for the *farthest-polygon Voronoi diagram* explored by Cheong *et al.* [208], even closed loop bisectors *and* disconnected regions may occur, while the diagram is still of linear complexity. In the Voronoi diagram of points with *multiplicative weights* both properties are violated as well, and its complexity may be quadratic; see Section 7.4.

7.5.3. *Algorithms and extensions*

Of the four algorithmic methods mentioned in Chapter 3 for computing the Euclidean Voronoi diagram, two have been adapted to abstract Voronoi diagrams.

How to generalize the *divide & conquer* approach has been shown in [461]. As in the case of nice metrics (Section 7.3) one has to assume that the index set, S, can recursively be split into subsets L and R such that the *merge chain* $B(L, R)$ contains no cycle, in order to find the starting edges of $B(L, R)$ at infinity. Once a starting edge has been found, $B(L, R)$ can be traced efficiently, because abstract Voronoi regions are pathwise connected; no stronger properties, like star-shapedness or convexity, are needed. This is illustrated in Figure 7.27, which should be compared to Figure 3.7.

It has been shown in Mehlhorn *et al.* [533] and in Klein *et al.* [466] that one can, without further assumptions, apply the *randomized incremental construction* technique to abstract Voronoi diagrams.

Theorem 7.11. *The abstract Voronoi diagram of a given admissible system* $\mathcal{J} = \{J(p, q) \mid p, q \in S \text{ and } p \neq q\}$, *where* $|S| = n$, *can be*

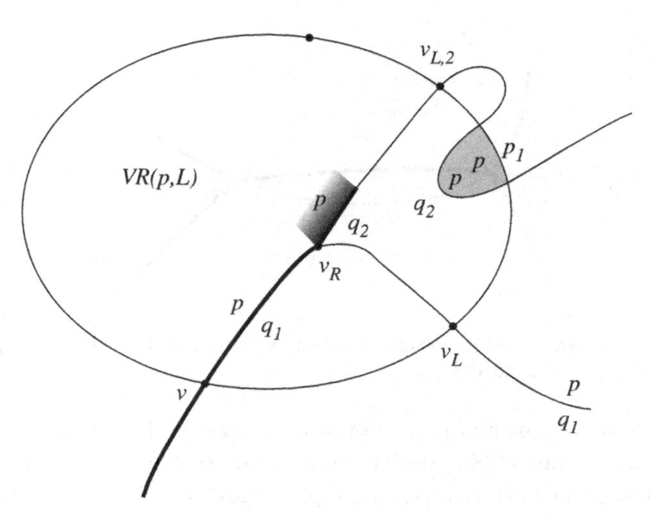

Figure 7.27. If $J(p, q_2)$ ran like this, the disconnected shaded areas would both belong to $\mathrm{VR}(p, \{p, p_1, q_2\})$, a contradiction. Therefore, the first point, $v_{L,2}$, on the boundary of $\mathrm{VR}(p, L)$ can still be found by scanning the boundary from v_L on counterclockwise.

constructed within expected $O(n \log n)$ many steps and expected space $O(n)$, by randomized incremental construction.

This algorithm provides a universal tool that can be used for computing all types of concrete Voronoi diagrams whose bisector curves have the properties required in Definition 7.7. To adapt the algorithm to a special type, one has to implement only certain elementary operations on bisector curves. For example, it is sufficient to have a subroutine that accepts five elements of S as input, and returns the graph structure of their abstract Voronoi diagram. All numerical computations can be kept to this subroutine.

It seems difficult to apply the *plane sweep approach* to abstract Voronoi diagrams. One reason is the absence of sites whose detection by the sweep line could indicate that a new region must be started.

Abstract Voronoi diagrams can also be thought of as *lower envelopes*, in the sense mentioned in Subsection 7.5.1. Namely, for each point x not situated on a bisecting curve, the relation

$$p <_x q \quad \text{iff } x \in D(p, q)$$

defines a total ordering on S, thanks to Lemma 7.7. If we construct a set of surfaces H_p, $p \in S$, in 3-space such that H_p is below H_q iff $p <_x q$ holds, then the projection of their lower envelope equals the abstract Voronoi diagram.

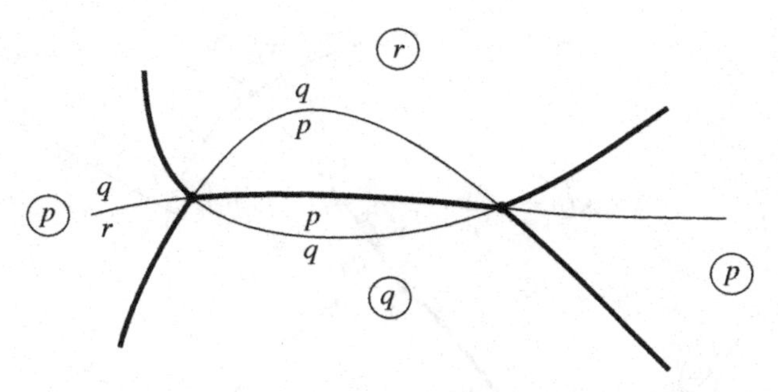

Figure 7.28. Abstract inverse Voronoi diagram for the bisectors of three sites p, q, and r. The region of p is not connected.

The concept of abstract Voronoi diagrams has been generalized to 3-space in Lê [488], under structural restrictions. For example, any two bisecting surfaces $J(p,q), J(p,r)$ must have an intersection that is homeomorphic to a line, or empty. These assumptions apply, e.g., to *convex distance functions* whose unit sphere is an *ellipsoid*; cf. Section 7.2.

Rasch [602] and Mehlhorn *et al.* [534] have studied *abstract inverse* (or *farthest-site*) *Voronoi diagrams* in the plane; cf. Section 6.5. Here the region of p is defined as the intersection of all open domains $D(q,p)$, where $q \in S \setminus \{p\}$. In contradistinction to the Euclidean case, the resulting regions are in general not pathwise connected; see Figure 7.28.

If the admissible system $\mathcal{J}$ is such that none of the regions in the (closest-site) abstract Voronoi diagram generated by $\mathcal{J}$ is empty, then the inverse diagram with respect to $\mathcal{J}$ is connected. It is a *tree* of complexity $O(n)$ and can be constructed within expected $O(n \log n)$ many steps and expected linear storage.

Bohler *et al.* [137] have recently introduced *abstract Voronoi diagrams* of arbitrary *order* k. For a subset M of size k of S, its abstract order-k region is defined as the intersection of all domains $D(p,q)$, where $p \in M$ and $q \in S \setminus M$; cf. Section 6.5. Such order-k regions need not be connected; even for $k = 2$ a region can consist of up to $n - 1$ faces.

Let us assume that the bisector system $\mathcal{J}$ is in general position, so that no more than three curves pass through one point and all intersections are transversal. Moreover, suppose that the standard order-1 abstract Voronoi regions are non-empty. Then the following upper bound can be shown [137].

Theorem 7.12. *Under the conditions above, an abstract order-k Voronoi diagram has at most $2k(n-k)$ many faces, and this bound is tight.*

This result generalizes and sharpens the $O(k(n-k))$ upper complexity bounds for the order-k diagrams of points and line segments; see Lee [495] and Papadopoulou and Zavershynskyi [584], respectively. Not covered is the case of point sites under the *geodesic distance* in a polygon with holes, which has recently been studied by Liu and Lee [510].

Testing order-1 abstract Voronoi regions for non-emptiness is facilitated by the following lemma complementing Theorem 7.10. If all bisecting curves $J(p, q)$ were lines, the claim would follow directly from *Helly's theorem* (see e.g. [371]) on intersections of convex sets.

Lemma 7.9. *In order to verify that all Voronoi regions $VR(p, S')$, where $S' \subseteq S$, are non-empty it is sufficient to consider the subsets S' of size 4.*

7.6. Time distances

7.6.1. *Weighted region problems*

Imagine a large modern-style city, with streets arranged in north–south and east–west directions. The city is equipped with a public transportation network like a subway or a bus system. Time is precious and people intend to follow the quickest path from their homes to their desired destinations, using the network whenever appropriate. For some people several facilities of the same kind are equally attractive (think of post offices or hospitals), and their wish is to find out which facility is reachable first. There also may be commercial interest (from real estate agents, or from a tourist office) to make visible the area which can be reached in, say one hour, from a given location in the city (the apartment for sale, or the recommended hotel).

Applications like this lead to Voronoi diagram models where distances are measured differently at certain *places* of the plane, rather than for different sites, as it is the case in the weighted model (in Section 7.4): Within the transportation network, movement is quicker, so distances are shorter.

In its general form, such a model is treated in Mitchell and Papadimitriou [548]. They consider a partition of the plane into *weighted regions*, the weight expressing internal travelling speed (or its reciprocal, the time per unit distance). As a basic task, for two given points p and q the quickest (Euclidean) path between p and q across the weighted regions is sought, entering and exiting regions at any place. The solution in [548] is based on *Snell's law of refraction*. The query time is $O(t^7 \cdot \ell)$, where t is the complexity of the polygonal partition and ℓ denotes the precision of the problem instance.

Allowing only weights 0 and 1 leads to the *shortest* (*obstacle-avoiding*) *path* problem (or geodesic distance problem), for which much more efficient solutions are known; see e.g. Mitchell [546], and Section 5.4 for the resulting *geodesic Voronoi diagram*.

A given *transportation network* T in the plane can be viewed as a one-dimensional instance of the weighted region problem. Choose a weight $v_e > 1$ for each network segment e, and weight 1 on all connected parts of the complement of T in the plane. This variant (and others) have been studied by Gewali *et al.* [360], who construct the quickest path between two given points p and q in time $O(t^2)$, where t denotes the complexity (number of edges) of the network.

The *air-lift distance* concept in Section 7.3 constitutes a zero-dimensional variant; distances are shorter only between points in a given finite set.

7.6.2. *City Voronoi diagram*

In the following, we will present the special one-dimensional case where the transportation network T is *isothetic* and of constant speed v, and where the underlying metric is the L_1*-norm*. This exactly models the practical situation mentioned at the beginning of this section, and has been investigated in Aichholzer *et al.* [38].

For two points x and y in the plane, let $Q_T(x, y)$ denote a *quickest path* between them, that is, a path minimizing the travel duration between x and y, using T. Let $d_T(x, y)$ be the time for traversing $Q_T(x, y)$. Note first that d_T induces a metric in the plane, which we shall call the *city metric*: d_T is non-negative, symmetric, and obeys the triangle inequality because, for each point $z \in Q_T(x, y)$, the concatenation of $Q_T(x, z)$ and $Q_T(z, y)$ gives $Q_T(x, y)$. Quickest paths are isothetic polygonal lines which, apart from special cases, are not unique. A single path may contain $\Theta(t)$ pieces on T plus $\Theta(t)$ shortcuts (foot walks outside T). The complexity of the city metric is also apparent from the shape of the induced circles

$$C(x, r) = \{y \mid d_T(x, y) = r\}.$$

Not only may $C(x, r)$ break into $\Theta(t)$ pieces, its shape also varies with its radius r, as well as with the relative position of its center x to T. Still, the 'disk' interior to $C(x, r)$ is a connected set, as being the union of quickest paths.

Let now S be a set of n point sites in the plane. The *city Voronoi diagram* $V_T(S)$ contains, for each site $s \in S$, the region of all points given by

$$\text{reg}(s) = \{x \mid d_T(x, s) < d_T(x, u), \forall u \in S \backslash \{s\}\}.$$

All regions are *path-connected* (and in fact, the diagram consists of *simply connected* regions), by the fact below.

Observation 7.1. reg(s) *is the union of all quickest paths from* $x \in$ reg(s) *to the site* s, *and equivalently, to the set* S.

Observation 7.1 does *not* imply that $V_T(S)$ is of size $O(n)$. Vertices of degree 2 may occur, whose number cannot be bounded by applying the *Eulerian theorem* to a planar graph with n faces. A trivial upper bound is $O(n \cdot t)$, where t counts the number of edges of the network T. The true complexity of $V_T(S)$ is only $O(n + t)$, as we shall see later.

Let us consider the *bisector* of two sites $s, u \in S$ in the city metric, which is the set

$$B_{su} = \{x \mid d_T(x, s) = d_T(x, u)\}.$$

The set B_{su} is connected by Observation 7.1. However, due to well-known L_1-specific degeneracies (see Section 7.2), B_{su} may fail to be one-dimensional. A canonical one-dimensional form of B_{su} results from the straight skeleton interpretation of $V_T(S)$ given next. In this form, B_{su} is a (possibly cyclic) polygonal curve with $\Theta(t)$ edges in the worst case. Unfortunately, cyclic and non-constant sized bisectors are known to be a main obstacle for efficiently applying the construction techniques divide & conquer (Section 3.3) and randomized incremental insertion (Section 3.2).

There is a useful link between the city Voronoi diagram and *straight skeletons* introduced in Section 5.3. Its implications are manyfold [38]. To aid the intuition, let us recall the interpretation of the standard *L_1-metric Voronoi diagram* in the *growth model.* Imagine each site in the given set S sends out a *wavefront,* in the shape of an expanding L_1 circle (called a *diamond*), at time 0 and with translational speed $1/\sqrt{2}$ for all its edges. Wavefront movement stops at every point x where two diamonds come into contact, and x is declared a point on the edge graph of the L_1-Voronoi diagram.

This expansion view actually shows that every L_1-Voronoi diagram for point sites is a straight skeleton, with these diamonds as its defining figures.

How does the network T influence this growth model? The circles expanding around the sites change their shapes, in a manner determined by T which we want to explore next. Let $s \in S$ be a fixed site, and recall that $C(s, r)$ denotes the circle of radius r around s in the city metric. Our aim is to interpret $C(s, r)$ as the common wavefront propagated from a certain set F_s of figures.

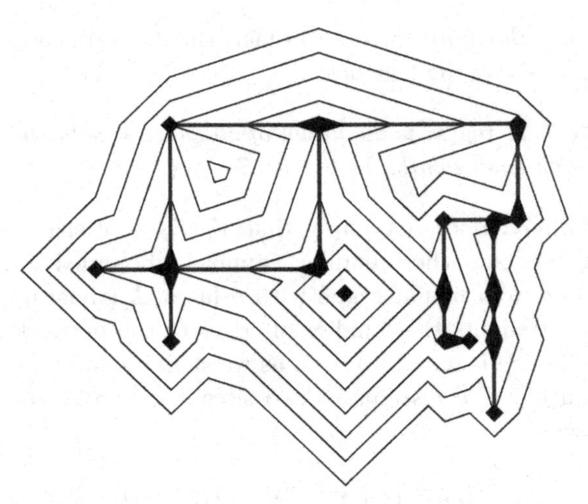

Figure 7.29. Co-centric d_T-circles around a site s. The small black polygons depict the generated figures in the set F_s (from [38]).

See Figure 7.29. For sufficiently small $r > 0$, $C(s, r)$ is a diamond with center s as before. Let F_s be a set of figures which initially contains this diamond, associated with its place s and time 0 of birth. Now consider all points $x \in T$ where either a vertex of $C(s, r)$ runs into a network segment, or an edge of $C(s, r)$ slides into a network node. At time $r = d_T(s, x)$, and at no other point in time, the circle $C(s, r)$ changes its combinatorial shape. Accordingly, we add to the set F_s an appropriate figure with time r and place x of birth, constructed as follows.

Observe that, at time $r + \varepsilon$, a diamond D with center x and diagonal length ε appears on $C(s, r)$, along with sharp-angled wedges tangent to D, whose peaks move at speed v in all possible directions on T that point to the exterior of $C(s, r)$. The figure we construct is simply the union of D and these wedges. Each wedge is an isosceles triangle with base b and height $\varepsilon \cdot v$, where b is a diagonal of D.

We categorize the figures in F_s by their number j of wedges and call them *j-needles*. Note that $0 \leq j \leq 3$ holds, because the degree of a network node is at most four, and at least one of the four isothetic directions points to the interior of $C(s, r)$; see Figure 7.29 again.

Let $W_r(F_s)$ denote the common wavefront of all the figures in F_s. By construction, $W_r(F_s)$ enjoys the following property: $W_r(F_s)$ coincides with $C(s, r)$ at any time r. This property has an obvious but important consequence: It draws the connection to straight skeletons.

Theorem 7.13. *Let $F = \bigcup_{s \in S} F_s$. Then the straight skeleton* SK(F) *of* F *contains the city Voronoi diagram* $\mathcal{V}_T(S)$ *as a subgraph.*

The set F in Theorem 7.13 contains *redundant* figures, i.e., figures which do not contribute to SK(F). Their wavefronts solely and always consist of points swept over earlier (or at the same time) by wavefronts of other figures. For example, for each site $s \in S$, all figures of F_s whose places of birth lie outside the region reg(s) in $V_C(S)$ are redundant in F. Edges of SK(F) which are not edges of $V_C(S)$ stem from figures born in the *same* region of $V_C(S)$. We will alternatively call SK(F) the *refined city Voronoi diagram* of S and C, and denote it with $\mathcal{V}_C(S)$. The two-dimensional components of $\mathcal{V}_C(S)$ are its *regions*, which are the regions of $V_C(S)$, and which are partitioned into *faces*, the faces of SK(F); see Figure 7.30.

The diagram $\mathcal{V}_T(S)$ = SK(F) is a *planar graph* whose faces are connected and whose vertices have degree at least three, as in every straight skeleton. The number of edges and vertices of $\mathcal{V}_T(S)$ therefore is linear in the number of its *faces*. This number, in turn, is bounded by the total number of *edges* of all figures which are non-redundant in the set F, because each such edge (possibly) sweeps out a single face of $\mathcal{V}_T(S)$.

How many figures are non-redundant? Clearly, each site $s \in S$ defines a non-redundant diamond with birth place s. Moreover, each node of C gives rise to a single non-redundant j-needle born at this node. This gives $n + t$ relevant figures so far. It remains to examine the 2-needles born in the interior of network segments. Their number is $\Theta(n \cdot t)$ in the worst case: each of the n sites may cause t such figures, born on t parallel segments of T; However, if f is some interior 2-needle on network segment σ, then the interior 2-needle f' born when the wavefront of f hits some segment parallel to σ is redundant in the set $\{f, f'\}$.

Redundant needles of the last kind indicate *unused portions* of T (dotted in Figure 7.30). They should not appear in an efficiently designed network.

An interior 2-needle never stems from a contact between a network segment and some sharp-angled vertex of a needle. Therefore, only the interior 2-needles that stem from contacts with vertices incident to some diamond edge are non-redundant in the set F. As such vertices disappear from the common wavefront of F after their first contact with T, we have at most $4(n + t)$ figures of this kind. In summary, F contains at most $5(n + t)$ non-redundant figures. Each figure is of constant complexity, which gives a total of $O(n + t)$ figure edges, and with it, skeleton faces.

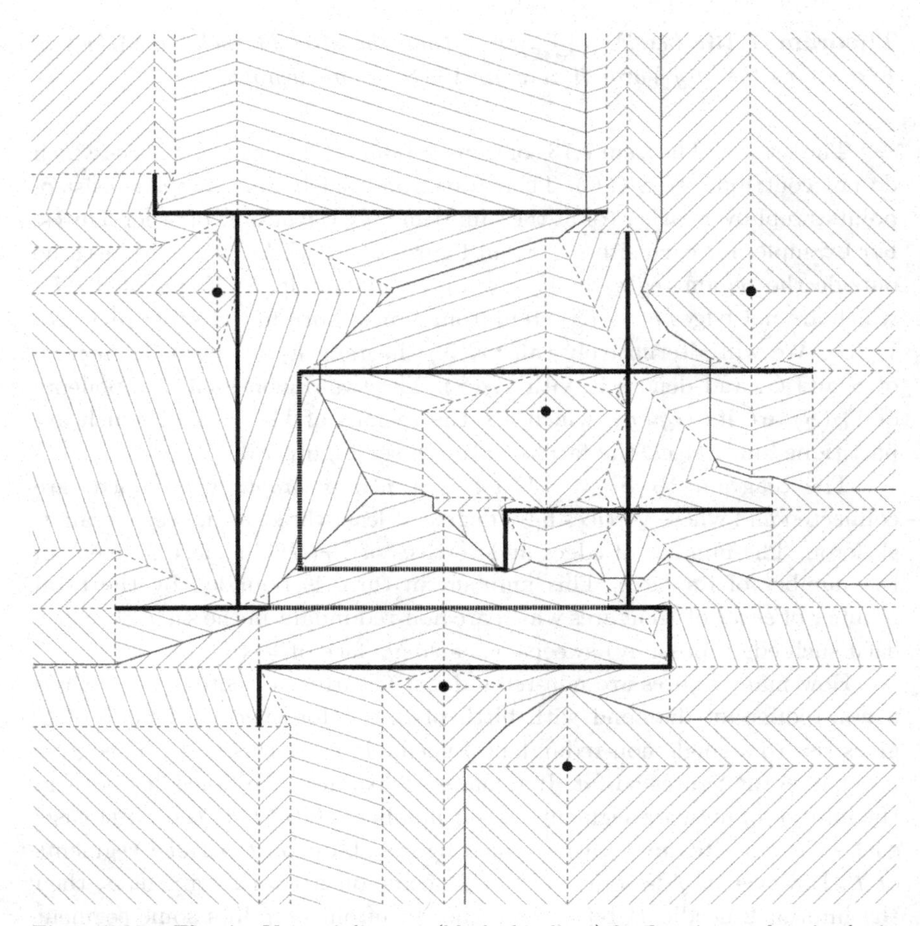

Figure 7.30. The city Voronoi diagram (black thin lines) for five sites and an isothetic transportation network (bold lines). The faces of the diagram are hatched by iso-contours for the city metric (gray lines); from [38].

Theorem 7.14. *The refined city Voronoi diagram $\mathcal{V}_T(S)$ consists of $O(n + t)$ faces, edges, and vertices.*

7.6.3. *Algorithm and variants*

On the algorithmic side, all non-redundant figures can be determined in time $O(n \log n + t^2 \log t)$, by exploring the isothetic grid defined by the vertices of T with the *continuous Dijkstra technique.* It remains to construct the straight skeleton of these figures in an efficient way.

Fortunately, the *abstract Voronoi diagram* machinery (Section 7.5) can be employed in this special case — and unlike for general straight skeletons — in spite of the fact that the bisector system defined by taking these figures as abstract sites is *not admissible*: Areas belonging to no abstract Voronoi region (no-man's lands) may arise, and bisectors may be closed curves. These shortcomings can be remedied, by modifying the figures so as to behave accordingly while leaving their straight skeleton unchanged; see [38] for details.

An abstract Voronoi diagram with m sites and constant bisector complexity (as it is the case for our figures) can be computed in time $O(m \log m)$. Particularly attractive is a *randomized incremental construction*; see Mehlhorn *et al.* [533] and Klein *et al.* [466]. To adapt this algorithm to our situation, it suffices to have a subroutine that accepts five figures as input and returns their straight skeleton. Any trivial skeleton algorithm may be implemented in this subroutine, because all input figures are convex, *12-oriented*, and have at most six edges.

In summary, since $m = O(n + t)$ holds by Theorem 7.14, the time for finding the figures dominates, and $O(n \log n + t^2 \log t)$ time (and optimal $O(n + t)$ space) is taken for constructing the city Voronoi diagram.

In the paper [374], Görke *et al.* showed how to find the figures more efficiently, in $O((n + t) \text{ polylog } (n + t))$ time, which is faster for large t.

Naturally, time distances for the *Euclidean metric* (rather than the L_1-metric) are of interest in several applications. For the basic case where the transportation network T is a single straight line (a highway, for example), the resulting Voronoi diagram was studied in Abellanas *et al.* [7]. Lee *et al.* [498] generalize to two or more straight lines, and Bae and Chwa [109] treat the general case, i.e., *non-isothetic* polygonal networks.

Unfortunately, any slight alteration of the city Voronoi diagram model, such as non-isothetic networks, individual travel speeds on network segments, or metrics different from L_1, may yield a diagram size of $\Theta(n \cdot t)$, see [38], and thus significantly increases the computation time and storage requirement, making the concept less useful in applications.

Time distances also have been used to generalize geometric structures different from Voronoi diagrams. The convex hull $\text{conv}(S)$ of a set S of n points in the plane is a noteworthy example.

As being the smallest convex set which contains S, $\text{conv}(S)$ can be (implicitly) defined as the minimal set X that contains S and all the line segments connecting two points in X. Given now, in addition, a straight line T in the plane (with fast travel speed), the so-called *highway hull*, $H(S,T)$, is the minimal set containing S as well as the quickest paths

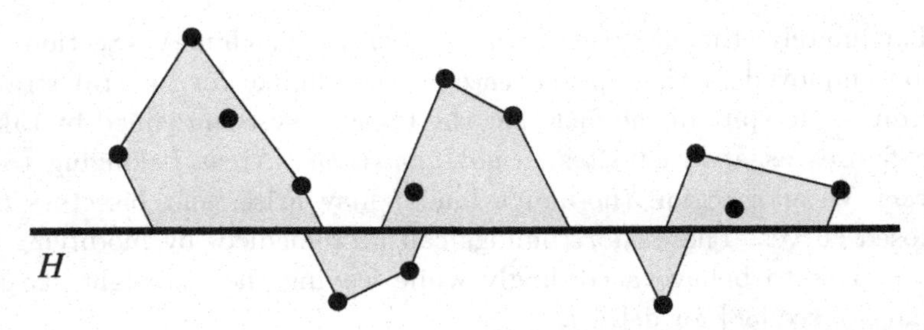

Figure 7.31. Highway hull for the Euclidean metric. It consists of convex polygons along the (horizontal) highway H. Boundary segments touching H are of slope k or $-k$, respectively, under which it is best to approach the highway. Slope k increases with the highway speed. (From [715].)

between all pairs of points in $H(S,T)$, using the highway T. See Figure 7.31 for an illustration.

Aloupis *et al.* [49] give an optimal $O(n \log n)$ algorithm to find the highway hull under the L_1-metric, as well as an $O(n \log^2 n)$ algorithm for the Euclidean case, improving over earlier $O(n^2)$ methods. They also identify the region a highway T must intersect such that the quickest path between at least one pair of points uses T. Ahn *et al.* [21] tackle the problem from the optimization point of view. Given a set S of point sites, and a travel speed v, find an axis-parallel line T that minimizes the maximum travel time over all pairs of sites in S. They show that $O(n)$ time algorithms exist, for both the L_1-metric and the Euclidean metric.

Chapter 8

APPLICATIONS AND RELATIVES

This chapter is devoted to the numerous applications of the Voronoi diagram $V(S)$ and its dual, the Delaunay triangulation $\mathrm{DT}(S)$, in solving geometric problems. Quite a few applications have already been mentioned in previous chapters, mostly along with generalizations of the classical type. Applications in relation to optimization properties of the Delaunay triangulation have been discussed in Section 4.2.

Whereas distance is implicitly involved in almost all applications of the Voronoi diagram, we start with some examples of 'pure' distance problems. These distance problems are, apart from their direct use in practical applications that will be mentioned occasionally, frequently arising subroutines in more involved geometric algorithms.

8.1. Distance problems

Unless otherwise stated, the following problems address point sites in the plane under the standard Euclidean distance. However, most solutions can be generalized, at least to convex distance functions (Section 7.2), and sometimes also to sites other than points (Chapter 5), and to dimensions higher than two (Chapter 6).

8.1.1. *Post office problem*

Our first example is the *post office* (or *nearest neighbor*) *problem* mentioned in the seminal paper [637] by Shamos and Hoey. Given a set of n sites (post offices) in the plane, determine, for an arbitrary query point x, the post office closest to x.

The Voronoi diagram of the post offices represents the *locus approach* to the post office problem: It partitions the plane into regions of equal answer, i.e., regions whose points x have identical nearest post offices; see Figure 8.1.

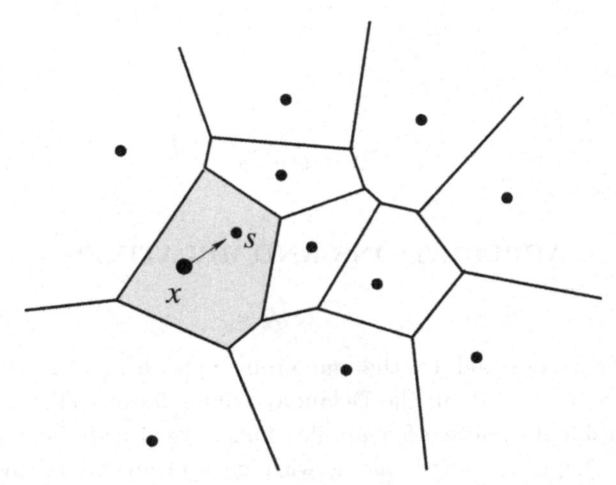

Figure 8.1. Locating a point x among the regions of a Voronoi diagram determines the post office s closest to x.

It remains to quickly determine the region that contains the query point, x. This *point-location problem* has received a lot of interest in computational geometry. Here we sketch an elementary solution first mentioned in Dobkin and Lipton [285], called the *slab method.*

Through each Voronoi vertex a horizontal line is drawn. These extra lines partition the Voronoi regions into triangles and trapezoids. By construction, no horizontal 'slab' between two consecutive lines contains a Voronoi vertex in its interior, so that all crossing Voronoi edges are ordered within the slab. To locate a query point, $x = (x_1, x_2)$, we use one binary search for x_2 among the slabs, and another one for x_1 among the edge segments of the slab found. This gives us $O(\log n)$ query time, at quadratic storage cost.

There are search structures like in Kirkpatrick [456], Edelsbrunner *et al.* [305], and Edelsbrunner and Maurer [308] that are equally efficient, but need only $O(n)$ storage and can be derived in $O(n)$ time from the Voronoi diagram. An early method for general subdivisions has been given in Maurer [523]. With any of the optimal Voronoi diagram algorithms in Chapter 3, this yields the following solution to the post office problem.

Theorem 8.1. *Given a set S of n point sites in the plane, one can, within $O(n \log n)$ time and $O(n)$ storage, construct a data structure that supports nearest neighbor queries: For an arbitrary query point x, some nearest neighbor in S can be found in time $O(\log n)$.*

Storing the *history* of the Voronoi diagram during *incremental insertion* (cf. Section 3.2) even obviates the need of processing the diagram for

point location. Guibas *et al.* [390] showed that nearest neighbor queries are supported in (expected) time $O(\log^2 n)$ by the resulting structure.

Analogously, the *order-k Voronoi diagram* can be used for finding the *k-nearest neighbors* in S of a query point x. Aurenhammer and Schwarzkopf [102] generalized the practical approach in [390], obtaining a structure with expected query time $O(k \log^2 n)$ that additionally allows insertions and deletions of sites, at a space requirement of $O(k(n-k))$; see Subsection 9.1.1 for more material on the *dynamic post office problem*. A sophisticated technique of *compacting* order-k Voronoi diagrams in Aggarwal *et al.* [18] achieves optimal query time $O(k + \log n)$ and space $O(n)$, but with high constants.

The *locus approach* also works for more general sites and distance measures. Let us assume that the site set, S, consists of k non-intersecting convex polygons with n edges in total. The bisector of two sites is composed of $O(n)$ many straight or parabolic segments; if we also count the endpoints of such segments as vertices (of degree 2), then the Voronoi diagram of the k convex polygons is of complexity $\Theta(n)$, as Lemma 5.1 shows. It could be constructed in time $O(n \log n)$ by first computing the *line segment Voronoi diagram* of the polygon edges (see Section 5.1), and then joining those regions that belong to parts of the same convex polygon.

A more efficient structure for solving the post office problem for k *disjoint convex polygons*, the so-called *compact Voronoi diagram*, has been introduced in McAllister *et al.* [527]. Using only $O(k)$ many line segments, it partitions the plane into regions bounded by at most six edges each. With each region, one or two sites are associated; for any point x in the region, one of them is the nearest neighbor of x in S.

This planar straight-line graph can be defined quite easily. Its vertices are the $O(k)$ original Voronoi vertices defined by the k polygonal sites, plus, for each Voronoi vertex v defined by three sites, the closest points to v on their respective boundaries. Straight-line edges run from each vertex v to its closest points and, around the boundary of each site, between the points closest to its Voronoi vertices; see Figure 8.2. For each site C in extreme position (i.e., appearing on the convex hull of the set of polygons), a halfline to infinity is added all of whose points have the same closest point on the boundary of C.

Theorem 8.2. *Given a set S of k disjoint convex polygonal sites in the plane with a total of n edges, one can, within $O(k \log n)$ time and $O(k)$ storage, construct a data structure that allows nearest neighbor queries to be answered within time $O(\log n)$.*

The essential task is in constructing the $O(k)$ many original Voronoi vertices of the k polygons, spending only $O(\log n)$ time on each vertex.

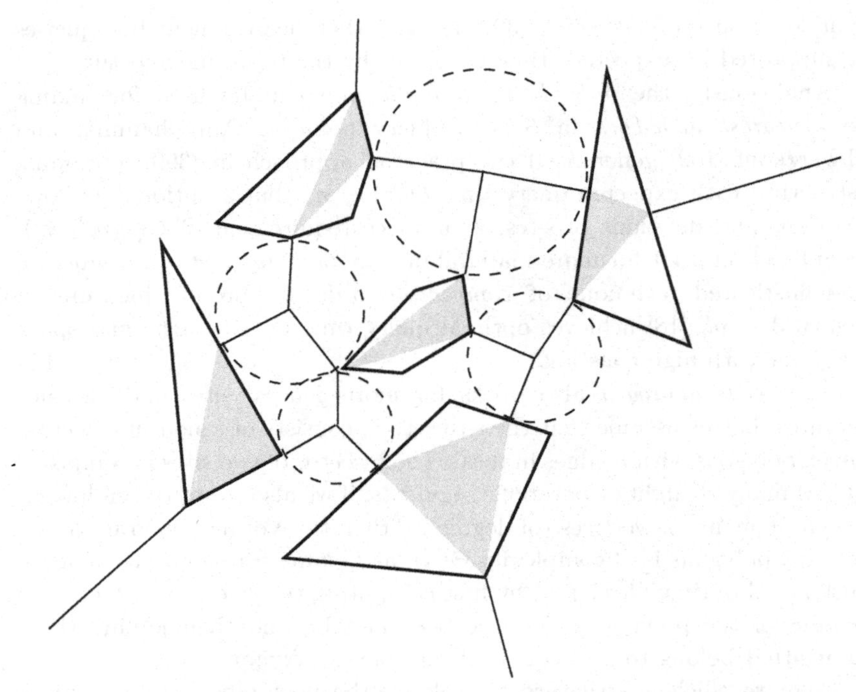

Figure 8.2. A piecewise linear subdivision of size $O(k)$ for solving the post office problem of convex polygons.

This can be achieved by a clever adaption of the *sweep line approach* described in Section 3.4, and by means of a subroutine that computes, in $O(\log n)$ time, the Voronoi vertex of three convex polygons. The subroutine uses the *tentative prune-and-search* technique described in Kirkpatrick and Snoeyink [458]. The same approach works for convex distance functions; see Section 7.2.

In general dimensions d, the post office problem (which is also called *proximity problem, similarity problem*, or *nearest neighbor search problem* in the literature) can be efficiently solved by methods different from the ones described in this subsection. Chapter 10 gives a detailed discussion.

8.1.2. *Nearest neighbors and the closest pair*

Another distance problem the Voronoi diagram directly solves is the *all nearest neighbor problem*: For each point site in the set S, a nearest neighbor in S is required. An example is shown in Figure 8.3; arrows are pointing from each site towards its nearest neighbor. For sites in *general position*, the resulting nearest neighbor graph is a *forest*, i.e., a vertex-disjoint collection

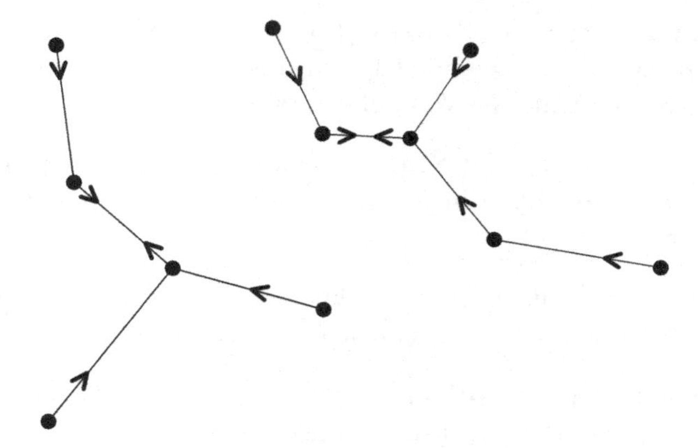

Figure 8.3. The nearest neighbor graph of 11 points.

of trees, each of whose vertices is of outdegree 1 and indegree at most 6. (The bound on the indegree follows from the containment of this forest in the *minimum spanning tree* of S; see Subsection 8.2.1.)

The solution offered by the Voronoi diagram is based on the following fact.

Lemma 8.1. *Let $S = P \cup Q$ be a disjoint decomposition of the point set S, and let $p_0 \in P$ and $q_0 \in Q$ be such that*

$$d(p_0, q_0) = \min_{p \in P,\ q \in Q} d(p, q).$$

Then the Voronoi regions of p_0 and q_0 are edge-adjacent in $V(S)$.

Proof. Otherwise, the line segment p_0q_0 contains a point z that belongs to the closure of some Voronoi region $\mathrm{VR}(r, S)$, where $r \neq p_0, q_0$. Let us assume that r belongs to Q; the case $r \in P$ is symmetric. From $z \in \mathrm{VR}(r, S)$ follows $d(z, r) \leq d(z, q_0)$, hence

$$\begin{aligned} d(p_0, r) &\leq d(p_0, z) + d(z, r) \\ &\leq d(p_0, z) + d(z, q_0) = d(p_0, q_0) \leq d(p_0, r) \end{aligned}$$

by minimality of $d(p_0, q_0)$. Hence, equality must hold, and we obtain $d(p_0, r) = d(p_0, q_0)$ and $d(z, r) = d(z, q_0)$.

But since p_0 lies on the bisector $B(q_0, r)$, each point z in the interior of $\overline{q_0p_0}$ is strictly closer to q_0 than to r, which is a contradiction. □

If we apply Lemma 8.1 to the subsets $\{p\}$ and $S \backslash \{p\}$, it follows that the nearest neighbor of p is sitting in a neighboring Voronoi region (i.e., it is a *Delaunay neighbor*). Therefore, it is sufficient to inspect, for each $p \in S$, all neighbors of p in the Voronoi diagram, and select the closest of them. This way, each edge of $V(S)$ will be accessed twice. Since their total number is linear, due to Lemma 2.3, we can solve the all nearest neighbor problem by constructing the Voronoi diagram.

Theorem 8.3. *Given a set S of n points in the plane, $O(n \log n)$ time and linear space are sufficient for determining, for each $p \in S$, a nearest neighbor in S.*

Once for each p its nearest neighbor in S is known, we can easily determine a pair of points whose distance is a minimum.

Corollary 8.1. The *closest pair* among n points in the plane can be determined within $O(n \log n)$ time and linear space.

Hinrichs *et al.* [407] have shown that it suffices to maintain certain parts of the Voronoi diagram by a *sweep algorithm* (cf. Section 3.4), in order to detect the closest pair.

Whereas finding the closest pair seems an easier task than constructing the Voronoi diagram, it still has an $\Omega(n \log n)$ lower bound, by reduction from the *ε-closeness problem*; cf. Theorem 3.1.

In dimension d, constructing the Voronoi diagram is no longer the method of choice for finding the closest pair, due to its exponentially increasing size; see Section 6.2. Already in 1976, Bentley and Shamos [122] provided an $O(n \log n)$ algorithm for arbitrary but *fixed* dimensions. Golin *et al.* [369] showed that randomization plus use of the floor function (which is forbidden in the algebraic decision-tree model of computation) allows the closest pair of n points in d-space to be found within expected $O(n)$ time, for any fixed d. Additional results on finding closest pairs in high dimensions, including *approximate solutions*, are described in Chapter 10.

More involved is the detection of the so-called *bichromatic closest pair* in a set S of n points. In this variant each point has a color, either red or blue, and one asks for the minimum distance between differently colored points. In the plane, a simple $O(n \log n)$ solution exists, by constructing the Voronoi diagram of, say, the r red points, in $O(r \log r)$ time, and performing *point location* for the b blue points among the red regions, in $O(\log r)$ time per step (with $r + b = n$); see Subsection 8.1.1.

Agarwal *et al.* [11] solve the bichromatic closest pair problem in general d-space, achieving a runtime of (roughly) $O((r \cdot b \cdot \log r \cdot \log b)^{2/3})$ in dimension 3.

The problem of finding the closest pair of n points can be generalized to reporting the *k closest pairs in S*, for some number $k \leq \binom{n}{2}$. A brute-force solution would sort all interpoint distances in time $O(n^2 \log n)$ and then report the k smallest. Time $O(n^2)$ is sufficient if one uses one of the well-known selection methods [234] for finding, in time $O(m)$, the kth smallest of m objects (i.e., their so-called *k-median*). Smid [658] has shown how to enumerate the $O(n^{2/3})$ smallest distances in time $O(n \log n)$ and linear space.

A simple yet elegant solution using the Delaunay triangulation DT(S) of S has been provided by Dickerson *et al.* [272]. Their approach is as follows.

The closest pair in S corresponds to the shortest edge, e_1, of the Delaunay triangulation DT(S), as Lemma 8.1 shows. The *second-closest pair*, e_2, must be in DT(S) as well, as both edges are, by Kruskal's construction, in the *minimum spanning tree* of S, which is a subgraph of DT(S); cf. Subsection 8.2.1. (Note that e_1 and e_2 cannot cross; otherwise, two of their endpoints would be closer to each other than e_2's length, by the *quadrangle inequality*. This inequality states that, in any convex quadrangle, the sum of two opposite sides lengths is strictly smaller than the sum of the two diagonal lengths.)

But the kth closest pair, (p, q), need not define an edge of DT(S), not even for small values of $k \geq 3$. However, there always exists a path of Delaunay edges connecting p and q which are of lengths smaller than $d(p, q)$ each.

This can be used for proving the following elegant algorithm correct. It maintains a *priority queue*, Q, of pairs of points by distance. Initially, Q contains all Delaunay neighbors. In the ith step, the closest pair (p, q) in Q is removed and reported ith closest. Then, for all Delaunay edges $\overline{op}$ satisfying $d(o, p) \leq d(p, q)$, the pair (o, q) is inserted into Q; similarly, each pair (p, r) is added to Q where $\overline{qr}$ is a Delaunay edge of length at most $d(p, q)$.

The performance of this algorithm is stated in the next theorem.

Theorem 8.4. *The k closest pairs of a set S of n points can be computed in time $O((n + k) \log n)$ and space $O(n + k)$.*

A similar approach even works efficiently in higher dimensions. Dickerson and Eppstein [273] showed that several *interdistance enumeration problems* for a point set S in d-space can be solved by means of DT(S). The prohibitive size of DT(S) in d-space is reduced by augmenting S to have a linear-size, *bounded-degree Delaunay triangulation*, with the method of Bern *et al.* [128].

8.1.3. *Largest empty and smallest enclosing circle*

Suppose someone wants to build a new residence within a given area, as far away as possible from each of n sources of disturbance — a typical *facility location problem*.

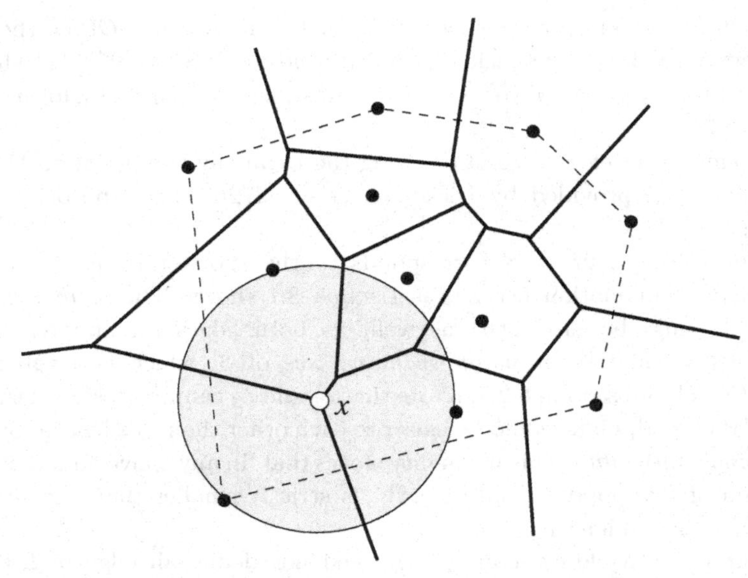

Figure 8.4. Largest empty circle with center inside the convex hull of the sites. In this example, the circle is centered at a Voronoi vertex.

If the area is modeled by a convex polygon A over m vertices, and the disturbing sites by a point set S, we are looking for the largest circle with center in A that does not contain a point of S. The task of determining this circle has been named the *largest empty circle problem.*

Shamos and Hoey [637] observed that this circle must have its center at a Voronoi vertex of $V(S)$, or at the intersection of a Voronoi edge with the boundary of A, or at a vertex of A. (See Figure 8.4, where the first case occurs.) If a circle C with center $x \in A$ contains no $p \in S$, not even on its boundary, we blow it up until it does. If its boundary now contains three sites then x is a Voronoi vertex; see Lemma 2.1. If there are two sites on its boundary, x lies on a Voronoi edge. In this case we move x along their bisector away from the two sites until the expanding circle hits a third site, or x hits the boundary of A. If, in the beginning, only one point site $p \in S$ lies on the boundary of C then we expand C, while keeping its boundary in contact with p, until x reaches a vertex of A, or one of the before-mentioned events occurs.

This observation leads to the following result.

Theorem 8.5. *The largest circle not containing a point of S, whose center lies inside a convex polygon A, can be found within $O(m+n\log n)$ time and linear space. Here, m denotes the number of vertices of A, and n is the size of S.*

Proof. We spend $O(n \log n)$ time on constructing the Voronoi diagram $V(S)$. The largest empty circles centered at the Voronoi vertices can be determined in constant time each. We can trace the convex boundary of A through the Voronoi diagram, in order to detect its intersections with the Voronoi edges, in time $O(n + m)$. Simultaneously we can find, for each vertex v of A, its nearest neighbor in S, by determining the Voronoi region containing v. $\square$

Based on the above characterization of largest empty circles, Kaplan and Sharir [443] give an efficient algorithm for preprocessing S, so that, for a given query point q, we can quickly report the largest (open) disk D with $q \in D$ but $D \cap S = \emptyset$. The storage required by their data structure is $O(n \log n)$, at a preprocessing cost of $O(n \log^2 n)$, and a query takes time $O(\log^2 n)$.

Queries of this kind occur when one wants to transmit a message over the largest possible area (which is a disk if propagation proceeds uniformly) such that q will receive the message, but none of the 'spies' sitting at the points of S. A more friendly interpretation is to position a sprinkler to water the largest possible area that includes plant q, without wettening any location in S.

Repeatedly selecting centers of largest empty circles (i.e., of optimally far away points) is a task which also arises in *quality mesh generation* within polygonal domains, as we briefly describe next.

Starting with the Delaunay triangulation, DT_0, of a convex polygon P, the triangulation can be refined by insertions of centers of largest empty circles. More precisely, DT_i is constructed from DT_{i-1} by inserting the center p_i of the largest empty circle for DT_{i-1} within P, for $i \geq 1$, until certain quality criteria are met. (Note that p_i might lie on the boundary of P, thus splitting an edge of P when being inserted.) This idea of *Delaunay refinement* (which is also called *canonical Voronoi insertion*) originates with Gonzalez [370] and Feder and Green [337], who developed it for approximating *optimal clusterings*. Strategies of this kind also play a role in *circle* and *sphere packing* problems; see the survey article by Fejes Tóth [339].

For mesh generation purposes, Delaunay refinement has been used, among others, by Chew [213], Ruppert [612], Shewchuk [649], and Aurenhammer *et al.* [99]. The last paper discusses the problem of approximating *uniform triangular meshes*, in the sense that the maximal edge length, or the maximal edge ratio or perimeter of a triangle, is to be minimized. See Figures 8.5 and 8.6 that illustrate a convex polygon, meshed with small edge ratio. Refinement also works for general *simple polygons*,

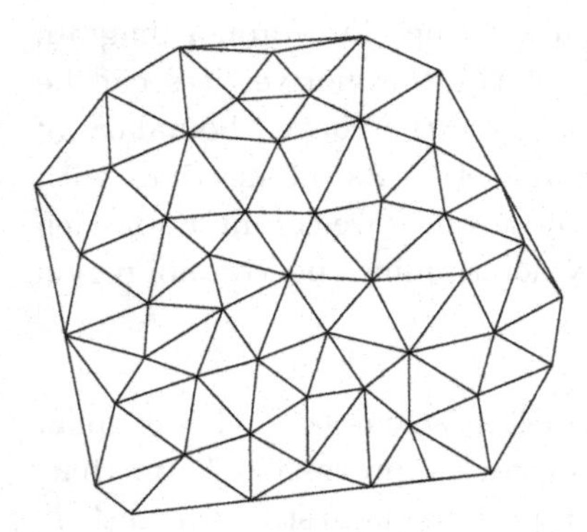

Figure 8.5. Delaunay-refined mesh with 50 points (from [99]).

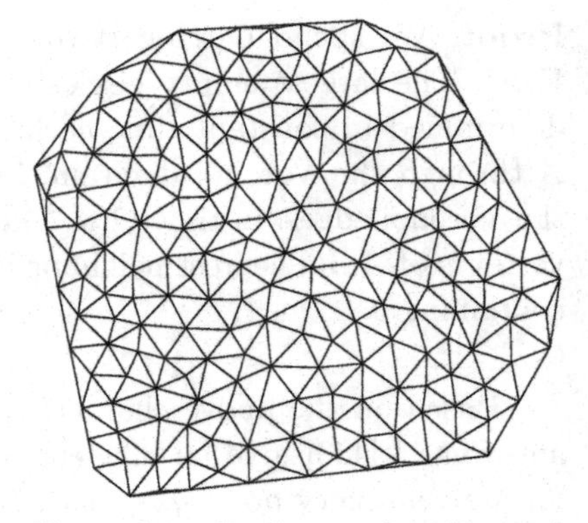

Figure 8.6. Further mesh refinement with 150 additional points.

where the *constrained Delaunay triangulation* is used (Section 5.4). The approximation bounds in [99] have been recently improved by Jiang [428], who confirmed an interesting structural result conjectured there:

For any two triangular meshes T_1 and T_2 within a fixed simple polygon P, which have the same number m (but *different* sets) of internal mesh vertices, the maximum edge length in T_1 is at least the minimum vertex distance in T_2.

An inverse problem, in some sense, is the *smallest enclosing circle problem*. It asks for the circle of minimum radius that encloses a given set S of n points in the plane. Applications are the placement of a least powerful transmitter station that can reach a given set of locations, or the placement of a service station that minimizes the maximum distance to the customers.

As was observed in Bhattacharya and Toussaint [130], there are two cases. Either the smallest enclosing circle contains three points of S on its boundary, or it runs through two antipodal points whose distance equals the *diameter* of S, that is, the longest distance spanned by S.

From the convex hull of S its diameter, together with a pair of points realizing it, can be derived in linear time; then it can be checked if the smallest circle through these points contains S.

Otherwise, the center v of the smallest enclosing circle C must be a vertex of the *farthest-site* (or *order* $n-1$) *Voronoi diagram*, $V_{n-1}(S)$, introduced in Section 6.5. Namely, the three sites on the boundary of C are farthest from v.

This yields an $O(n \log n)$ algorithm for computing the smallest enclosing circle. The *linear programming* technique by Megiddo [531] is more efficient: For fixed dimension d, it allows the *smallest enclosing sphere* for n points in d-space to be constructed in time $O(n)$. A more practical way of achieving linear (*randomized*) time is the elegant *minidisk algorithm* by Welzl [702]; here the constant in O has been shown to be a subexponential function of d by Matoušek *et al.* [522]. The minidisk algorithm also works for smallest enclosing *ellipsoids*.

The smallest enclosing circle problem is the simplest instance of a clustering problem called *k-center problem*. This problem asks for a set C of k centers such that S is covered by k disks centered at C, with the largest arising disk radius δ getting minimized; see Subsection 8.4.1. In other words, a partition of S into k subsets (clusters) and a center for each cluster are sought, such that the maximum distance, δ, of a cluster point to its center is minimized. Clearly, δ is the radius of the smallest enclosing circle for the respective cluster.

Efficient algorithms for the k-center problem are partially based on the observation that clusters are linearly separable hence hull-disjoint. If for two clusters one point each would lie on the 'wrong' side of a straight line, then re-assignment of points to centers would decrease (or leave equal) the critical radius δ. Particular attention has been paid to the *2-center problem*, for which various superquadratic algorithms were proposed until Sharir [640] and Eppstein [322] succeeded in achieving near-linear runtimes; $O(n \log^c n)$ for $c = O(1)$, and $O(n \log^2 n)$, respectively.

Recently, a *coloring version* of the 2-center problem has been investigated in Arkin *et al.* [65]. The points in S are to be colored (pairwise) either red or blue. Thereby, a given function $f(\delta_R, \delta_B)$ of the radii of two disks $D_R = (c_R, \delta_R)$ and $D_B = (c_B, \delta_B)$ is to be minimized, such that D_R covers all the red points, and D_B covers all the blue ones. Unlike for the bichromatic closest-pair problem (in Subsection 8.1.2), the coloring is *not* given but part of the output. This *bichromatic 2-center problem* is motivated by an application in air traffic management, where a flight corridor (the line segment $\overline{c_R c_B}$) should be designed such that the traffic outside the corridor is minimized.

For f being the maximum or the sum of radii, polynomial-time solutions of $O(n^3 \log^2 n)$ and $O(n^5 \text{ polylog } n)$, respectively, are given in [65]. They also obtain much more efficient algorithms for approximating the optimum value of f, and for the L_∞-metric.

We have seen that, for the largest empty circle problem, the closest-site Voronoi diagram $V(S)$ of the input point set S is of help, whereas for the smallest enclosing circle problem, the farthest-site Voronoi diagram $V_{n-1}(S)$ plays this role. In solving the *smallest annulus problem* for S, information from both diagrams is used. Here, two co-centric disks D_1 and D_2 of radii $r_1 > r_2$ are sought, such that D_1 covers S and D_2 avoids S, and the width

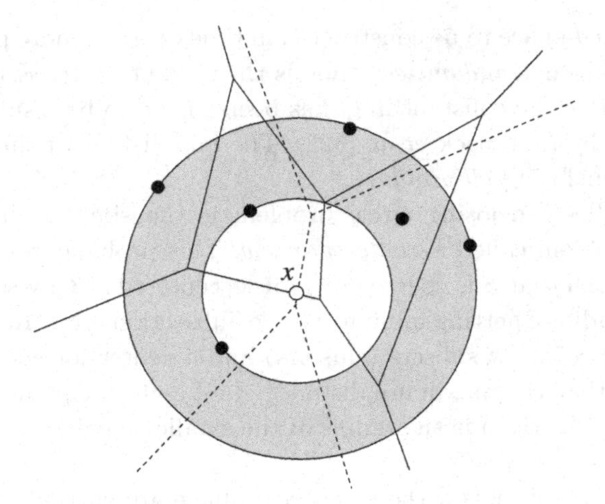

Figure 8.7. Minimum-width annulus covering a set of six points. Its center, x, is located at a crossing of their closest-site and farthest-site Voronoi diagrams.

$r_1 - r_2$ of the *annulus* $D_1 \backslash D_2$ they define is minimized; see Figure 8.7. Applications stem from testing the *roundness* of a point set, used as a tolerance measure in mechanical design; see Le and Lee [489], Swanson *et al.* [672], and Smid and Janardan [659].

The center of the smallest annulus is a vertex of $V(S)$, or a vertex of $V_{n-1}(S)$, or the intersection point of two edges in these diagrams. This gives an $O(n^2)$ algorithm for its construction, as was first observed in Ebara *et al.* [299]. Using different (and more complex) techniques, Agarwal and Sharir improved this to $O(n^{\frac{3}{2}+\varepsilon})$; we refer the interested reader to their survey article [16] on geometric optimization algorithms. Moreover, a $(1+\varepsilon)$-approximation can be obtained in time $O(n+\varepsilon^{-c})$, for a suitable constant $c > 1$, using the *linear programming* methods in Chan [188].

If roundness of a convex or simple polygon P is to be checked, then the *medial axis* of P (Sections 5.1 and 5.2), and the farthest-site Voronoi diagram of P's vertices, have to be taken for characterizing the location of the center of the minimum-width annulus that covers P; see [489, 672].

8.2. Subgraphs of Delaunay triangulations

The Delaunay triangulation DT(S) of a set S of n point sites contains, as subgraphs, various structures with diverse applications. One example, the *all nearest neighbor graph* of S, has already been discussed in Subsection 8.1.2. Efficient algorithms for computing these structures follow from the fact that DT(S) has size $O(n)$, and can be constructed in $O(n \log n)$ time, in the plane. This positive effect is partially lost in higher dimensions, as DT(S) may be the *complete graph* on S already in 3-space. Alternative methods for computing subgraphs of Delaunay triangulations in d dimensions, and for approximating them, are discussed in Chapter 10.

8.2.1. *Minimum spanning trees and cycles*

A *minimum spanning tree*, MST(S), of S (sometimes also called *shortest connection network for* S) is a planar straight line graph on S which is connected and has minimum total edge length. Such a graph is necessarily acyclic and thus is a tree. Minimum spanning trees play an important role, for instance, in transportation problems, pattern recognition, and clustering. Shamos and Hoey [637] first observed the connection to Delaunay triangulations.

Lemma 8.2. *MST(S) is a subgraph of DT(S).*

Proof. Let e be an edge of MST(S), and see Figure 8.8. Removal of e splits MST(S) into two subtrees, and S into two subsets S_1 and S_2. Clearly, e is the shortest edge connecting S_1 and S_2, because a shorter edge, if existent, would lead to a spanning tree shorter than MST(S). It follows that the circle C with diameter e is empty of sites in S: A site enclosed by C would have to belong to either S_1 or S_2, leading to a connection between S_1 and S_2 shorter than e. By Definition 2.2 in Chapter 2, C proves e Delaunay. □

We thus can select the $n - 1$ edges which build up MST(S) from the at most $3n - 6$ edges of DT(S), rather than from all the $\binom{n}{2}$ edges spanned

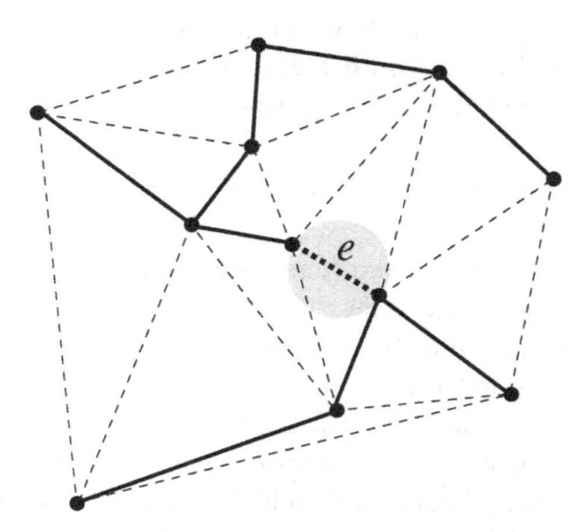

Figure 8.8. Delaunay triangulation, and minimum spanning tree (in bold style).

by S, by standard application of *Kruskal's* [475] or *Prim's* [597] *'greedy' algorithms.* Edges are considered in increasing length order, and an edge is classified as a tree edge if it does not violate acyclicity.

In Kruskal's version, to detect if a considered edge $e = (p, q)$ yields a cycle, the *forest* of subtrees of MST(S) constructed so far (which initially consists of n singleton vertices, the sites in S) is stored in a *union-find data structure*; see e.g. [234, 538]. Performing FIND(p) $= i$ and FIND(q) $= j$ returns the indices of the sets M_i and M_j that store the vertex sets (subsets of S) of the current subtrees T_i and T_j that contain p and q, respectively. Now, if $i \neq j$ then edge e does not close a cycle but rather joins the subtrees T_i and T_j into a larger one. In order to reflect this in the data structure, we join the sets M_i and M_j with UNION(M_i, M_j).

We choose an implementation that supports the FIND operation in $O(1)$ time, and all $n-1$ UNION operations in total time $O(n \log n)$. For each set M_i, we just link all its members (the vertices of T_i) to the set index, i. When joining two sets, the links for the smaller set are redirected to the larger set. This way, each fixed member changes the set which contains it at most $O(\log n)$ times. An overall runtime of $O(n \log n)$ for constructing MST(S) is obtained, including the computation of DT(S).

Lemma 8.2 implies that MST(S) is a *crossing-free* tree, i.e., its edges do not intersect in their interiors. This is because the edges in any triangulation do not cross.

There is a generalization of this lemma, for constrained Delaunay triangulations and spanning trees. Consider an arbitrary crossing-free spanning tree, T, of the point set S. We define the *minimum spanning tree of S constrained by T*, for short $\mathrm{MST}|_T(S)$, as the shortest spanning tree for S that does not cross (but may contain) edges of T. It is easy to see that $\mathrm{MST}|_T(S)$ is a subgraph of the *constrained Delaunay triangulation* CDT(S, T) discussed in Section 5.4. Exploiting this containment relation, Aichholzer *et al.* [36] showed that the sequence of spanning trees

$$T_0, T_1, \ldots, T_k, \quad \text{with } T_0 = T \quad \text{and} \quad T_i = \mathrm{MST}|_{T_{i-1}}(S), \quad \text{for } 1 \leq i \leq k$$

terminates at $T_k = \mathrm{MST}(S)$, the (unconstrained) minimum spanning tree of S, after $k = O(\log n)$ iterations. This so-called *fixed tree theorem* allows us to gradually though quickly transform two arbitrary spanning trees of a given point set into each other — a result with possible applications to *morphing* of skeletons and shapes in the plane.

Lemma 8.2 also generalizes to metrics different from the Euclidean (for example L_1, see Hwang [416]), and to higher dimensions. When staying in the plane, equally efficient construction algorithms result.

In higher dimensions, however, DT(S) may not be of much use for the construction of MST(S), because of the possibly quadratic size. Subquadratic worst-case time algorithms for computing the Euclidean minimum spanning tree in d-space exist, for example by Yao [710] and Agarwal *et al.* [11]. The latter algorithm runs in randomized time $O((n \log n)^{4/3})$ for dimension 3, and uses the computation of a *bichromatic closest pair* in S as a subroutine; see Subsection 8.1.2.

It still remains open whether an $O(n \log n)$ time algorithm can be developed, at least for 3-space. The problem is apparently easier for the L_1*-metric* and the L_∞*-metric.* The minimum spanning tree can be computed in $O(n \log n)$ time for these metrics, in fixed dimensions d; see Krznaric [476].

The minimum spanning tree MST(S) should not be confused with the (Euclidean) *minimal Steiner tree* of a point set S, which is the *shortest connection network* for S when the usage of *extraneous* points is allowed for length reduction. Indeed, already a three-point example shows that adding a fourth point can reduce the length of the resulting tree structure. In particular, the *Fermat point* of a triangle with angles less than $\frac{2\pi}{3}$, which is the point minimizing the total distance to the triangle's vertices, has this property. (By contrast, the minimum spanning tree would consist of the two shortest sides of the triangle.) Minimal Steiner trees may use $O(n)$ Fermat points as extraneous points. Finding a minimal Steiner tree is an *NP-hard* problem; see Garey and Johnson [352].

Another related tree concept is the *minimum diameter spanning tree* of S, where the Euclidean length of the longest path in the tree is to be kept to minimum. Figure 8.9 displays an example. This tree is either a

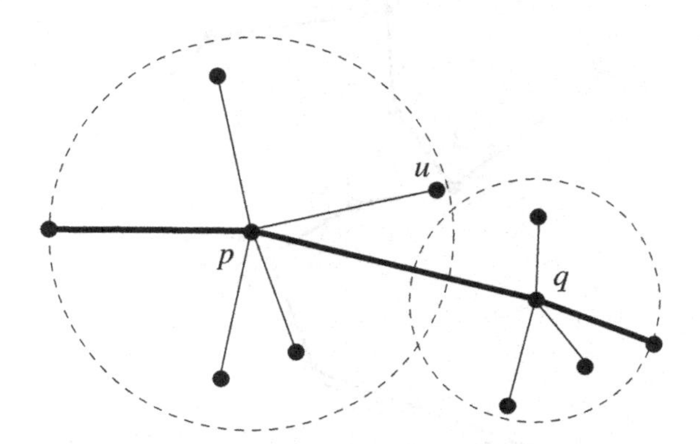

Figure 8.9. The Euclidean minimum diameter spanning tree is a double-star in this case. Its diameter is realized by a path of 3 edges, drawn in bold style. Note that point u is connected to p and not to q, whereas $d(u,p) > d(u,q)$.

star or a *double-star* and — by exploiting its simple topology — can be found in $O(n^3)$ time using the farthest-site and the order-$(n-2)$ Voronoi diagrams of S (Section 6.5). Consult Ho *et al.* [408], and Chan [189] for an $O(n^{17/6+\varepsilon})$ improvement. One can compute $(1+\varepsilon)$-length approximations much faster, in $O(n+\varepsilon^{-3})$ time, with the grid-oriented methods in Spriggs *et al.* [662]. Some *bi-criteria* optimization problems based on minimum diameter spanning trees have been shown to be *NP-hard*, in Seo *et al.* [634].

A *travelling salesman tour*, TST(S), for a set S of point sites in the plane is a minimum length *spanning cycle* (cycle passing through all the sites). It is known that TST(S) is not a subgraph of DT(S), in general. Dillencourt [276] showed that DT(S) need not even contain *any* spanning cycle (a so-called *Hamiltonian cycle*), and that finding Hamiltonian cycles in Delaunay triangulations is *NP-hard* [279]. He also gives a partial explanation for the fact that DT(S) is Hamiltonian in most cases [277].

A related optimization graph which is known to be not part of DT(S) is a *minimum length matching* for S; see Akl [46]. However, DT(S) is guaranteed to contain some (*perfect*) *matching* (i.e., a set of edges that pairs up all the points in S except possibly one, and hence is of cardinality $\lfloor \frac{n}{2} \rfloor$); see Dillencourt [277]. This property is not shared by all triangulations of S.

Finding a travelling salesman tour has been shown to be *NP-hard* in Papadimitriou [578], but a *factor-2* length *approximation* of TST(S) can be found easily, using MST(S). Let $A(S)$ be a cycle through S that results from traversing MST(S) in *preorder*; consult Figure 8.10. Rosenkrantz *et al.* [606]

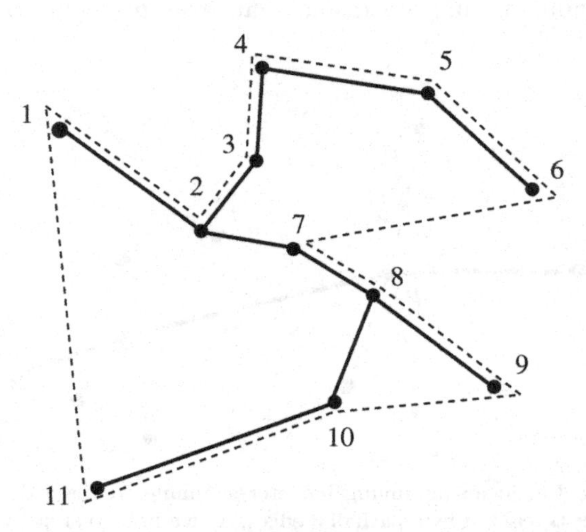

Figure 8.10. Minimum spanning tree, and short cycle (in dashed style).

observed the following. Let $|X|$ denote the total length of a set X of edges.

Lemma 8.3. $|A(S)| < 2 \cdot |\mathrm{TST}(S)|$.

Proof. When traversing each edge of MST(S) twice, a tour longer than $A(S)$ is obtained. Hence $|A(S)| < 2 \cdot |\mathrm{MST}(S)|$ holds. To see $|\mathrm{MST}(S)| < |\mathrm{TST}(S)|$, note that removing an edge from TST(S) leaves a path which is some, but not neccessarily the minimum, spanning tree of S. □

$A(S)$ can be constructed in linear time from MST(S). Note that tree traversal in preorder is facilitated by the fact that the *maximum vertex degree* in MST(S) is at most six: Consecutive edges around each tree vertex form an angle of at least 60°, to let them be the two shortest edges of the spanned triangle.

A more sophisticated construction of a *spanning cycle* by means of MST(S) is given in Christofides [222]. An approximation factor of 1.5 is achieved, at an expense of roughly $O(n^2\sqrt{n})$ in construction time.

Another *NP-hard* problem, which has a *factor-2 approximation* by means of MST(S), is the construction of optimal radii graphs; see Chen and Huang [200]. Let R be a real-valued vector that associates each site $p \in S$ with an individual radius $r_p > 0$. The corresponding *radii graph*, $G_R(S)$, contains an edge between two sites $p, q \in S$ iff $d(p,q) \leq \min\{r_p, r_q\}$. The optimization problem now asks for a radii vector R^* such that $G_{R^*}(S)$ is connected and $|R^*| = \sum_{p \in S} r_p^*$ is minimum. Radii graphs have applications, among others, in the design of strongly connected *radio* and *sensor networks*.

Consider the vector R that takes, as a radius for each site $p \in S$, the length of the longest edge of MST(S) incident to p. Then $G_R(S)$ is connected, as it has to contain MST(S) as a subgraph. Moreover, we have the following quality guarantee.

Lemma 8.4. $|R| < 2 \cdot |R^*|$.

Proof. Let $e = (p,q)$ be some edge of MST(S). Then e appears at most twice as a radius in R, namely if e is the longest edge incident to p, and the longest edge incident to q. This gives $|R| \leq 2 \cdot |\mathrm{MST}(S)|$.

To verify $|\mathrm{MST}(S)| < |R^*|$, let us consider $G_{R^*}(S)$. As being connected, this graph contains some spanning tree, T. We may orient the edges of T such that each site $p \in S$ (except one) has a unique edge e_p of T pointing at it. As e_p is also an edge of $G_{R^*}(S)$, we have $r_p^* \geq |e_p|$. Hence $|R^*| > |T| \geq |\mathrm{MST}(S)|$. □

Lemma 8.4 remains true for more general measures $|R| = \sum_{p \in S} f(r_p)$, for any increasing function f. This is because MST(S) also minimizes $\sum_{e \in T} f(|e|)$, for all spanning trees T of S. The case $f(r) = r^2$ is relevant to the application mentioned above, as the received power of a radio terminal decreases with the square of the distance.

8.2.2. *α-shapes and shape recovery*

Extracting the shape of a given set S of point sites in the plane is a problem that arises, for example, in *data visualization*, *pattern recognition*, and in particular, in *curve reconstruction*. To some extent, the shape of S is reflected by the convex hull of S. The edges of the convex hull are part of the Delaunay triangulation DT(S). The concept of *α-shape*, introduced in Edelsbrunner *et al.* [306, 300], generalizes the convex hull of S for the sake of better shape approximation, while still remaining a subgraph of DT(S).

For $\alpha > 0$, the *α-shape* $\alpha(S)$ of S is defined to contain an edge between sites $p, q \in S$ iff there is some (open) disk of radius α that avoids S and has p and q on its boundary. By Definition 2.2 in Chapter 2, $\alpha(S)$ is always part of DT(S) and thus has only $O(n)$ edges.

For α being sufficiently large, the *convex hull* of S is obtained. The smaller is the value of α, the finer is the level of resolution of the shape of S; see Figure 8.11. The graph $\alpha(S)$ contains no edges (only singleton points) if the disk diameter, 2α, is smaller than the distance of the *closest pair* in S.

The definition of $\alpha(S)$ can be extended to negative values of α, by requiring that the disks of radius $-\alpha$ fully contain S. In this case, approximations of the shape of S more crude than the convex hull are obtained, and the edges of $\alpha(S)$ appear in the dual triangulation of the *farthest-site Voronoi diagram* of S, that is, in the *farthest-site Delaunay triangulation* of S; see Section 6.5.

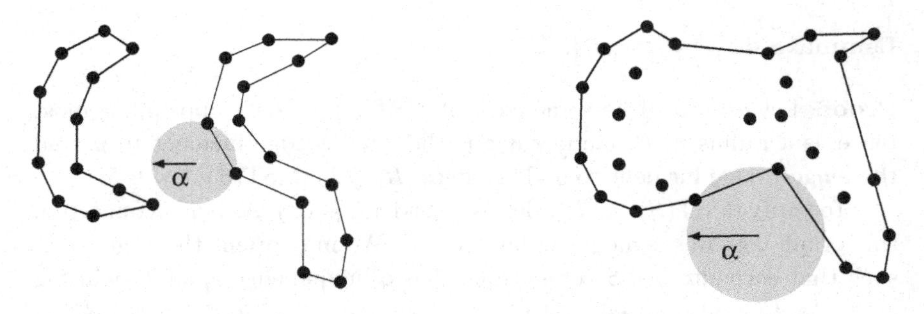

Figure 8.11. Two α-shapes of a point set for different values of parameter α.

It follows that all the edges of the whole family of α-shapes for S, for $-\infty < \alpha < \infty$, are contained in two triangulations of size $O(n)$. For each triangulation edge e, an interval of activity can be specified, containing all values of α such that $\alpha(S)$ contains e. These activity intervals can be computed in $O(n \log n)$ time, and allow us to extract $\alpha(S)$ for an input value α in $O(n)$ time; see [306].

Another nice feature of α-shapes is their flexibility. The notion of α-shape, along with its relationship to Delaunay triangulations, nicely generalizes to higher dimensions. α-shapes in 3-space are of particular interest, and implementations have been used in several areas of science and engineering; see Edelsbrunner and Mücke [307].

Also, the concept of α-shape can be generalized to represent different levels of detail at different parts of space; see Edelsbrunner [302]. This is achieved by weighting the sites in the given set S individually. The corresponding *weighted α-shapes* are subgraphs of the *regular triangulation* for the weighted set S which, in turn, is the dual of the *power diagram* for this set. The combinatorial and algorithmic properties of weighted α-shapes are similar to that of their unweighted counterparts. Interpreting weighted sites as spheres leads to a compact representation of the respective *union of balls* by their weighted α-shapes. This is discussed in Subsection 6.6.2.

Different approaches for *reconstructing shapes and surfaces* in 3-space by means of Delaunay triangulations have been pursued. Boissonnat and Geiger [143] exploit additional information on the sites available in certain applications, namely that S is contained in k parallel planes (corresponding to cross sections of the object to be reconstructed, e.g., in computer tomography). The Delaunay edges connecting sites in neighboring planes can be computed in an *output-sensitive* manner; see Boissonnat *et al.* [139]. The edges describing the final shape are selected according to several criteria. Geiger [355] reports that satisfactory shapes are produced even for complicated medical images.

A particularly useful concept for reconstructing an object from a set S of *sample points* on its boundary is the so-called *crust* of S, introduced in Amenta and Bern [58]. As does the α-shape, also the crust consists of faces of the Delaunay triangulation of S. In the two-dimensional case, all edges are included in the crust that can be enclosed by a circle empty of sites *and* of all vertices of the Voronoi diagram of S. In three dimensions, the (sphere) emptiness condition on Voronoi vertices has to be mildened for certain reasons, to so-called *pole vertices*, which are the two vertices per region (if they exist) that are farthest from the object boundary on either side. If S fulfills specific sampling conditions (namely, being an *r-sample*, a condition based on the *medial axis* of the object), then the crust indeed connects S by edges (triangles) in a topologically correct way.

Surface recovery can also be based, as done with advantage in Boissonnat and Cazals [140], on the *natural neighbor interpolants* described in Section 4.2. Manifold reconstruction in high dimensions is a demanding problem. Various *abstract simplicial complexes* have been defined and utilized for this purpose; see Section 10.5 for a short discussion of these structures, and their relationship to Delaunay simplicial complexes.

An extensive study of curve and surface reconstruction techniques is provided in the book by Boissonnat and Teillaud [149].

8.2.3. β-skeletons and relatives

Another family of graphs, defined by empty disks and thus consisting of subgraphs of $\mathrm{DT}(S)$, is the family of *β-skeletons* of a set S of point sites. β-skeletons have been introduced in Kirkpatrick and Radke [457] as a class of *empty neighborhood graphs* and have later received increased attention, not least because of their relation to minimum-weight triangulations. In [457] both a *circle-based* and a *lune-based* version of β-skeletons are proposed. Here we will mainly consider the former, as defined below.

In contrast to α-shapes, the radii of the empty disks defining a β-skeleton are not fixed but depend on the interpoint distances in S. Let $p, q \in S$. For $\beta \geq 1$, an edge between p and q is included in the *β-skeleton* $\beta(S)$ iff the two disks of diameter $\beta \cdot d(p,q)$ which have the segment $\overline{pq}$ as a chord are empty of sites in S. See Figure 8.12.

Note that $\beta(S)$ is a subgraph of $\beta'(S)$ if $\beta > \beta'$. For the minimum value, $\beta = 1$, the two disks above coincide, and $\overline{pq}$ becomes their diametrical chord. The so-called *Gabriel graph* of S is obtained, which thus contains all possible circle-based β-skeletons for S. Whereas $\beta(S)$ may be disconnected for any $\beta > 1$, the Gabriel graph is always connected, as it contains the *minimum spanning tree* of S as a subgraph; cf. the proof of Lemma 8.2. It is easy to see that the Gabriel graph just consists of those edges of $\mathrm{DT}(S)$ that cross their dual Voronoi edges. This observation yields an $O(n \log n)$ algorithm for its construction. In fact, also general β-skeletons can be constructed from $\mathrm{DT}(S)$ in linear time; see Jaromczyk *et al.* [427] and Lingas [508]. Gabriel graphs have proved useful in the processing of geographical data, see Gabriel and Sokal [346] and Matula and Sokal [519], and more recently, in graph drawing questions and for routing in mobile networks, see Section 8.3 and Subsection 9.4.2.

It has been first observed by Keil [449] that, for β large enough, $\beta(S)$ is contained in the *minimum-weight triangulation*, $\mathrm{MWT}(S)$, of S, which

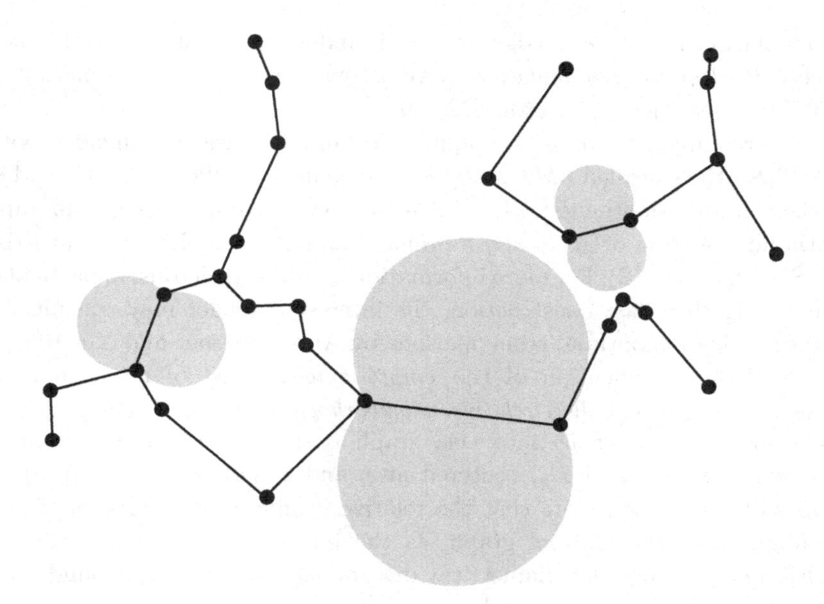

Figure 8.12. Disconnected (circle-based) β-skeleton for $\beta = \frac{6}{5}$.

is defined as a triangulation of S having minimum total edge length; see also Section 4.2. The original bound $\beta \geq \sqrt{2}$ in [449] has been improved in Cheng and Xu [206] to $\beta > 1.1768$, which is close to the value $2/\sqrt{3}$ for which a counterexample is known. Apart from the edges of the convex hull of S, only the shortest edge in S, as well as so-called *unavoidable edges* have been known to be part of MWT(S) before. (An edge interior to the convex hull of S is called unavoidable if no other edge spanned by S crosses it. Unavoidable edges have to appear in every triangulation of S.)

One of the heuristics for the intriguing problem of constructing MWT(S) first computes $\beta(S)$ for the smallest admissible value of β. The obtained k connected components of $\beta(S)$ then can be completed length-optimally to a triangulation, using *dynamic programming*, in $O(n^{k+2})$ time; see Cheng *et al.* [204]. Thus, if $\beta(S)$ is connected ($k = 1$) or has only a constant number of components, then MWT(S) can be constructed in *polynomial time*. Unfortunately, k tends to be linear in n.

Let us mention that there are more effective approaches to constructing MWT(S). One is based on the notion of a *light edge*, which is an edge e spanned by S such that all other spanned edges that cross e are longer. If the set of light edges for S happens to form a full triangulation of S, then it has to be the minimum-weight triangulation and, at the same time, the *greedy triangulation* (Section 4.2). Otherwise, a partition of the greedy triangulation into $k \geq 2$ edge subsets is induced, and if $k = O(1)$ then such *k-level greedy triangulations* give a constant-factor approximation of MWT(S); see Aichholzer *et al.* [32, 39].

Interestingly, there is a graph that 'almost always' coincides with MWT(S): the so-called *LMT-skeleton* defined in Belleville *et al.* [119] and in Dickerson and Montague [274]. This is an easy-to-compute large subgraph of the intersection of all *locally minimal triangulations* (LMTs) that exist for S (cf. Section 4.2). For more information on LMT-skeletons, in particular concerning their fast construction, the interested reader may consult the survey article on optimal triangulations by Aurenhammer and Xu [106].

A slight modification of the *empty neighborhood* of an edge in S gives rise to the so-called *relative neighborhood graph* of S. For $p, q \in S$, inclusion of the edge $\overline{pq}$ into this graph is done iff the intersection of the two disks D_p and D_q, centered at p and q and of radius $d(p,q)$, is empty of other sites. Note that the relative neighborhood graph is always a subgraph of the Gabriel graph, as the intersection $D_p \cap D_q$ (the so-called *lune*) covers the diametrical disk of $\overline{pq}$. On the other hand, this graph is still connected, as it has to contain the minimum spanning tree of S: Every edge e of MST(S) has an empty lune $L(e)$, because $s \in L(e)$ for another site s would mean that e is the longest side of the triangle spanned by s and e, which cannot be in MST(S). Applications in *pattern recognition* are reported in Toussaint [686]. Relative neighborhood graphs are constructible in $O(n \log n)$ time by exploiting — once more — their containment in DT(S); see Supowit [670] and Kirkpatrick and Radke [457].

When we let the disks D_p and D_q move on the straight line through p and q, such that their radii enlarge to $\beta \cdot d(p.q)$ for $\beta > 1$ while their boundaries still touch the segment $\overline{pq}$ at one endpoint, we obtain the spectrum of *lune-based* β*-skeletons* mentioned at the beginning of this subsection.

Again, these skeletons get sparser with increasing values of β; they are all part of the relative neighborhood graph. For $\beta = \infty$, the lune $D_p \cap D_q$ degenerates to a strip orthogonal to the segment $\overline{pq}$, and the ∞-skeleton is acyclic and consists of a forest of *monotone paths.* In spite of being subgraphs of DT(S) — as are their circle-based analogs — lune-based β-skeletons are less easy to deal with algorithmically. For $1 < \beta \leq 2$, a runtime of $O(n \log n)$ can still be achieved; see e.g. Lingas [508]. Also, Majewska and Kowaluk [515] showed recently that the entire 'β-spectrum' within this range is computable in $O(n \log^2 n)$ time. The best known construction algorithm for $2 < \beta \leq \infty$ is by Kowaluk [474] and runs in time $O(n^{3/2} \log^{1/2} n)$, which slightly improves over an earlier solution by Rao and Mukhopadhyay [601].

Versions of β-skeletons for values $\beta < 1$ exist, too. These graphs are non-planar in general, and get arbitrarily dense in the extreme case: The complete graph K_n is obtained, once β falls short of half of the minimum interpoint distance in S. Hurtado *et al.* [413] gave worst-case optimal $O(n^2)$ time algorithms for these structures.

8.2.4. *Paths and spanners*

Let S be a set of n point sites in the plane. A (connected) straight line graph G on S is said to have *dilation* t if, for any $p, q \in S$, the length of the *shortest path* in G between p and q is at most $t \cdot d(p, q)$, where d stands for the Euclidean distance. In this case, G is also called a t-*spanner* (of the *complete Euclidean graph*) for S.

Intuitively speaking, spanners are graphs on the point set S which contain shortest paths not much longer than the direct distance between source and target, and the objective is to keep their number of edges small, i.e., significantly subquadratic. *Sparse* t-spanners (i.e., which have only $O(n)$ edges) are of special interest in various applications, including the field of robotics and the study of wireless ad hoc networks. The dilation t is also referred to as the *spanning ratio* or the *stretch factor* of a t-spanner in the literature.

The *minimum spanning tree* MST(S), though being optimally sparse, may have dilation $\Theta(n)$. Also the (circle-based) β-*skeleton* can have unbounded dilation, including the special case of *Gabriel graphs*; see Eppstein [323]. By contrast, DT(S) is sparse but 'sufficiently connected' to exhibit constant dilation, independently from the size n of S. Consequently, good spanners for S can be constructed in $O(n \log n)$ time. A dilation of $t \approx 5$ for DT(S) is proved in Dobkin *et al.* [283], a result which has been strengthened to $t \approx 2.5$ in Keil and Gutwin [450] and, very recently, to $t < 1.998$ in Xia [708]. An easy lower bound is $t \geq \pi/2$. However, as Bose *et al.* [159] have shown, the dilation of DT(S) is almost always strictly larger than $\frac{\pi}{2}$. In certain applications, sparse spanner graphs where the *degree* of every vertex is bounded by a constant are sought. In fact, DT(S) contains such a graph; see, e.g. Li and Wang [506] and Bose, Smid, and Xu [161].

Surprisingly, competing upper bounds on the dilation t can be achieved by taking the Delaunay dual of generalized Voronoi diagrams.

The convex distance function whose unit circle is an equilateral triangle leads to a *shape Delaunay tessellation* with $t = 2$. (See Section 7.2 for convex distance functions, and Subsection 7.2.2 for their dual shape tessellations.) This bound was shown in Chew [212], along with the following result concerning the *constrained Delaunay triangulation* (Section 5.4) for the

L_1-metric: There is always a path between two given sites, whose length is at most $\sqrt{10}$ times the *geodesic distance* with respect to the constraining line segments. This yields an $O(n \log n)$ time algorithm for computing constant approximations of optimal paths in the presence of polygonal obstacles. It can be shown that shape Delaunay tessellations always exhibit a constant dilation, that depends only on the underlying convex shape; see Bose *et al.* [156].

Path lengths can be reduced by overlaying several graphs that show a small dilation in pre-specified directions. This idea has been pursued by Abam *et al.* [1], who utilize shape Delaunay tessellations induced by rotational copies of a sharp diamond whose smaller angle is a function of some (small) value $\varepsilon > 0$ (Figure 8.13). For directions close to parallel to the diameter of the diamond, the dilation then is only $1 + \varepsilon$. Now, using $O(1/\varepsilon^2)$ symmetrically rotated diamond copies, the same dilation can be maintained overall, at a price of increasing the size of the spanner graph to $O(n/\varepsilon^2)$. Moreover, a *kinetic* $(1 + \varepsilon)$*-spanner* is obtained in this way, which yields roughly $O(n^2/\varepsilon^2)$ events in the worst case, and has only $O(1)$ response time, assuming that the trajectories of the points in S can be described by reasonably simple curves; cf. Section 9.1.

Alternative triangulations of S are known to exhibit good spanner properties. Examples are the *greedy triangulation* and the *minimum-weight triangulation*; see Das and Joseph [238]. But only for DT(S) and its convex shape variants there are algorithms available that are worst-case efficient and easy to implement. However, graphs based on certain *edge exclusion* properties (which are shared, for example, by the minimum-weight triangulation) lead to good and efficiently computable spanners as well [161]. Observe that Delaunay triangulations, by contrast, are defined via an *edge inclusion* property, namely, the empty-circle (or empty-shape) property. Sharpening this inclusion property led to various subgraphs of DT(S), like the Gabriel graph and the nearest neighborhood graph, in Subsection 8.2.3.

In higher dimensions, Delaunay tessellations may lose their sparseness and thus are of minor interest for computing spanners. Different techniques have been used with success; see e.g. Vaidya [688], Chandra *et al.* [192], and the methods described in Chapter 10. An extensive treatment of spanner graphs is presented in the book by Narasimhan and Smid [561].

Let us mention that, in principle, any *convex partition* with vertex set S (i.e., any partition of the convex hull of S into convex polygons whose edges cover the points in S) is a candidate for a spanner graph — for instance, for the purpose of compass routing of messages, where network links are followed so as to get closer to the final recipient at each particular step.

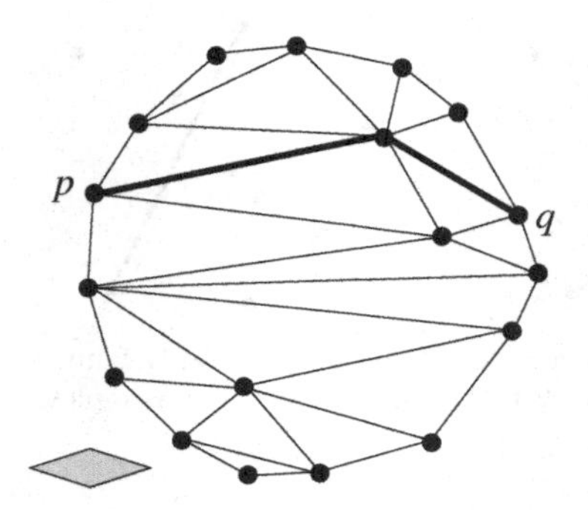

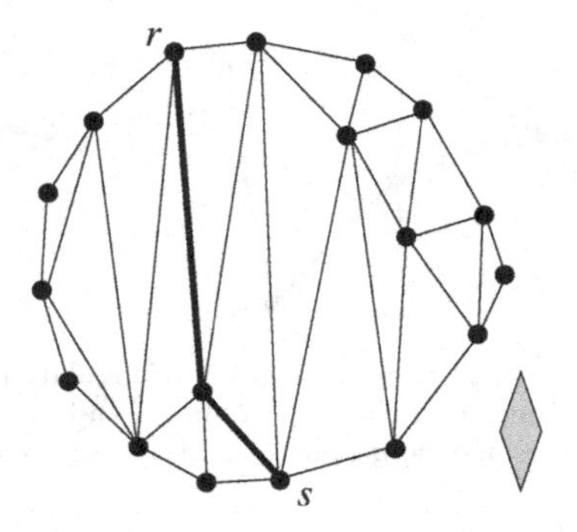

Figure 8.13. Shape Delaunay tessellations for a fixed set of sites and two different diamond shapes. The graphs allow for short paths in the horizontal and vertical direction, respectively.

Convexity ensures that messages can never get stuck at any node of the network.

There exists a considerable literature on *minimum convex partitions*, concerning their size (number of faces), weight (total edge length), and other criteria to be optimized, also in higher dimensions. We refer to Dumitrescu *et al.* [291], Dumitrescu and Tóth [292], and the references given there. In particular, minimum-weight convex *Steiner partitions* in the plane, where extraneous points are admitted to reduce weight, and points of S might lie *on* edges of the partition, are of relevance to spanner graph extraction. In [292] it is shown that such a partition can never be longer than $O(W \cdot \log n/ \log\log n)$, where W denotes the weight of a minimum spanning tree of S.

8.3. Supergraphs of Delaunay triangulations

The Delaunay triangulation is a planar graph and thus has a linear number of edges and triangles. In spite of these desirable features, more dense (non-planar) graphs which still exhibit some of the Delaunay properties, like well-shaped triangles, are sought in certain applications. In particular, several useful *supergraphs* of $\mathrm{DT}(S)$ have been investigated in the literature.

8.3.1. *Higher-order Delaunay graphs*

The Delaunay triangulation can be extended to higher order, in a way quite similar to the Voronoi diagram (Section 6.5). For $k \geq 0$, Gudmundsson *et al.* [386] define the *order-k Delaunay graph*, k-$\mathrm{DT}(S)$, of S to contain an edge between two sites $p, q \in S$ if and only if there exists a circle passing through p and q that encloses at most k sites of S; see Figure 8.14. Equivalently, one can define the *scope* of a triangle spanned by S as the

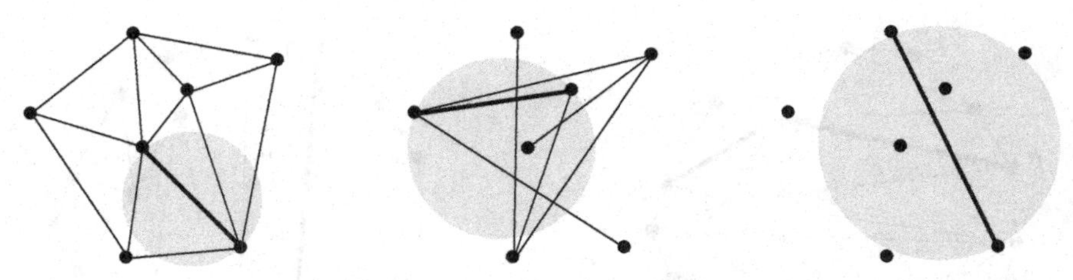

Figure 8.14. Delaunay triangulation DT(S) (left), and additional edges of the Delaunay graphs 1-DT(S) (middle) and 2-DT(S) (right). 2-DT(S) is already the complete geometric graph K_7 on these seven sites.

number of sites enclosed by its *circumcircle*. The graph k-DT(S) then contains all triangles of scope at most k.

Clearly, 0-DT(S) is just the classical Delaunay triangulation DT(S), and we have the inclusion property k-DT(S) $\subset$ $(k+1)$-DT(S). The motivation behind the generalization to higher order stems from the wish of gaining more flexibilty in choosing triangles, to build triangulations that model *elevation terrains* in geographic information systems applications. However, there exist edges and triangles in the graph k-DT(S) which cannot be used in *any* triangulation of S built from triangles of scope k. In [386] it is shown how to compute all the (in this sense) useful triangles in time $O(nk^2 + n \log n)$. They also prove that every triangulation of S with scope-1 triangles can be obtained from DT(S) by flipping *independent edges*, i.e., edges that do not belong to the same triangle.

Triangulations of S which are solely composed of triangles of scope $\leq k$ are called *order-k Delaunay triangulations*. The family of order-1 Delaunay triangulations is a suitable class where optimization regarding additional criteria can be done efficiently. Assume that some third coordinate is given for each point site in S, and consider the resulting *triangular surface* in three-space where sites are elevated accordingly. Then an order-1 triangulation that minimizes the number of local extrema in its terrain can be computed in $O(n \log n)$ time. Also, shapes of triangles and degrees of vertices can be optimized efficiently within that class. The maximal size of this class is 2^{n-3}; see Mitsche *et al.* [545], who also stress the fact that arbitrary large point sets may have only a single order-k Delaunay triangulation, even for high k (which then is just DT(S)).

Of particular interest in terrain modeling are *constrained* versions of *low-order Delaunay triangulations*, studied in Gudmundsson *et al.* [387]. (See Section 5.4 for the concept of a constrained Delaunay triangulation.) In this setting, a constraining set L of edges is prescribed, and a Delaunay triangulation of lowest order is sought which includes L. Note that, by contrast, the (classical) constrained Delaunay triangulation for S and L may contain triangles of very high scope. Define the *useful order* of an edge $e \in L$ as the lowest order of a triangulation of S that includes e. Then, if all edges in L have a useful order $k \leq 3$, the lowest-order completion to a triangulation which includes L can be computed in $O(n \log n)$ time. For $k \geq 4$, no *polynomial-time* algorithm is known for that problem. Several related variants of constrained order-k Delaunay triangulations are discussed in Silveira and van Krefeld [655].

On the combinatorial side, Abellanas *et al.* [4] give an upper bound of $3(k+1)(n-k-2)$ on the number of edges of k-DT(S). They also study some graph-theoretic properties of k-DT(S), proving, among others, that the order-15 Delaunay graph always has to contain some *Hamiltonian cycle*. Recall from Subsection 8.2.1 that, by contrast, DT(S) is not Hamiltonian, in general. The conjecture is that order 1 already suffices to restore this property.

Moreover, they show that any two order-1 Delaunay triangulations of a given point set S can be transformed into each other by flips that generate only triangulations in that class. That is, the *flip graph* of all order-1 Delaunay triangulations of S is connected; cf. Section 6.3. Connectivity is lost for $k \geq 2$, in general, unless the point set S is in *convex position*.

More results on k-DT(S) and its subgraph, the *order-k Gabriel graph*, are given in Bose *et al.* [157], concerning diameter, connectivity, dilation, and other properties. For example, the *dilation* of the latter graph is $\Theta(\sqrt{\frac{n}{k}})$ for $k \geq 1$ in the worst case, which generalizes the bound of $\Theta(\sqrt{n})$ for the classical (order-0) Gabriel graph (cf. Subsections 8.2.3 and 8.2.4).

Ábrego *et al.* [8] provide bounds on the number of crossings, both for the minimum and maximum number attained when all sets S of n points are considered. k-DT(S) realizes $\Omega(k^3 n)$ crossings for every point set S in general position, and $O(k^2 n^2)$ crossings in the worst case, thus relatively few compared to $\Theta(n^4)$ for the complete graph K_n. These numbers, and corresponding (smaller) numbers for several subgraphs of k-DT(S) given in [8], might be interesting from the point of view of *graph drawing* when proximity information is to be preserved.

Notice that an edge, $\overline{pq}$, is in the graph k-DT(S) iff there are two adjacent regions R and R' in the *order-*$(k+1)$ *Voronoi diagram* $V_{k+1}(S)$ of S, such that p is among the $k+1$ sites defining R, and q is among the $k+1$ sites defining R': Each vertex of $V_{k+1}(S)$ is the center of a circle through three sites and with k or $k-1$ sites inside, hence the spanned triangle is of scope at most k. Still, the dual of $V_{k+1}(S)$ for $k \geq 1$ is *not* k-DT(S), but rather the centroid-based convex hull projection in Subsection 6.5.3, which is a planar graph.

8.3.2. *Witness Delaunay graphs*

Another type of graphs related to DT(S) has been investigated recently. The so-called *witness Delaunay graph* of S with respect to a (witness) point set W, for short DT(S, W), includes an edge between two points $p, q \in S$ whenever there exists a circle that passes through p and q but does not enclose any point from W.

Figure 8.15 gives an illustration. If $W = \emptyset$, then DT(S, W) is just the *complete graph* on S. On the other hand, if $W = S$, then we have DT(S, W) $=$ DT(S). So the point set W can be used to control the inclusion of edges into this graph. Similar concepts, though based on interpoint distances rather than on empty circle (or sphere) properties, are the *witness complexes* discussed in Section 10.5. These are abstract simplicial complexes, which are related to *Delaunay simplicial complexes* in a certain sense, and have been applied to manifold reconstruction problems in high dimensions.

Dillencourt [278] proved that all maximal *outerplanar graphs* (i.e., planar graphs all of whose bounded faces are adjacent to the unbounded face) are realizable as subgraphs of Delaunay triangulations. Inspired by that result, Aronov *et al.* [70] study witness Delaunay graphs concerning their ability of realizing other graph classes. For example, all *trees* can be realized, whereas non-planar *bipartite graphs* cannot, regardless of the choice of S and W. (A graph is called *bipartite* if its vertex set can be partitioned into two subsets such that no edges run within either subset.)

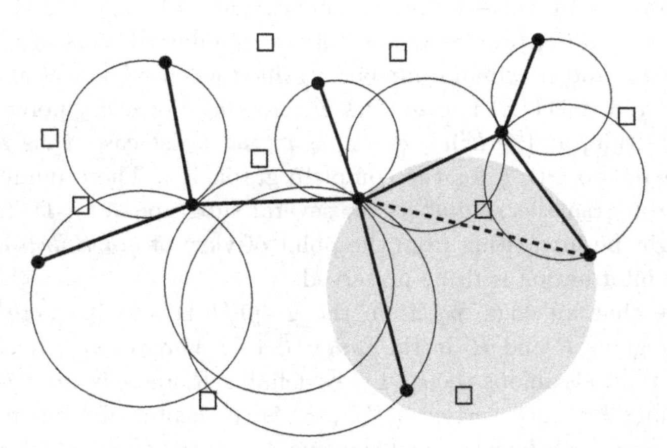

Figure 8.15. A tree realized as a witness Delaunay graph. For all non-tree edges (dashed) between sites (•), there is no enclosing circle that does not contain some witness point (□).

They also show that $\mathrm{DT}(S, W)$ can be computed in time $O(n \log^2 n + k \log n)$, where k denotes the (potentially superlinear) number of edges of $\mathrm{DT}(S, W)$, and $|S| = |W| = n$.

In *graph drawing* applications, witness-controlled versions of the *Gabriel graph* (Subsection 8.2.3) have been considered. The *witness Gabriel graph* of S and W consists of all edges spanned by S whose diametrical circle encloses no points from W.

Whereas all biconnected *outerplanar graphs* admit drawings as (classical) Gabriel graphs, *trees* that have vertices of degree higher than five cannot be drawn in that way; see Lenhart and Liotta [503] and Bose *et al.* [160], respectively. However, all trees become drawable when witness Gabriel graphs are used. This was shown by Aronov *et al.* [71] who also discuss several other properties of this class of geometric graphs.

8.4. Geometric clustering

Clustering a set of data means finding subsets (called *clusters*) whose in-class members are similar, and whose cross-class members are dissimilar, according to a predefined similarity measure. Data clustering is important in diverse areas of science, and various techniques have been developed; see e.g. Hartigan [397]. In many situations, similarity of objects has a geometric interpretation in terms of distances between points in d-space.

The role of Voronoi diagrams in the context of clustering is manifold. For certain applications, the relevant cluster structure among the objects is well reflected, in a direct manner, by the structure of the Voronoi diagram of the corresponding point sites; see e.g. Ahuja [24]. For instance, dense subsets of sites give rise to Voronoi regions of small area (or volume). Regions of sites in a homogeneous cluster will have similar shape. For clusters having a direction-sensitive density, the regions will exhibit an extreme diameter in the corresponding direction.

Perhaps more important is the fact that numerous types of *optimal clusterings* are induced by Voronoi diagrams, and/or can be computed by first computing a Voronoi diagram, for a certain set of sites. This set need not coincide with the set of points to be clustered.

A general distinction is between *offline* and *online* clustering methods. The former methods assume the availability of the whole data set before starting, whereas the latter cluster the data upon arrival, according to an online classification rule. Though Voronoi diagrams play a role in online clustering, too, see e.g. Hertz *et al.* [405], we restrict attention to offline methods below.

8.4.1. *Partitional clustering*

Let X be a set of n points in d-space. A *k-clustering* of X is a partition of X into k subsets $C_1, \ldots, C_k$, called clusters. The dissimilarity of a single cluster C is measured by an appropriate *intra-cluster* criterion $\mu(C)$, which may be variance, diameter, radius, etc. The dissimilarity of the whole clustering, in turn, is expressed as a (usually monotone) function f of $\mu(C_1), \ldots, \mu(C_k)$, called the *inter-cluster* criterion. Common examples for f are the maximum or the sum. $C_1, \ldots, C_k$ is called *optimal* if $f(\mu(C_1), \ldots, \mu(C_k))$ is minimal for all possible k-clusterings of X.

If k is part of the input, the problem of finding an optimal k-clustering is *NP-hard* in general, even in the plane; see Capoyleas *et al.* [182] for references. For fixed k, *polynomial-time* algorithms are known for various criteria. This is due to the fact that optimal clusters are separable in a certain sense, and in several cases are induced by the regions of (generalized) Voronoi diagrams. This approach was first systematically used in [182].

A common intra-cluster criterion is *variance*,

$$\mu(C) = \frac{1}{|C|} \sum_{p,q \in C} d(p,q)^2.$$

Variance can be rewritten as $\sum_{p \in C} d(p, s(C))^2$, with $s(C)$ being the *centroid* (center of mass) of C. Note that $s(C)$ is the point s^* in d-space that minimizes the sum of squared distances, $\sum_{p \in C} d(p, s^*)^2$, to C. Hence, if the inter-cluster criterion f is sum, the clustering problem is equivalent to the *k-centroid problem*: Given a set X of points to be clustered, find a set S of k sites (cluster centroids) that minimizes $\sum_{p \in X} d(p, S)^2$, where $d(p, S)$ is the distance from p to the closest site in S. This implies that the optimal k-clustering $C_1^*, \ldots, C_k^*$ (which is also called *k-means clustering* in the literature) has the following nice property; see Boros and Hammer [155]: For $1 \le i \le k$, cluster C_i^* is contained in the region of $s(C_i^*)$ in the Voronoi diagram of $S = \{s(C_1^*), \ldots, s(C_k^*)\}$.

The candidates for an optimal k-clustering of X thus are the possible partitions of X induced by a Voronoi diagram of k point sites. The number of such partitions is $n^{O(dk)}$ in d-space; see Inaba *et al.* [421]. This implies a *polynomial-time algorithm* for computing an optimal k-clustering for fixed k, which proceeds by enumerating all candidate solutions and comparing their dissimilarities.

The problem becomes considerably easier when the set S of k 'cluster centers' is *fixed* in advance. A clustering of X that minimizes the sum of the squared distances of the clusters to their centers is easily found by just constructing the Voronoi diagram of S; its regions will induce the desired optimal partition of X. Of more interest is the case where the *sizes* of the k clusters are prescribed too. Now, a corresponding *least-squares clustering* is induced by a *power diagram* of S (Section 6.4), a fact which leads to an algorithm with roughly $O(k^2n)$ runtime, if X is a set of n points in the plane; see Aurenhammer *et al.* [97].

In Inaba *et al.* [421], optimal k-clusterings for the intra-cluster measure

$$\mu(C) = \sum_{p,q \in C} d(p,q)^2$$

(like variance, but without division by $|C|$) are considered. Again, these clusterings are induced by a kind of power diagram of the cluster centroids, weighted additively by $|C|$ and multiplicatively by $\mu(C)$; cf. Section 7.4 for *mixed weighted distances.*

The reason why sums of *squared* distances, rather than just sums of distances, are used in the definition of the above intra-cluster measures partially lies in the intrinsic difficulty of solving the following, seemingly easy, question: Given a cluster C of size $m \geq 4$, find a point (cluster center) F that minimizes

$$\sum_{p \in C} d(p, F).$$

F is called the *Fermat point* (or *Fermat-Weber point*) of C. Its exact location is the solution to a high-degree polynomial in the coordinates, and can be computed only by numerical algorithms; see Bajaj [110]. For $m = 3$, a classical exact construction for F exists.

Another popular measure $\mu(C)$ is the radius of the smallest sphere enclosing a cluster C. If maximum is taken as the inter-cluster criterion, the *k-center problem* is obtained: Find a set S of k centers for X, such that

$$r_S = \max_{p \in X} d(p, S)$$

is minimized. In other words, choose a set S of centers such that X can be covered by k congruent spheres of minimum radius r_S. (This simplifies to the *smallest enclosing sphere problem* for $k = 1$; see Subsection 8.1.3). It is easy to see that the Voronoi diagram of S gives rise to optimal clusters. Capoyleas *et al.* [182] observed that an optimal k-clustering for μ, for any monotone increasing inter-cluster criterion f, is induced by the *power diagram* of the enclosing spheres.

For both intra-cluster measures μ above (but not for the Fermat point case), polynomial-time algorithms for fixed k result from considering the $n^{O(dk)}$ candidate clusterings.

For $k = 2$, the 'Voronoi property' of the 2-centroid or 2-center problem just means *linear separability* of the two optimal clusters. (Note that linear separability of two sets is equivalent to their (*convex*) *hull-disjointness.*) However, linear separability for all pairs of clusters in a given clustering does not always imply that this clustering has a realization by means of Voronoi or power diagrams. An example is the intra-cluster criterion diameter in 2-space [182].

The problem of constructing linearly (or *circularly*) separable clusterings in the plane, where each cluster C of X is separable by a straight line (or circle) from $X \setminus C$, is addressed in Dehne and Noltemeier [251] and Heusinger and Noltemeier [406]. They give a polynomial-time algorithm for deciding the existence, and finding an optimal clustering, with prescribed cluster sizes. A correspondence between separable clusters and regions of *higher-order* Voronoi diagrams for X (Section 6.5) is exploited.

The order-k Voronoi diagram, $V_k(X)$, of X is also useful for selecting from X a k-sized cluster C^* of minimal dissimilarity $\mu(C^*)$. This approach is pursued in Boyce *et al.* [162] and in Aggarwal *et al.* [19]. For instance, if μ is variance, C^* has a non-empty region in $V_k(X)$. If μ is diameter, C^* is contained in a subset of X having a non-empty region in $V_{3k-3}(X)$. These properties hold in arbitrary d-space, and lead to efficient *cluster selection* algorithms in the plane. Moreover, if $\mu(C)$ measures the perimeter of the axis-aligned enclosing square, or rectangle, of a cluster C, higher-order Voronoi diagrams in the L_∞-*metric* (cf. Sections 6.5 and 7.2) yield improved solutions. Except for the diameter case, the running time of all these algorithms is dominated by the cost of computing an order-k diagram.

8.4.2. *Hierarchical clustering*

Hierarchical methods are based solely on a given *inter-cluster* distance δ. They cluster a set S of n points as follows. Initially, each point is considered to be a cluster itself. As long as there are two or more clusters, a pair C, C' of clusters is joined into one cluster if $\delta(C, C')$ is minimum for all cluster pairs.

A frequently used inter-cluster distance is the *single-linkage distance*

$$\delta(C, C') = \min\{d(p, q) \mid p \in C, q \in C'\}.$$

As was pointed out in Shamos and Hoey [637], constructing the single-linkage cluster hierarchy for S just means simulating the *greedy algorithm of Kruskal* [475] for computing the *minimum spanning tree* MST(S) of S. We thus obtain a time bound of $O(n \log n)$ in the plane, and subquadratic bounds in d-space; see Subsection 8.2.1.

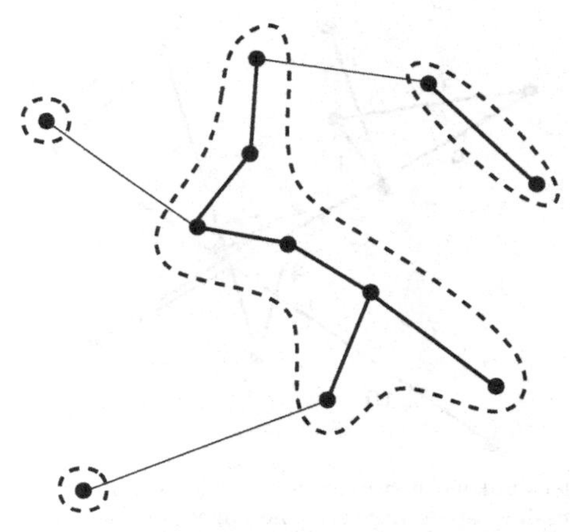

Figure 8.16. Removing the three longest edges of MST(S) leaves a 4-clustering of S where the minimum distance arising between any two clusters is maximized.

In fact, after having performed $n-k$ steps in constructing the hierarchy (that is, after having added the $n-k$ shortest edges of MST(S)), the intermediate k-clustering $C_1, \ldots, C_k$ is optimal in the following sense; see Asano *et al.* [75]: It maximizes the minimum single-linkage distance between the clusters, for all possible k-clusterings of S. Figure 8.16 illustrates these observations. Note further that $C_1, \ldots, C_k$ are *hull-disjoint*, by construction.

The practical value of *single-linkage clustering* is, however, restricted by the fact that the produced clusters tend to exhibit large dissimilarity in terms of variance, radius, and diameter. To remedy this deficiency, other inter-cluster distances have been used. Among them is the *complete-linkage distance*

$$\delta(C, C') = \max\{d(p,q) \mid p \in C, q \in C'\}.$$

Constructing the *complete-linkage hierarchy* for S efficiently is much more elusive. Figure 8.17 depicts a planar example. A direct approach leads to an $O(n^3)$ time and $O(n)$ space algorithm in general d-space. (The frequently cited $O(n^2)$ time *insertion algorithm* in Defays [246] only approximates the hierarchy. Its output depends on the insertion order.)

In the plane, it seems of advantage to use, as an auxiliary structure, the *Voronoi diagram* of the intermediate clusters $C_1, \ldots, C_k$, defined by the *Hausdorff distance* $h(x, C) = \max\{d(x,p) \mid p \in C\}$ of a point x to a cluster C; see Section 6.5. Unfortunately, the closest pair of clusters does not necessarily yield adjacent regions in this diagram. Moreover, there exist examples where the clusters violate *hull-disjointness*.

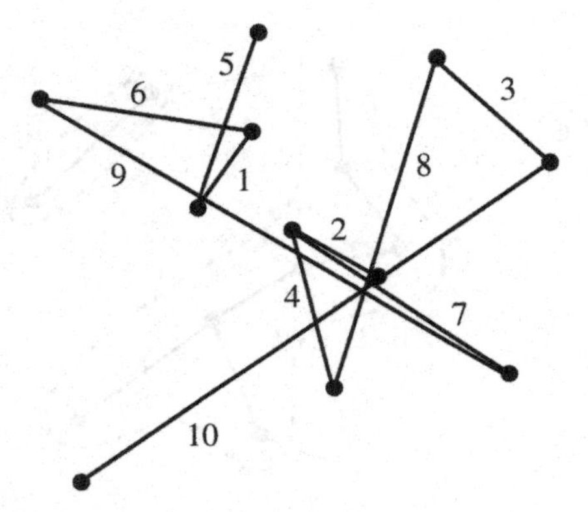

Figure 8.17. Complete-linkage hierarchy for the same point set as in Figure 8.16. In the resulting (self-crossing) spanning tree, the edges joining the clusters are numbered in their order of generation. Note that the tree is *pointed*, in the sense that the edges per vertex span an angle less than π (in fact, less than $\frac{\pi}{2}$).

An $O(n \log^2 n)$ time complete-linkage clustering algorithm for a set S of n points in the plane has been given in Krznaric and Levcopoulos [477]. Among other structures, they use the *Delaunay triangulation* DT(S), and the *farthest-site Voronoi diagram* (Section 6.5) of the points in the intermediate clusters C_i. They also show that *approximations* of the complete-linkage hierarchy for S can be obtained from DT(S) in $O(n)$ time.

Clustering data in higher dimensions is intricate. An interesting and effective approach is so-called *topological clustering*, based on *abstract simplicial complexes* related to Delaunay complexes; see Section 10.5. In particular, the results in Ghrist and Muhammad [362] and de Silva and Carlsson [259] are relevant to applications like data visualization, sensor networks, and others.

8.5. Motion planning

The general *motion planning problem* (also called *path planning problem*) is in finding a trajectory for a robot, from a given start configuration to a target configuration, that avoids collision with a set of obstacles and possibly satisfies additional requirements concerning cost, speed, or other constraints. Applications range from industrial robots to autonomous vehicles, and on to character motions in game design.

There is a rich literature on both the algorithmic aspects of this task and their practical realization. The reader may consult, e.g., the early surveys by Alt and Yap [52, 53], Yap [711], the handbook chapter by Sharir [641], or the monographs by LaValle [485] and Latombe [484].

In the following we will present some aspects of the general motion planning problem that involve Voronoi diagrams and employ methods interesting in their own right.

Of relevance in this context are also the *collision detection* methods in Kirkpatrick *et al.* [459], Abam *et al.* [2], and others; consult the survey article by Lin and Manocha [507]. The first paper uses *pseudo-triangulations* (see Section 6.3) of free space as an underlying data structure. As these methods are not primarily based on Voronoi diagrams, we refrained from their detailed description, though.

8.5.1. *Retraction*

Suppose that for a disk-shaped robot centered at some start point, s, a motion to some target point, t, must be planned in the presence of n line segments as obstacles. We assume that the line segments are pairwise disjoint, and that there are four line segments enclosing the scene (Figure 8.18).

While the robot is navigating through a gap between two line segments, ℓ_1 and ℓ_2, at each position x its '*clearance*', i.e., its distance

$$d(x, \ell_i) = \min\{d(x, y) \mid y \in \ell_i\}$$

to the obstacles, should be maximized. This goal is achieved if the robot maintains the same distance to either segment. In other words, the robot should follow the bisector $B(\ell_1, \ell_2)$ of the line segments ℓ_1 and ℓ_2 until its distance to another obstacle gets smaller than $d(x, \ell_i)$.

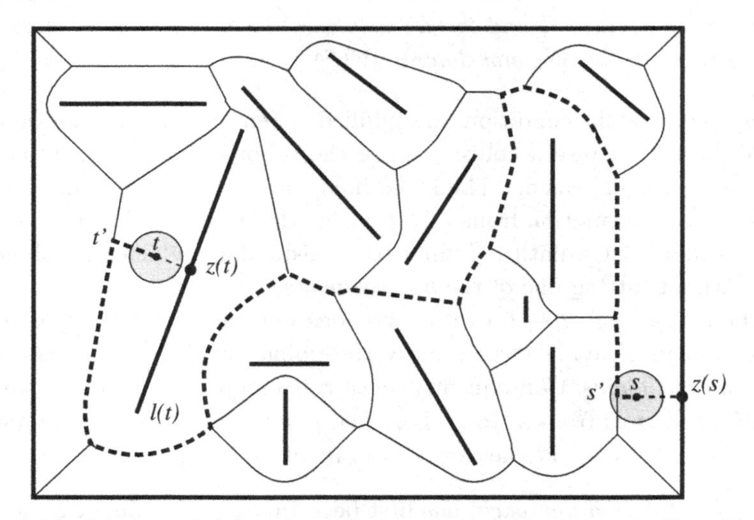

Figure 8.18. Moving a disk from s to t in the presence of barriers is facilitated by their Voronoi diagram.

Roughly, this observation implies that the robot should walk along the edges of the *Voronoi diagram* $V(S)$ *of the line segments* in $S = \{\ell_1, \ldots, \ell_n\}$ (see Sections 5.1 and 5.5). This diagram is connected, due to the four surrounding line segments.

If start and target point are both lying on Voronoi edges, the motion planning task immediately reduces to a *discrete graph problem*: After labeling each edge of $V(S)$ with its minimum distance to its two sites, and adding s and t as new vertices to $V(S)$, a breadth first search from s can find, within $O(n)$ time, a *collision-free path* to t in $V(S)$ if one exists. (Those edges whose clearance is less than the robot's radius are ignored during the search.)

If the target point, t, does not lie on any edge of $V(S)$, we first determine the line segment $\ell(t)$ whose Voronoi region contains t; to this end, *point-location* techniques as mentioned in Subsection 8.1.1 can be applied. Next, we find the point $z(t)$ on $\ell(t)$ that is closest to t; see Figure 8.18. If its distance to t is less than the robot's radius then the robot cannot be placed at t and no motion from s to t exists. Otherwise, we consider the ray from $z(t)$ through t. It hits a point t' on $V(S)$, which serves as an intermediate target point.

Similarly a point s' can be defined if the original start point, s, does not lie on an edge of $V(S)$. Now the formerly infinite problem of finding a collision-free path has become finite.

Theorem 8.6. *The robot can move from s to t without collision iff the following conditions hold: Its radius does not exceed one of the distances $d(z(t), t)$ and $d(z(s), s)$, and there exists a collision-free motion from s' to t' along edges of the Voronoi diagram $V(S)$.*

Proof. Suppose the conditions are fulfilled. Then the straight motion from s to s' does not cause a collision since the robot's distance to its closest obstacle is ever increasing. The same holds for t and t'. Combining these pieces with a safe motion from s' to t' yields the desired result. Conversely, let us assume that a path π from s to r exists along which the robot can move without hitting one of the line segments.

The mapping $f : t \mapsto t'$ can be extended not only to s but to all points x of the scene (leaving fixed exactly the points of $V(S)$). One can show that f is *continuous*. Consequently, path π is mapped by f onto a path π' in $V(S)$ that runs from s' to t'. For each point x on π, its corresponding point x' on π' is even farther away from its closest obstacle. □

This '*retraction approach*' has first been taken by Ó'Dúnlaing and Yap [569]. The Voronoi diagram of n line segments (Section 5.1) plus a structure

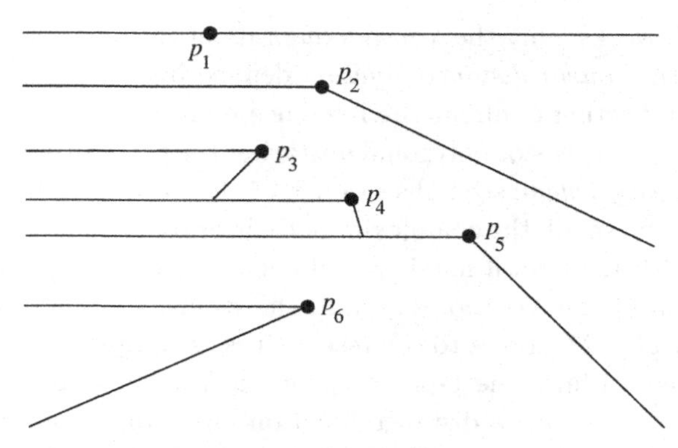

Figure 8.19. The polar Voronoi diagram for six point sites. Each region $\mathrm{VR}(p_i)$ is bounded by two rays at p_i : The negative horizontal ray for p_i, and a ray defined by p_i and the unique site $p_j \neq p_i$ with $\mathrm{VR}(p_j) \ni p_i$. An exception is the region of the topmost site, which is bounded by a single horizontal line.

suitable for point location in $V(S)$ (Subsection 8.1.1) can be preprocessed in time $O(n \log n)$. Afterwards, for each pair (s, t) of start and target points, the above method can find a safe motion from s to t within time $O(n)$, if it exists, or report otherwise.

Several variants of the retraction approach are meaningful. For example, *convex distance functions* (Section 7.2) can be used instead of the Euclidean distance, to adapt to the shape of the robot.

We mention another variant, as it uses a nice type of Voronoi diagram, introduced in Grima *et al.* [379, 380]. The *polar angle* of a point x with respect to a point site p is the angle formed by the positive horizontal ray through x, and the ray from x to p. We require x to have smaller y-coordinate than p, so the polar angle is smaller than π. The *polar Voronoi diagram* of a set of point sites allots to each site p the region of all points x yielding the smallest polar angle; see Figure 8.19. Motion planning in a specific direction now can be done by finding a chain of points such that each of them can see its successor, using the polar Voronoi diagram in that direction (not necessarily the horizontal one).

The polar Voronoi diagram is different from the *visibility-constrained Voronoi diagram* in Section 5.4, where individual visibility angles are attached to sites.

8.5.2. *Translating polygonal robots*

If approximating the robot's shape by a disk is not accurate enough, a compact convex set C can be used, with some fixed reference point in its interior. As long as only *translational motions* are considered, the retraction approach still works; only the Voronoi diagram of the obstacles should now be based on the *convex distance function* defined by the convex set C' that results from reflecting C about the reference point; see Section 7.2.

If the scene consists of polygonal obstacles, the same situation occurs as in the *post office problem*; see Subsection 8.1.1: Even though the number k of obstacles may be small, the complexity of their Voronoi diagram increases in proportion with their total number n of edges. Again, the piecewise-linear approximation of the Voronoi diagram, the *compact Voronoi diagram* by McAllister *et al.* [527] comes to the rescue; it is of complexity $O(k)$ and can be constructed within time $O(k \log n)$ for the Euclidean distance, and in time $O(k \log n \log m)$ for a distance function based on a convex m-gon. In the latter setting, the diagram applies to solving the translational motion planning problem of a convex robot with m edges.

Alternatively, planning translational motions for a convex robot R can be reduced to motion planning for a *point*, using an observation by Lozano-Pérez and Wesley [512] described below.

Supposing that the origin o is a vertex of R, we replace each obstacle P_i by the *Minkowski difference*

$$P_i \ominus R = \{p - r \mid p \in P_i,\ r \in R\}.$$

Geometrically, this amounts to moving the reflected image, R', of R around P_i in such a way that o traces the boundary of P_i. In the resulting *configuration space*, each point x not contained in any enlarged obstacle $P_i \ominus R$ corresponds to a collision-free placement of R in workspace. (This space is also called the *free space* of the robot R.) As a consequence we may, without loss of generality, assume that the robot is a point.

If the robot is also allowed *rotational motions*, Voronoi diagrams can still be applied in some cases. Ó'Dúnlaing *et al.* [567, 568] have studied the problem of moving a line segment amidst polygonal obstacles.

8.5.3. *Clearance and path length*

As mentioned earlier, a robot moving along the edges of the Voronoi diagram maintains maximal clearance of the obstacles; however, the resulting path may be longer than necessary. The *shortest* possible path, on the other hand, has clearance zero because it does, in general, touch obstacle boundaries. It can be obtained in time $O(n^2)$ by running *Dijkstra's algorithm* on the *visibility graph* of the n obstacle vertices; see Overmars and Welzl [575] or Ghosh and Mount [361]. In this graph, two vertices

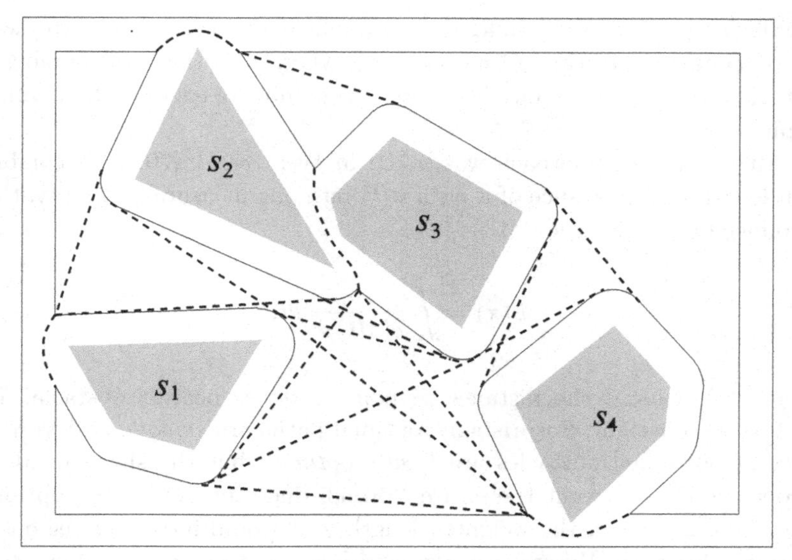

Figure 8.20. Visibility-Voronoi complex (dashed style) for the convex polygonal obstacles $s_1, \ldots, s_4$. Its straight edges form the visibility graph.

v, w are connected by an edge iff they 'see' each other, that is, if the line segment $\overline{vw}$ does not intersect the interior of any obstacle polygon.

Wein *et al.* [700, 701] have presented different methods to obtain both, sufficient clearance *and* short paths. In [700] a constant $c \geq 0$ can be specified, requiring that the robot must stay away a distance of at least c from each obstacle at all times. Each obstacle s_i is enlarged by a belt of width c, by forming the *Minkowski sum* $s_i \oplus C$ with a circle C of radius c, that is, the *c-offset* of C. Consult Figure 8.20.

By construction, any shortest path in the visibility graph of the grown obstacles is an optimum trajectory of the required clearance. Moreover, the path is *smooth*, because obstacle vertices have been replaced with circular arcs of radius c. But grown obstacles may overlap, thus closing a narrow passage that existed before. In such a case, the user might rather accept a smaller clearance than not find a collision-free path at all. To this end, the Voronoi diagram of the original obstacles is added to all belt intersections. Its edges may be assigned a weight higher than their Euclidean length, as a penalty for their lack of clearance.

As the parameter c grows from 0 to ∞, the hybrid structure just introduced (called the *visibility-Voronoi complex*) changes from the visibility graph of the original obstacles into their Voronoi diagram. After

$O(n^2 \log n)$ preprocessing time, the optimum path for a given clearance c can be computed in time $O(n \log n + k)$, where n is the total number of obstacle vertices and $k = O(n^2)$ denotes the number of edges in the visibility graph.

An alternative approach was taken in Wein *et al.* [701]. To combine both length and clearance of a path $\pi(t)$ into one measure of quality, they introduced a weighted length

$$L(\pi) = \int_\pi \frac{1}{c(\pi(t))^\delta} \, dt,$$

where $c(z)$ denotes the distance of point z to its nearest obstacle, and $\delta > 0$ is some weight. Surprisingly, optimal paths are *smooth*, and Voronoi edges of polygonal obstacles are *locally optimal*; but the sharp bends at Voronoi vertices prevent the entire Voronoi diagram from being optimal. In a discretization of the weighted length, one would introduce regions of constant clearance. Where two regions of different clearance values share an edge, *Snell's law of refraction* applies to optimal paths.

For $\delta = 1$ one can, in $O(\frac{\Lambda^2}{\varepsilon^2} n)$ time, construct a path whose weighted length is by at most $O(n) \cdot \varepsilon$ larger than optimal. Again, n is the total number of all obstacle vertices, and Λ equals the total weighted length of a substructure of the Voronoi diagram.

8.5.4. *Roadmaps and corridors*

Instead of using the Voronoi diagram of the obstacles as a retract of the *free space* where a robot may be placed, *randomized roadmaps* were studied, e.g., in Overmars [574] and in Amato and Wu [56]. These structures can be built incrementally by drawing points at random. If the new sample point lies in free space, one tries to connect it by a straight edge to an existing point of the roadmap. If this edge does not intersect any obstacle, it is added to the structure, otherwise the edges (or the point) are discarded. In this way a graph results that can be used for path planning.

Clearly, the *sampling technique* is critical. One would hope to reduce the number of edges that must be discarded, and to increase the sampling density near narrow passages. Yershova *et al.* [714] suggest to maintain, for each vertex v of the structure already computed, a visible Voronoi region containing all points of free space that are closest to, and visible from, v. New sample points are chosen from a uniform distribution over these regions. The appropriate structure reflecting this distance information is the *visibility-constrained Voronoi diagram* in Section 5.4.

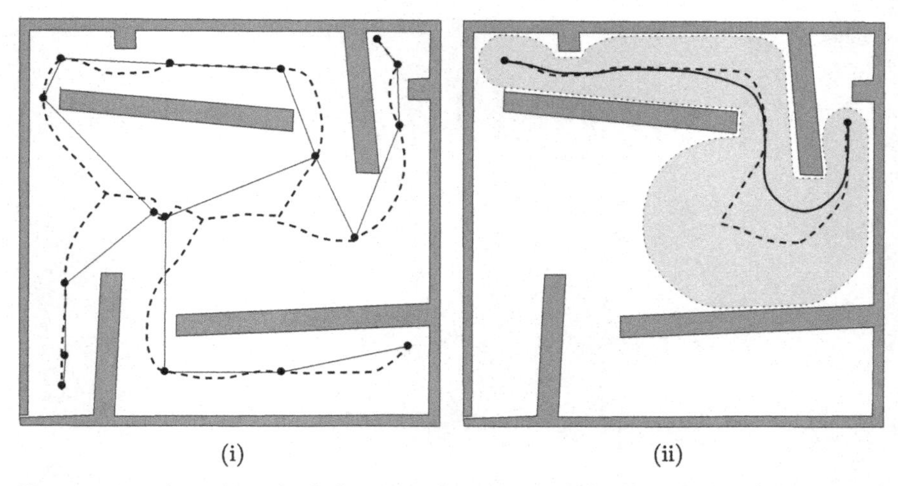

Figure 8.21. A roadmap built from sampling points of the obstacles' medial axis, and the medial axis segments connecting them (i). The corridor defined by two sample points and its backbone path (ii). In the absence of further obstacles, the robot might follow the solid path which results from a potential field approach. But the corridor is large enough for the robot to locally avoid smaller fixed or mobile obstacles; see [359].

Geraerts and Overmars [359] suggest choosing sampling points from the *medial axis* (see Chapter 5) of the large, static obstacles of a scene. Figure 8.21 gives an illustration. They use them to build backbone paths of *corridors.* A corridor is the union of all disks whose centers z lie on the backbone and whose radii equal the clearance $c(z)$ of z, up to a global upper bound. Agents move in corridors, but need not follow fixed paths. Instead, the corridor's width provides some flexibility that can be used for coordinating motions of groups, or for local obstacle avoidance. To this end, *potential-field methods* can be employed. Wein *et al.* [701] proved that their method for computing (near) optimal clearance-weighted paths mentioned in Subsection 8.5.3 also applies to constructing optimal corridors.

There are several other applications of Voronoi diagrams in motion planning. We refrain from more details here and refer the interested reader, e.g., to the work by Canny [180, 181] and Choset and Burdick [220, 221].

Chapter 9

MISCELLANEA

9.1. Voronoi diagram of changing sites

Constructing the Voronoi diagram of a fixed set S of sites is an algorithmic problem which is inherently *static*. Several applications require data structures which can represent proximity information that changes over time, however. To name a few, one might need to preserve an optimal cluster structure in a database for multivariate data that undergoes insertions and deletions (Section 8.4), maintain clearance for a robot moving among non-stationary obstacles (Section 8.5), or simply wish to visualize the influence a shopping market would exert on its competitors when being closed down and re-opened at a promising location (Section 9.2).

Basically there are two — algorithmically quite different — possibilities of altering the underlying set S of sites, which we will review in this section.

9.1.1. *Dynamization*

Discrete changes involve insertions and/or deletions of sites, and require a *dynamization* of the data structure for Voronoi diagrams, like dynamic *search trees* are used for storing and maintaining univariate (one-dimensional) data [234, 538].

Usually, the sequence of insertions and deletions is not given in advance. Updates have to be performed online, without information about future requests. Note that deleting a site at location x and re-inserting it at location y may simulate 'jumping' of a site from x to y.

We have seen ways of dynamizing the planar Voronoi diagram in Sections 3.2 and 3.3 and also in Subsection 6.5.3 — based on the algorithmic paradigms incremental insertion and divide & conquer — where the *construction history* of the diagram, or of its dual Delaunay triangulation, is stored for later use as a search structure. Dynamic geometric data structures are well understood, and efficient solutions are available. In particular, efficiency can be gained by using diverse *randomization methods.*

In our context, we refer to Devillers *et al.* [264], Schwarzkopf [624], and the monograph by Mulmuley [556], for handling randomized update sequences of sites in $O(\log n)$ expected time per operation. The two latter references also treat convex hull maintenance in general dimensions, further investigated in Clarkson *et al.* [228] and in Chan [190] — results of direct relevance for dynamizing higher-dimensional power diagrams, and Voronoi diagrams as a special case, by Theorem 6.2.

Also in [556], the problem of *dynamic point-location* in a planar Voronoi diagram is discussed in detail. This gives a solution to the *post-office problem* (Subsection 8.1.1) for a dynamically changing set S of sites. Among several methods, the powerful technique of *random sampling*, introduced to computational geometry by Clarkson [224], is applied to a triangulation of the Voronoi regions radially from the sites. In a nutshell, this technique randomly (and recursively) chooses a subset S_R of S of suitable size, constructs the desired data structure for S_R, and bases the structure for S on that substructure. The resulting hierarchy has a depth of $O(\log n)$ and a total size of $O(n)$, provided a constant fraction of sites is eliminated at each step.

Random sampling techniques lead to many efficient algorithms in computational geometry but are usually difficult to analyze, and touch upon complex concepts from the theory of *ε-nets* and *discrepancy*; see e.g. the monographs by Matoušek [520] and Chazelle [195].

In our case, an update time of $O(\log n)$ and a point location time of roughly $O(\log^2 n)$ can be achieved for the post office problem, in the expected sense.

A simpler but slightly less efficient approach, which also allows for *k-nearest neighbor searching*, has been taken in Aurenhammer and Schwarzkopf [102]. It is based on the history of randomized incrementally constructing the *order-k Voronoi diagram*, and yields expected update time $O(k^2 \log n + k \log^2 n)$ and expected query time $O(k \log^2 n)$. The space consumption is the same as for the diagram, $O(k(n-k))$; see Subsection 6.5.3 for details.

More methods for dynamic k-nearest neighbor searching (and also for dynamic point location in general 'non-Voronoi' planar subdivisions) can be found in Mulmuley's book [556], which offers a careful treatment of geometric algorithms based on various randomization techniques.

9.1.2. *Kinetic Voronoi diagrams*

Continuous changes in S need to be modeled in various applications as well. The sites now move steadily on predefined curves, their trajectories.

Such *Voronoi diagrams of moving sites* are so-called *kinetic* data structures, a concept that has received considerable attention in the past fifteen years; see e.g. Basch *et al.* [117, 118], Guibas [388], and Kirkpatrick *et al.* [459].

In the *kinetic Voronoi diagram*, regions change shape continuously, whereas the Delaunay triangulation DT(S) changes only in a discrete way from time to time, when an *edge flip* (Section 3.2) has to restore local Delaunayhood at the moment of cocircularity of four moving sites. For this reason, implementations prefer to maintain the graph DT(S), and derive the primal structure $V(S)$ only when demanded.

The geometric aspect of *kinetic Delaunay triangulations* is almost trivial. Delaunay edges admit local *certificates* of their validity, due to their *empty-circle property*: By *Thales' theorem*, an (interior) edge e in the moving triangulation stays locally Delaunay (hence globally Delaunay by Theorem 4.2), until the sum of the two angles opposite to e in the incident triangles exceeds π. This can be calculated in $O(1)$ time under reasonable assumptions on the trajectories. A *priority queue* of 'lifespans' of Delaunay edges with at most $3n - 6$ entries can be kept, in order to predict the next edge to be flipped, and to store the lifespan of the newly flipped-in edge, in $O(\log n)$ time.

In its combinatorial aspects, the setting is surprisingly complex. Several authors considered the kinetic Voronoi (or Delaunay) problem, even prior to its first attack by computational geometers, in Guibas *et al.* [391]. How many Delaunay flips may occur in the worst case? In the paper above, a nearly cubic upper bound of $O(n^2\lambda_s(n))$ is shown. Here $\lambda_s(n)$ is the maximum length of an (n, s)-*Davenport–Schinzel sequence*, and s is a constant depending on the algebraic degree of the trajectories of the point sites. The function $\lambda_s(n)$ is superlinear in n for $s \geq 3$, see Sharir and Aggarwal [642], which includes all cases of interest here. Still, for constant speed on straight lines, the bound above was improved to $O(n^3)$. A quick lower bound — valid even for that simplest case — is $\Omega(n^2)$, the number of combinatorial changes of the convex hull of S; see Agarwal *et al.* [13]. In comparison, the naive upper bound for Delaunay flips is $O(n^4)$.

Albers *et al.* [47] generalize the bound in [642] to higher dimensions, and also consider the case where only k sites are moving and $n - k$ stay fixed, a scenario sensitive to the 'fluctuation' in S. See also Huttenlocher *et al.* [414], who analyze the case where several groups of sites move rigidly in the plane.

Very recent progress was made in Rubin [611], who proved what has been conjectured for a long time (and what is true in the L_1-metric; see Chew [214]): Only $O(n^{2+\varepsilon})$ flips can occur, under certain restrictions on the cocircularities and collinearities that may arise from choosing the sites' trajectories and velocities. Still, the most basic case of constant speed on straight lines is not yet covered, and remains a challenging open question.

Agarwal *et al.* [12] proposed a way to relax the problem. They restrict attention to a so-called *stable subgraph* of DT(S) which is less volatile. It includes each Delaunay edge whose endpoints see the dual Voronoi edge at a (necessarily equal) angle of at least α, for some appropriate value of α. This graph need not contain any of the subgraphs of DT(S) discussed in Section 8.2, but rather consists of certain edges of the *shape Delaunay tessellation* (Subsection 7.2.2) for a regular polygonal shape. The stable Delaunay subgraph undergoes only a near-quadratic number of changes, each of cost $O(1)$ in the implementation. The hidden constant depends on parameter α.

We conclude this section with a few closely related results.

Fan *et al.* [334] consider kinetic variants of the *Voronoi diagram for graphs*, which decomposes a given network into subparts owned by the individual sites; see Subsection 9.4.1. Sites are moving continuously along the network, and nearest-neighbor queries (i.e. the *post office problem*) and k-nearest neighbor queries are to be responded to.

Of kinetic nature are also certain algorithms for constructing (static) Voronoi diagrams. For instance, Gold *et al.* [366, 367] compute the Voronoi diagram for line segments (Section 5.1) by *growing points* into line segments, after having precomputed the classical Voronoi diagram for one segment endpoint each.

Similarly, Kim and Kim [453] let sphere radii, i.e., additive site weights, grow one by one, in order to obtain the Voronoi diagram of spheres in 3-space (Subsection 6.6.3).

Finally, Shewchuk [650] considers the problem of '*repairing*' Delaunay tessellations and regular complexes (Section 6.3), which arises in mesh generation applications. He proves that, when such a simplicial complex is distorted to non-Delaunay or non-regular, but only in a measurably moderate way, then $O(n)$ bistellar flips will restore the property.

9.2. Voronoi region placement

Let S denote a set of n points in the plane representing supermarkets. If products are identically priced at each market, customers usually shop at the supermarkets nearest to their residences. Thus, the Voronoi diagram $V(S)$ shows the market areas of the individual shops. (Were products sold at different prices, we could model the situation by an *additively* or *multiplicatively weighted* Voronoi diagram, as studied in Section 7.4.)

If we assume that customers are uniformly distributed, a market's rate of sale is proportional to the area of its Voronoi region. Hence, many important problems in economy translate into geometric questions concerning Voronoi diagrams and related structures. The same is true for diverse applications in the natural sciences, where Voronoi regions may model, e.g., the extensions of habitats of biological species, or the structural behavior of crystalline solids.

9.2.1. *Maximizing a region*

A typical *facility location problem* is how to place a new supermarket where it wins over as many customers as possible from the competitors that already exist. This task amounts to finding a position p for the new market such that the area of its Voronoi region $\mathrm{VR}(p, S \cup \{p\})$ against its competitors in S is *maximized.*

Two aspects of this problem statement need to be clarified. First, if the new position p were chosen outside the *convex hull* of S then its Voronoi region would be unbounded. There are several ways to deal with this problem. One could require that p must be chosen 'within city limits', that is, within some area F contained in the interior of the convex hull of S. Or one could use a non-uniform customer distribution to the same extent. Another option is to study the problem on a bounded surface, like, e.g., the sphere or the torus; see Section 7.1.

Second, the Voronoi region of p is formally undefined if p coincides with an existing location q in S. For our maximization problem, we could assume that p moves towards q along a halfline ℓ ending in q. Then the bisector $B(p, q)$ of p and q converges to the line through q and orthogonal to ℓ. It cuts the former Voronoi region $\mathrm{VR}(q, S)$ into two parts. The part intersected by ℓ would become the Voronoi region $\mathrm{VR}(p, S \cup \{p\})$ of p. Even if it were not allowed to place p exactly onto an existing point, a close approximation would be possible.

Clearly, p can steal (close to) half of the area of $\mathrm{VR}(q, S \cup \{p\})$, by choosing a suitable 'line of attack' ℓ against q, and even more than half if $\mathrm{VR}(q, S \cup \{p\})$ is not *point-symmetric* about q. In total time $O(n)$ one can determine the largest region that p can obtain by attacking a point $q \in S$ in this way.

An even larger region might be secured by placing p at some point not in S; see Figure 9.1. Let $A(p)$ denote the area of $\mathrm{VR}(p, S \cup \{p\})$ as a function of p. As long as p stays within the same cell of the arrangement of all circumcircles of the Delaunay triangles in $\mathrm{DT}(S)$ (see Theorem 2.1), the combinatorial structure of the Voronoi diagram does not change. In

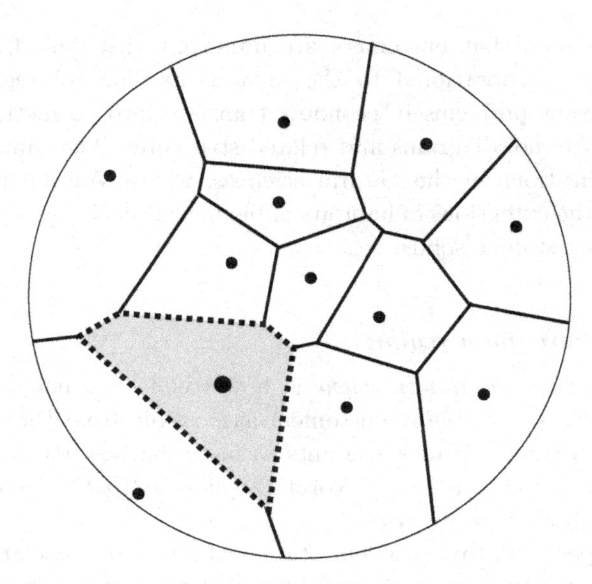

Figure 9.1. Placing a large Voronoi region within a bounding circle.

particular, the Voronoi neighbors of p, that is, the points $p_i \in S$ for which $\mathrm{VR}(p, S \cup \{p\})$ and $\mathrm{VR}(p_i, S \cup \{p\})$ share a Voronoi edge, are stable.

On such an arrangement cell, area $A(p)$ can be expressed by the Cartesian coordinates of the Voronoi vertices v_i situated on the boundary of $\mathrm{VR}(p, S \cup \{p\})$. If $v_i = (x_i, y_i)$, where $0 \leq i \leq m$ in counterclockwise order, then

$$A(p) = \frac{1}{2} \sum_{i=0}^{m-1} (x_i y_{i+1} - x_{i+1} y_i)$$

holds, reading indices modulo m. The Voronoi vertices can be obtained as intersections of bisector lines. As they depend on $p = (p_x, p_y)$, their coordinates x_i and y_i are rational functions, whose numerators are polynomials containing linear and quadratic terms in p_x and p_y, while the denominators are linear in either variable; see Okabe *et al.* [571]. The area function $A(p)$ can be shown to be continuously differentiable for $p \notin S$, even when the Voronoi diagram undergoes combinatorial changes due to the motion of p; see [571] and the references provided there. But despite this favorable property, it is not known how to compute the exact maximum of $A(p)$.

Dehne *et al.* [249] proved the following partial result. Let N denote the set of Voronoi neighbors of the new point p in S, and let $P(N)$ be the *star-shaped* polygon around p with vertex set N. Moreover, let L_N be the locus of all positions p that have N as their neighbor set.

Theorem 9.1. *Suppose that the Voronoi neighbor set N of p is in* convex position. *Then the area function $A(p)$ attains at most one local maximum on the polygon $P(N) \cap L_N$.*

Proof. (Sketch) It is sufficient to show that $A(p)$ has at most one maximum on each line, G. As $P(N)$ is convex, area $A(p)$ can be expressed in terms of $P(N)$ and the triangles formed by p and by its Voronoi neighbors p_i, rather than by the Voronoi vertices that vary with p. The sum of the (signed) triangle areas has poles wherever line G crosses a supporting line through neighboring points p_i, p_{i+1} of $P(N)$, because here the Voronoi vertex associated with p_i, p_{i+1} and p is at infinity. By convexity of $P(N)$, line G can pass only two supporting lines *between* their defining points p_i and p_{i+1}. Let l and r denote these crossing points. Since no other crossings occur on G between l and r, one can show by studying second-order derivatives, that $A(p)$ attains at most one maximum in the interval $[l, r]$. $\square$

An *approximate* solution to the Voronoi area maximization problem was provided by Cheong *et al.* [207]. They assume that the existing locations in S are situated in the unit square under *torus topology*, such that all Voronoi regions are bounded.

Let b denote the maximum distance between a location $p_i \in S$ and a point v on its region boundary in $V(S)$. If we place the new market at position v, its distance to any site in S must be at least b, because p_i is a nearest neighbor of v. This implies

$$\max_p A(p) \geq A(v) \geq \frac{\pi b^2}{4},$$

the latter inequality holding because $\mathrm{VR}(v, S \cup \{v\})$ contains an empty circle of radius $\frac{b}{2}$ centered at v. Moreover, each point of $\mathrm{VR}(p, S \cup \{p\})$ is at most a distance b away from its nearest neighbor in S, implying a distance $\leq b$ from p. Each candidate region $\mathrm{VR}(p, S \cup \{p\})$ is, therefore, of diameter $\leq 2b$.

The main idea is to approximate the real area $A(p)$ by the number of vertices of an ε-by-ε grid Γ_ε that are contained in $\mathrm{VR}(p, S \cup \{p\})$, times ε^2. By the diameter bound, the approximation error cannot exceed $8\, b\, \varepsilon$.

Let p_{opt} be a point maximizing $A(p)$, and let p_{app} be a point whose Voronoi region contains a maximum number μ_{app} of grid points of Γ_ε. Let

μ_{opt} denote the number of grid points in the region of p_{opt}. Then,

$$\begin{aligned} A(p_{\text{app}}) &\geq \mu_{\text{app}}\varepsilon^2 - 8\,b\,\varepsilon \geq \mu_{\text{opt}}\varepsilon^2 - 8\,b\,\varepsilon \\ &\geq A(p_{\text{opt}}) - 16\,b\,\varepsilon \geq A(p_{\text{opt}}) - 16\,b\,\varepsilon \cdot \frac{4}{\pi b^2} \cdot \frac{\pi b^2}{4} \\ &\geq \left(1 - \frac{64\,\varepsilon}{\pi b}\right) \cdot A(p_{\text{opt}}). \end{aligned}$$

Here the approximation bound has been used twice. The last line follows from the above estimate for the region area. It remains to determine the point p_{app}.

Theorem 9.2. *A point p_{app} approximating the maximum Voronoi area to a factor of $1 - \delta$ can be found in time $O(\frac{n}{\delta^4} + n \log n)$.*

Proof. In time $O(n \log n)$ we construct the Voronoi diagram $V(S)$ and determine the value of b. For the sake of accounting, a grid Γ_b of square cells of size b is superimposed to Γ_ε. The complexity of Γ_b is $O(n)$, since each cell is within distance $2b$ of a site in S, by definition of b. If p is a candidate point in some cell Q of Γ_b, its Voronoi region and its Voronoi neighbors are contained in Q and its 24 neighboring cells, by the diameter bound on $\text{VR}(p, S \cup \{p\})$. Let S_Q denote the set of points of S situated in these 25 cells.

For a fixed cell Q of Γ_b and an arbitrary grid point g of Γ_ε in Q, let

$$W_g = \{p \in Q \mid g \in \text{VR}(p, S \cup \{p\})\}.$$

The set W_g is the largest disk centered at g that contains no interior points of S, intersected with Q. For each of the $N = b^2/\varepsilon^2$ grid points g in Q we determine W_g by checking the distance from g to the elements of S_Q. Next, we construct the arrangement of all shapes W_g in time $O(N^2)$, and find a cell of this arrangement contained in a maximum number of these sets. A point p_Q of such a cell contains a maximum number of grid points in its Voronoi region, over all points in Q. Point p_{app} is obtained by maximizing over all cells Q. The running time is $O(n \log n + nN + N^2)$, and for $\varepsilon = \frac{1}{64}(\pi\, b\, \delta)$ the claim follows. □

9.2.2. *Voronoi game*

In Subsection 9.2.1 we have studied the situation where the area of *one* Voronoi region was to be maximized. There the difficulty of an exact solution lies with the analytical complexity of the area function. Now we will consider problems involving several regions that have a game theoretic flavor, too.

Suppose two players, White and Black, can place n point sites each in some bounded area. Their game proceeds in rounds. In each round, first White and then Black are allowed to place a certain number of their points, subject to specific rules. Points must not lie upon each other. When all points have been positioned, the Voronoi diagram of the $2n$ points is computed, and the total areas of all Voronoi regions owned by White and Black, respectively. The player wins whose regions occupy the larger total area. This example of *competitive facility location* is called a *Voronoi game.*

If we assume for a moment that only the vertices of a very fine grid are legal site locations, each player would have only a finite number of moves when his turn comes. As in all finite zero-sum games of two players with complete information, either White has a winning strategy, or Black does, or both can force a draw. For general Voronoi games in two-dimensional domains, it seems not to be known which case applies (a situation as for chess). But there are interesting results for special cases.

Ahn *et al.* [23] have studied the one-dimensional case where players compete for a circle or for a line segment of unit length. They proved the following.

Theorem 9.3. *On a circle or line segment, Black has a winning strategy for the Voronoi game of $n > 1$ rounds. But against any strategy of Black, White can secure an area of $\frac{1}{2} - \varepsilon$, for any given $\varepsilon > 0$.*

Proof. (Sketch) In the analysis, a set of n equidistant 'key' locations plays an important role. To win on a circle, Black plays onto empty key points as long as possible. Then, he keeps splitting largest intervals that have two white endpoints. With his last point, if only one such white interval, of length ℓ, is left, Black plays into a bichromatic interval, less than $\frac{1}{n} - \ell$ away from its white endpoint.

To defend himself, White can place his points at distance $\frac{\varepsilon}{2n}$ from the key points. In this way, each white interval has length $\geq \frac{1-\varepsilon}{n}$, and each black interval is of length $\leq \frac{1+\varepsilon}{n}$. Since there are as many white as black intervals, and because bichromatic intervals are fairly split among the two players, Black's excess is at most $\frac{2\varepsilon(n-1)}{n}$. □

One should observe that Theorem 9.3 does not hold for one-round games. On the circle, White can win by placing his points *exactly* onto the n equidistant key points. Afterwards, Black can form at most $\lfloor \frac{n}{2} \rfloor$ black intervals of length $< \frac{1}{n}$ each. A similar approach works for the line segment.

Cheong *et al.* [209] were the first to investigate the *one-round Voronoi game* in dimension two and higher. They proved the following result which implies that Black has a winning strategy on the square.

Theorem 9.4. *For any set W of $n \geq n_0$ points in the unit square, there exists a set B of n points not in W such that, in the diagram $V(W \cup B)$, the Voronoi regions of the points in B have total area at least $\frac{1}{2} + c$. Here, n_0 and $c > 0$ are independent constants.*

Proof. (Sketch) If there are $\leq \lfloor \frac{n}{2} \rfloor$ regions in $V(W)$ covering an area of $> \frac{1}{2} + c$ in total, Black can take over nearly all of their area by attacking each $w \in W$ by two of his own points from either side, in the way described in Section 9.2. Otherwise, most regions in $V(W)$ have about the same size. In this case, one can make use of the fact that even a point s chosen at random has a Voronoi region $\mathrm{VR}(s, W \cup \{s\})$ of expected area $\frac{1+\beta}{2n}$, for some constant $\beta > 0$. This fact can be generalized to more than one point. □

Since Black has a winning strategy for the one-round Voronoi game on a square while White can always win on a line segment, the question arises where the situation changes as the square is squeezed into a segment. Fekete and Meijer [338] gave a surprising answer. Let $\rho \leq 1$ denote the *aspect ratio* of the rectangle in which the game is played (that is, the ratio of the lengths of its shorter and longer edges).

Theorem 9.5. *On a rectangle of aspect ratio ρ, Black has a winning strategy for the one-round game with n points if $n \geq 3$ and $\rho > \frac{\sqrt{2}}{n}$ holds, or if $n = 2$ and $\rho > \frac{\sqrt{3}}{2}$. Otherwise, White has a winning strategy.*

The proof of Theorem 9.5 is based on the following observations. If the Voronoi diagram of the white points, $V(W)$, contains a region which is not *point-symmetric* about its site, then Black can win by attacking the points in W; cf. Subsection 9.2. One can conclude that White's only chance for winning is to play his points in a regular grid. Afterwards, Black can win iff he is able to place one point p that conquers an area strictly larger than $\frac{1}{2n}$ in $V(W \cup \{p\})$. Clearly, this condition is necessary. It is also sufficient, because after placing p, Black can still steal close to half of its former region area from any point in $w \in W$, by placing a black point next to w on the opposite side of p. Since the regions in $V(W)$ were all of area $\frac{1}{n}$, the claim follows.

Fekete and Meijer [338] also considered the one-round Voronoi game in a polygon with holes. They proved that Black's optimization problem is *NP-hard* once White has played his points.

9.2.3. *Hotelling game*

In the Voronoi game, the position of a point site is fixed once the point has been placed. A scenario allowing for relocation of sites in some bounded area is known as the *Hotelling game*, named after the economist H. Hotelling.

Here each single point site p_i in S is owned by a different player P_i, for $1 \leq i \leq n$. In each round, each of the players $P_1, P_2, \ldots, P_n$ can, in this order, move his site to a new location that maximizes the area of its Voronoi region, while all other sites remain fixed. The question is if an *equilibrium* will be reached where no player can gain by moving his site. Also a characterization of all equilibrial configurations would be of interest.

For the one-dimensional case, some of these questions can be answered. On a line segment or on a circle, a point site p with two neighboring sites q, q' has a Voronoi region of size $d(q, q')/2$, independent of its precise location between q and q'. Here d denotes the length of the (straight or circular) segment connecting q and q' that contains p. For the circle, this implies the following.

Observation 9.1. *Let S be a set of n points on the circle, labeled $p_0, p_1, \ldots, p_{n-1}$ in counterclockwise order. Then S is at equilibrium iff*

$$d(p_j, p_{j+1}) \leq d(p_{i-1}, p_i) + d(p_i, p_{i+1})$$

holds for all indices i and j modulo n.

Indeed, if this condition were violated, it would be profitable to relocate p_i to the circular segment from p_j to p_{j+1}. On the line segment, the situation is more complicated because, for the two outermost sites, it always pays to attack their neighbors. Eaton and Lipsey [298] obtained the following result.

Theorem 9.6. *For $n \in \{2, 4, 5\}$ point sites on a line segment, there are unique equilibrial states. For $n = 3$, no equilibrium exists. For each $n \geq 6$, there is an infinite number of equilibrial configurations.*

See Figure 9.2. More illustrations and many further results, also of experimental nature, can be found in Okabe *et al.* [571]. Given the difficulty of computing the maximum area attainable in a single move (see Subsection 9.2.1), it may not seem surprising that a general solution to the spatial Hotelling problem in dimensions two and higher has not yet been discovered.

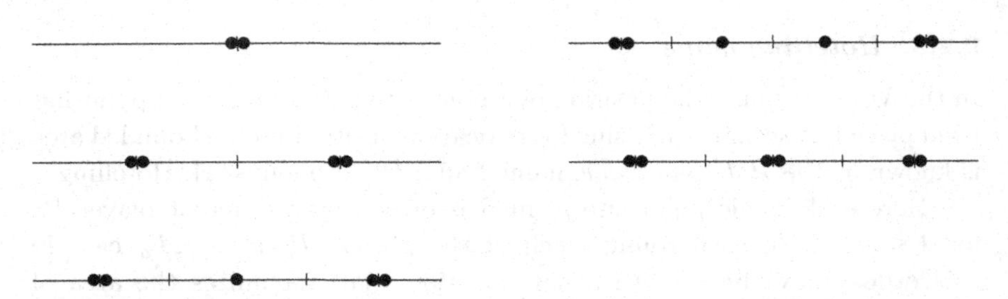

Figure 9.2. In the Hotelling game on a line segment, the equilibria for 2, 4, and 5 points are unique. For six points, the distance between points 3 and 4 can range from 0 to $\frac{1}{4}$; the extremes are shown on the right-hand side. It is interesting to observe that points at equilibrium may have Voronoi regions of different size. See [298] for details.

9.2.4. *Separating regions*

The problem of optimally placing Voronoi regions becomes easier in its combinatorial and topological setting, rather than in its geometric formulation where region areas etc. have to be considered.

Imagine an agent network represented by n black point sites in the plane. The goal is to disconnect the network as much as possible, in the sense that black Voronoi regions become non-adjacent when white point sites are being placed. Another motivation for questions of this kind arises in economics, where new shops are to be built so as to interfere with existing (and competing) shops as much as possible.

Several variants of such *separating region problems* have been studied in the literature, in their original form as well as in their equivalent *dual* form: Given the set B of black sites, place a set W of white sites such that, in the Delaunay triangulation $\mathrm{DT}(B \cup W)$, the number of black connected components is large. Ahmed *et al.* [20] report that, with n white sites, Black can be made less connected than White, i.e., the number of black components in $\mathrm{DT}(B \cup W)$ is larger than the number of white ones. Similarly, with only $\frac{2n+2}{3}$ sites, White can get all black sites as his neighbors, if $n \geq 12$. The intention behind the last task is to try to entice customers away from existing sites (shops) with better service.

Maximizing the number of black components, such that each of them consists of just an isolated black region (see Figure 9.3) results in the following *stabbing problem* for Delaunay triangulations: Find a preferably small point set W that 'stabs' all the triangles' circumcircles in the black triangulation $\mathrm{DT}(B)$, that is, each such circumcircle has to enclose at least one point in W. Aronov *et al.* [70] give an upper bound of $2n-2$ on $|W|$, in their paper studying the concept of *witness Delaunay graphs*; see Subsection 8.2.3. Ahmed *et al.* [20] offer an alternative, $5\,|M|$, where M is

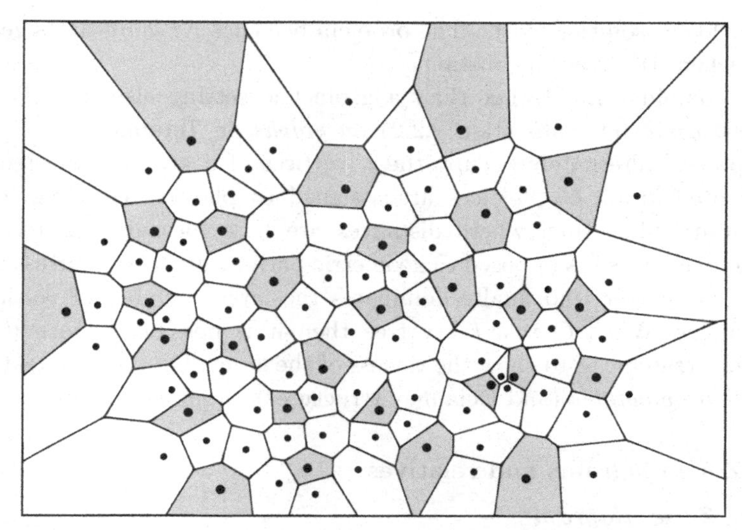

Figure 9.3. The presence of the white regions prevents the black regions from being neighbors in this Voronoi diagram (from [70]).

a *perfect matching* in DT(B). It is known that the Delaunay triangulation is guaranteed to contain some perfect matching; see Subsection 8.2.1.

The best upper bounds so far have been proven in Aichholzer *et al.* [41]. Their geometric constructions yield the following:

Lemma 9.1. *$|W| \leq \frac{3n}{2}$ can be achieved, and can be improved to $|W| \leq \frac{5n}{4}$ if the point set B is in* convex position.

They also argue that $n-1$ white points are always necessary to 'block' the black regions, i.e., to completely isolate them. For, if W blocks B, then DT($B \cup W$) does not contain any edge connecting two black points. That is, B is a so-called *independent vertex set* in this triangulation. But DT($B \cup W$) has to contain some perfect matching M, having $\lfloor \frac{n+|W|+1}{2} \rfloor$ components (one of the components might be a vertex rather than an edge). Since at most one point per component in M can be black, it follows that $n \leq \lfloor \frac{n+|W|+1}{2} \rfloor$, and thus $|W| \geq n-1$.

A kind of reverse problem, namely, the problem of getting all the white Voronoi regions *connected* by removing some black sites, has been studied in de Berg *et al.* [241]. More precisely, given sets W and B of white and black sites, respectively, how to find a minimum-sized subset $B' \subset B$ such that the white regions form a single connected component in the Voronoi diagram of $W \cup (B \setminus B')$? If there are only two white sites, $|W| = 2$, then an $O(n \log n)$ solution exists. The problem becomes *NP-hard* in its general form, when $|W|$ is not a constant.

A combinatorial rather than a geometric setting also underlies the *Voronoi game* (cf. Subsection 9.2.2) *on graphs*, in Teramoto *et al.* [682]. Two players alternately occupy the n vertices of a given undirected and unweighted graph G. Vertices are assigned to players according to the nearest neighbor rule, where distances are here measured in terms of numbers of edges (as opposed to geometric path lengths) on shortest paths in G. The player who finally dominates the larger number of vertices is the winner. When G is a k-ary tree then it is possible to formulate a winning strategy, or to show the tieness of the game. For a short treatment of *Voronoi diagrams* (and Delaunay structures) *on graphs*, see Section 9.4.

9.3. Zone diagrams and relatives

9.3.1. *Zone diagram*

As a fundamental property of Voronoi diagrams, they induce a partition of their underlying space: Voronoi regions are (interior) disjoint by definition, so the regions define a *packing* in space. On the other hand, for each point there exists at least one closest site, which implies that the (closures of the) regions yield a *covering* of the space. The concept of *zone diagrams*, introduced in Asano *et al.* [76] and to be discussed now, lacks the covering property, thus leaving a 'neutral zone' not being owned by any site.

Let S be a set of n point sites in the Euclidean plane $\mathbf{R}^2$. The *zone*, $Z(p)$, of a site $p \in S$ is defined to consist of all points in $\mathbf{R}^2$ closer to p than to the zone of any other site in S. More formally, for a subset $Z \subset \mathbf{R}^2$, let

$$\mathrm{dom}(p, Z) = \{x \in \mathbf{R}^2 \mid d(x,p) \leq d(x, Z)\},$$

where $d(x, Z) = \inf_{z \in Z} d(x, z)$. Clearly, $\mathrm{dom}(p, Z)$ is a convex set that necessarily contains p, as it can be expressed as an intersection of halfplanes which contain p. The zones are now required to fulfill

$$Z(p) = \mathrm{dom}\left(p, \bigcup_{q \in S \setminus \{p\}} Z(q)\right), \quad \text{for all } p \in S.$$

Observe that the description of this system of zones is implicit, as each zone is defined in terms of the remaining ones. So, the issues of existence and uniqueness of such a system arise.

The following *equilibrium interpretation* of a zone diagram gives some intuition about this structure. Citing [76], "...the notion of the zone diagram can be illustrated by a story about equilibrium in an age of wars. There are n

mutually hostile kingdoms. The ith kingdom has a castle at the site p_i and a territory $Z(p_i)$ around it. The territories are separated by a no-man's land, the neutral zone. If the territory $Z(p_i)$ is attacked from another kingdom, an army departs from the castle p_i to intercept the attack. The interception succeeds if and only if the defending army arrives at the attacking point on the boundary of $Z(p_i)$ sooner than the enemy. However, the attacker can secretly move his troops inside his territory, and the defense army can start from its castle only when the attacker leaves his territory. The zone diagram is an equilibrium configuration of the territories, such that every kingdom can guard its territory and no kingdom can grow without risk of invasion by other kingdoms."

Let us mention some basic facts about the behavior of zone diagrams first. Obviously, $p_i \in Z(p_i)$ holds, and $Z(p_i)$ is a convex set that is contained in p_i's region $\mathrm{VR}(p_i)$ in the classical Voronoi diagram of S. See Figures 9.4 and 9.5, where the Voronoi diagram for five sites and the corresponding zone diagram are shown. The distance of a zone boundary point x to the respective site equals the distance of x to the closest different zone. However, we face a peculiar behavior of zones, which is counter-intuitive for Voronoi diagram-like structures (though not for straight skeletons, which are not defined via distances; see Section 5.3): Zones can *gain* territory when new sites are added. For example, placing a new site very close to an existing one may make other neighboring zones expand. This phenomenon is quite realistic in the above 'kingdom' interpretation: The two close sites, as being 'enemies', weaken each other, and the surrounding zones get relatively stronger. As another unusual property, all zones may be *bounded* sets, which does not happen for 'normal' Voronoi diagrams.

In the simplest case, we have only two sites p and q, and the two resulting zones are bordered by what is called the *trisector curves* C_1 and

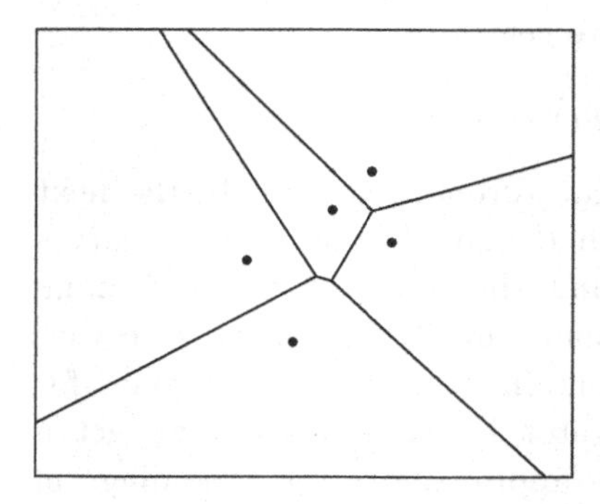

Figure 9.4. Voronoi diagram.

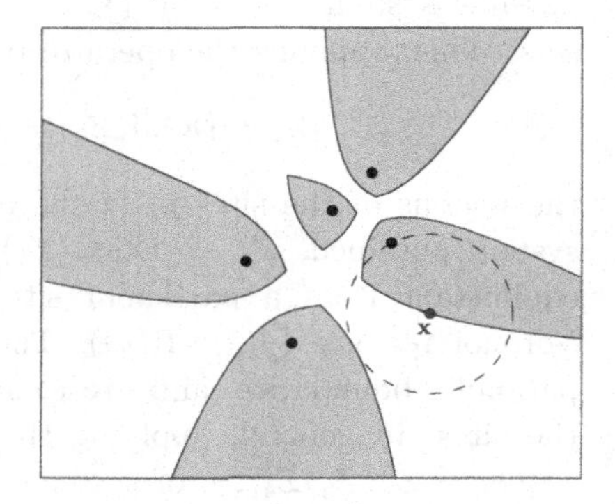

Figure 9.5. ... and zone diagram [76].

C_2 of p and q. These curves are defined such that every point on C_1 has equal distance to q and C_2, and every point on C_2 has equal distance to p and C_1. Asano *et al.* [77] prove that trisector curves exist and are unique analytic curves. It is not known whether these curves are algebraic or even expressible by elementary functions. Interestingly, the intersection of C_1 (or of C_2, its mirror image) with a given line parallel to the segment $\overline{pq}$ can be computed, to any desired precision, in time polynomial in the number of required digits.

In the case of more than two sites, things are even more complicated. An intuitive argument for the existence and uniqueness of the zone diagram could be based on the following *growth model*. From each site, a crystal is growing, everywhere on its boundary and at unit speed. As soon as the distance of a boundary point x of site p's crystal $K(p)$ to some other crystal $K(p')$ becomes as large as $d(x, p)$, the growth of $K(p)$ stops at x. The set of stopping points resulting from the overall growth process should now delineate the zone diagram. (Note the difference between this growth process and the process defining the *crystal growth Voronoi diagram*, in Subsection 7.4.3.)

Theoretically easier to deal with, Asano *et al.* [77] prove existence by iteratively *approximating* the zones, as follows. Consider an (ordered) system $\mathcal{A} = (A_1, \ldots, A_n)$ of non-empty subsets of $\mathbf{R}^2$, such that each A_i covers exactly one site p_i. Define

$$\mathrm{DOM}(\mathcal{A}) = (B_1, \ldots, B_n) \quad \text{with} \quad B_i = \mathrm{dom}\left(p_i, \bigcup_{j \neq i} A_j\right), \quad \text{for all } i.$$

Then, the system $\mathcal{A}$ describes the zone diagram if $\mathcal{A} = \mathrm{DOM}(\mathcal{A})$, that is, if $\mathcal{A}$ is a *fixed point* of the operator DOM.

To approximate the zone diagram, we now start with the smallest possible system, $\mathcal{Z}_0 = (\{p_1\}, \ldots, \{p_n\})$, which consists just of singleton sets. When applying the operator once, we get

$$\mathcal{Z}_1 = \mathrm{DOM}(\mathcal{Z}_0) = (\mathrm{VR}(p_1), \ldots, \mathrm{VR}(p_n)),$$

the regions of the sites p_i in the classical Voronoi diagram. In the next system obtained, $\mathcal{Z}_2 = \mathrm{DOM}(\mathcal{Z}_1)$, each domain is bounded by curves equidistant from a particular site p_i and the union of the remaining Voronoi regions, $\bigcup_{j \neq i} \mathrm{VR}(p_j)$. The domains in $\mathcal{Z}_2$ thus have piecewise parabolic boundaries, and are clearly contained in the desired zones for the sites. In general, applying the operator DOM iteratively, we get a sequence $\mathcal{Z}_0, \mathcal{Z}_2, \mathcal{Z}_4, \ldots$ of systems which approximates the zone diagram from inside, and another sequence $\mathcal{Z}_1, \mathcal{Z}_3, \mathcal{Z}_5, \ldots$ which approximates the zone diagram from outside. Both sequences converge to the same limit, the system $\mathcal{Z} = (Z(p_1), \ldots, Z(p_n))$ of unique zones for these sites.

This has been shown in [77], by applying *Schauder's fixed-point theorem.* They also outline an algorithm that constructs polygonal approximations of the various obtained domains. No estimates on the convergence rate of this algorithm are known, however. Alternative and simpler proofs of existence are provided in Kopecká *et al.* [473] and in Imai *et al.* [419]. The latter paper also establishes the existence of *k-sector curves* for general $k \geq 2$. In this context, we refer to an interesting paper by Reem [603] that addresses several existence and stability questions for general Voronoi diagrams.

The combinatorial size of the zone diagram is, of course, an interesting quantity. By analogy with Voronoi diagrams, we can define a *vertex* of a zone $Z(p_i)$ as a point on the boundary of $Z(p_i)$ with at least two distinct closest points in $\bigcup_{j \neq i} Z(p_j)$. *Edges* are boundary pieces that connect two consecutive vertices of a zone. An edge e of $Z(p_i)$ is always part of the bisector of p_i and some fixed zone $Z(p_j)$. The total number of edges and vertices can be shown to be $O(n)$. Still, this seemingly harmless complexity is misleading, as e might have a complicated structure stemming from boundary parts of $Z(p_j)$ that contain many vertices.

Generalizations of the zone diagram have been considered in Kawamura *et al.* [448]. They prove uniqueness for disjoint and compact sites of general shape in d-space, and also in certain normed spaces where unit balls are *smooth* and *strictly convex.* (Compare Section 7.2 on *convex distance functions.*) On the other hand, uniqueness is already lost, for example, for the L_1-*norm* in the plane, because its unit circles are not smooth.

9.3.2. *Territory diagram*

A modification of different flavor, called territory diagrams, is studied in de Biasi *et al.* [258]. Compared to the definition of a zone diagram, set equality is now replaced by set inclusion. More specifically, given point sites $p_1, \ldots, p_n$, any system of territories $T_1, \ldots, T_n$ in $\mathbf{R}^2$ with $p_i \in T_i$ is called a *territory diagram* (for these sites) if it fulfills the condition

$$T(p_i) \subseteq \operatorname{dom}\left(p_i, \bigcup_{j \neq i} T(p_j)\right), \quad \text{for } i = 1, \ldots, n.$$

Territory diagrams trivially arise whenever $T(p_i) \subseteq Z(p_i)$ holds, with $Z(p_i)$ being the zone of p_i. In particular, any zone diagram is a territory diagram. Also, in the sequence of systems $\mathcal{Z}_0, \mathcal{Z}_2, \mathcal{Z}_4, \ldots$ used above to (convexly) approximate zones from inside, each system $\mathcal{Z}_i$ constitutes a trivial instance. Valid territories need not be convex, however. In fact, *non-trivial* territory diagrams exist, as a simple example with two sites shows, where $T(p_1) \subset Z(p_1)$ but $T(p_2)$ is *not* contained in $Z(p_2)$.

Territory diagrams are 'mollified version' of zone diagrams, best explained in the 'kingdom' interpretation given before. Suppose that all the kings in the country have come to an agreement that, after one of them claims new territory, this will not be contested by others — if it does not lie closer to any's territory location $x \in T(p_i)$ as is the distance from x to p_i (since that would allow the respective king to successfully defend x from his castle, p_i). This regulation enables a particular king to conquer parts of the country larger than delineated by its zone $Z(p_i)$ in the zone diagram, by claiming territory sooner than his antagonists.

A territory diagram is called *maximal* if none of its territories can be expanded without violating the definition. It is clear that zone diagrams are maximal territory diagrams. The converse, however, is not true, and territories in a maximal territory diagram may still be nonconvex. The existence of maximal territory diagrams can be established via *Zorn's lemma* on partial orders, which leads to a somewhat simpler existence proof for zone diagrams.

9.3.3. *Root finding diagram*

A structure apparently similar to planar Euclidean zone diagrams arises in Kalantari [438], in the context of *polynomial root finding.* Let us view a point $p_j = (x_j, y_j)$ in $\mathbf{R}^2$ as a complex number

$$z_j = x_j + \mathbf{i} \cdot y_j, \quad \text{for } \mathbf{i} = \sqrt{-1}.$$

We can now describe a set of n point sites in the plane as the roots of a degree-n polynomial $Q(z)$ in the complex numbers,

$$Q(z) = \prod_{j=1}^{n} (z - z_j).$$

Conversely, given a polynomial $Q(z)$, root finding in the complex plane can be done by starting with some initial value z_0 and applying one of the known iterative methods. For suitable choices of z_0, convergence to some root z_j of $Q(z)$ will be achieved. That is, the corresponding point p_0 will be allotted to the site p_j. The set of all points in $\mathbf{R}^2$ whose orbit (under the iterations of a particular root finding method) converges to p_j is called the *basin of attraction*, $A(p_j)$, of p_j. Depending on the method used, $A(p_j)$ is a more or less accurate approximation of the Voronoi region $\mathrm{VR}(p_j)$. Unfortunately, $A(p_j)$ typically has a complicated structure, being disconnected, and even of *fractal* nature at its boundary. The connected component of $A(p_j)$ that contains p_j is called the *immediate* basin of attraction of p_j.

It is challenging to design root finding methods whose immediate basins of attraction resemble the shapes of Voronoi regions as much as possible. This goal is pursued in [438], by constructing a sequence $B_m(z)$, $m \geq 2$, of iteration functions whose first two members, $B_2(z)$ and $B_3(z)$, are well known as *Newton's method* and *Halley's method*, respectively. For $m \to \infty$, the basins of attraction of $B_m(z)$ uniformly approximate the Voronoi diagram of the roots from inside, to within any prescribed tolerance. That is, the 'chaotic parts' will lie in a controllable neighborhood of the Voronoi edges. On the other hand, it is shown that no rational iteration function is capable of achieving the same goal.

Using the sequence $B_m(z)$, one can define certain 'safe' convex subregions of the immediate basins of attraction, ordered by inclusion as m grows and the uncovered parts of the plane shrink. These regions exhibit a shape very similar to those in the zone diagram of the sites; see Figure 9.6.

Figure 9.6. Layering of convex subregions of basins of attraction (from [438]).

Apart from the visual similarity, however, a formal link between the two concepts — basins of attraction and zone diagrams — is still missing.

Each complex polynomial with n single roots will draw its individual layering of safe subregions, a fact used in what is called *polynomiography* in Kalantari [437]. This field of study is based on the computer visualization of the process of solving polynomial equations, and has applications in art, education, and science.

We remark at this place that Voronoi diagrams did not go unnoticed in the world of art and design. For example, Kaplan [442] creates attractive ornamental designs using Voronoi regions, Dobashi *et al.* [282] generate mosaic images for a non-photorealistic rendering method that show an artistic effect, and Vanhoutte explores the expressional power of fractal Voronoi diagrams. An internet search for *Voronoi art* will reveal numerous pleasing and colorful demonstrations of Voronoi diagrams as an artistic tool.

9.3.4. *Centroidal Voronoi diagram*

In the iterative process for approximating zone diagrams, described in Subsection 9.3.1, the sites remain fixed whereas the regions converge to the desired zones. There is quite a different type of Voronoi diagram where, in contrast to zone diagrams, the *sites* are allowed to move in the iteration process until they reach 'stable' positions.

A so-called *centroidal Voronoi diagram* is a (Euclidean) Voronoi diagram for point sites in a given convex domain $\Omega \subset \mathbf{R}^d$, such that each site p_i represents the *centroid* (center of mass) of its Voronoi region. (Usually, the region centroid can be located quite far away from the defining site.) Although probably not related to zone diagram structures, we briefly discuss this interesting special instance of classical Voronoi diagrams at this place, as it can be seen as the fixed point of a certain operator on Voronoi diagrams as well.

Sometimes this concept has been also called *self-centered Voronoi diagram* in the literature. This should not be confused with the notions of self-centered tetrahedral (or simplicial) complexes in Section 6.1, and of self-centered power diagrams in Section 10.5.

Centroidal Voronoi diagrams are usually defined for a given density function ρ on Ω, that is, *weighted centroids* of the regions are considered. The main question that arises is to find sets of sites, of given cardinality k, such that their Voronoi diagram is centroidal with respect to ρ. Figures 9.7–9.9 (from [435]) depict three planar examples, where different density functions are taken in the square $[-1, 1]^2$.

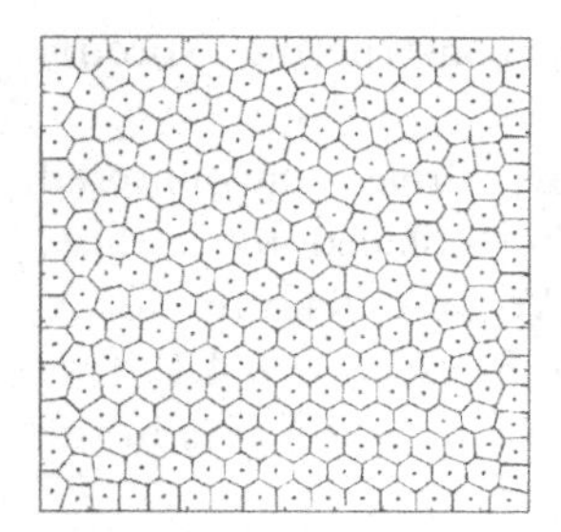

Figure 9.7. Centroidal Voronoi diagram, for $\rho(x,y) = 1$ (area).

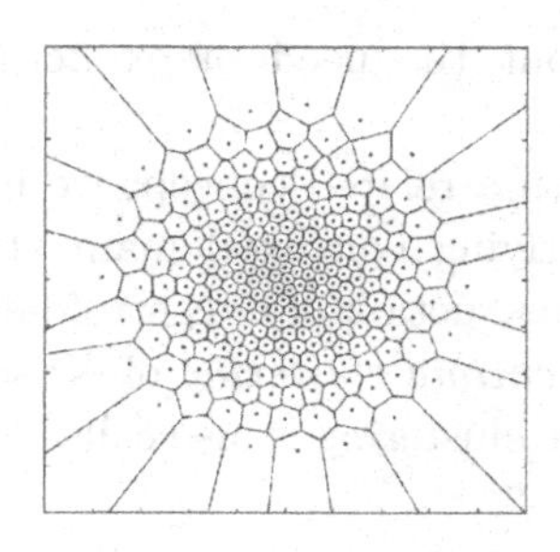

Figure 9.8. Centroidal Voronoi diagram, for density function $\rho(x,y) = e^{-10(x^2+y^2)}$.

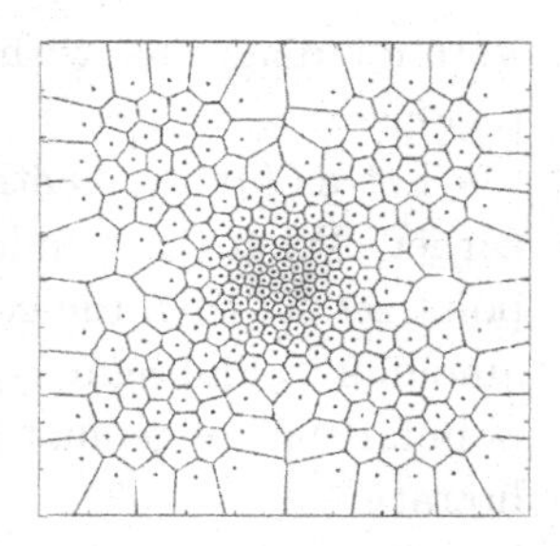

Figure 9.9. Centroidal Voronoi diagram, for $\rho(x,y) = e^{-20(x^2+y^2)} + \frac{1}{20}\sin^2(\pi x)\sin^2(\pi y)$.

Centroidal Voronoi diagrams do always exist, for any number k of sites. However, they are ambiguous structures, even in their simplest form where $\rho(x) = 1$ for all $x \in \Omega$ (uniform distribution), as a simple example with two sites in $\mathbf{R}^2$ shows.

In their survey paper, Du *et al.* [289] discuss various applications of this appealing diagram in diverse areas. Among others, they include data compression in image processing (where ρ represents the color density), estimation of data (for example in meteorology, where the sites represent optimal locations for precipitation measurement), facility location (sites indicate the optimal placement of resources), and territorial behavior in biology (where synchronous settlement of animals tends to take place at the sites of a centroidal Voronoi diagram).

An iterative construction algorithm, known as *Lloyd's method* [511] in the electrical engineering literature, alternates between constructing Voronoi regions and (weighted) centroids. Let S_0 be a set of k point sites in $\mathbf{R}^d$, possibly initially generated by a Monte Carlo method. For $i \geq 0$, we write $S_{i+1} = \mathrm{OP}(S_i)$, where the operator OP constructs the Voronoi diagram of S_i and then assigns to S_{i+1} the centroids of its regions. Clearly, any *fixed point* of the operator OP corresponds to a centroidal Voronoi diagram for k sites.

Several variants of Lloyd's method, as well as other iteration schemes for *approximating* centroidal Voronoi diagrams, are described in Du *et al.* [289] and Du and Wang [290]. Computation costs are usually high, as they may involve the construction of higher-dimensional Voronoi regions and the evaluation of multi-dimensional integrals.

Alternative construction methods that avoid these drawbacks are given, e.g., in Ju *et al.* [435]. Their probabilistic method, which is also suited to *parallelization*, obtains accurate and efficient approximations of centroidal

Voronoi diagrams without the need of explicitly constructing Voronoi regions.

Let us finally mention a related concept we have already encountered earlier. When the underlying convex domain Ω is replaced by a *finite* point set X, then the construction of a 'centroidal Voronoi diagram' for X amounts to finding a *k-centroid clustering* of X; see Subsection 8.4.1. This popular type of geometric clustering is also called *k-means clustering* in the literature.

9.4. Proximity structures on graphs

In certain applications, it is not appropriate to model distances between locations in the entire plane, but rather on a given connectivity structure. In fact, this is a standard framework in the area of facility location, where clients and suppliers lie on a connectivity network, usually a *planar graph* or even a *tree*. The network may have to be specified as an abstract weighted graph (where the triangle inequality does not hold, in general), or by some geometric graph, usually with straight edges in the plane.

9.4.1. *Voronoi diagrams on graphs*

Graph Voronoi diagrams have first been considered in Mehlhorn [532], for finding approximations of *minimal Steiner trees* in graphs. The underlying model is an edge-weighted graph $G = (V, E)$, where each edge has assigned a positive weight (or cost). The distance, $d(a, b)$, between two vertices a and b is defined as the sum of edge weights on a *shortest path* in G between a and b. For geometric graphs, weight can be just the Euclidean length. Distances between any two points placed on edges of G, not just between vertices, are meaningful then.

Given a subset $S = \{s_1, \ldots, s_n\} \subset V$ of distinguished vertices (the sites), the *Voronoi diagram of* G *and* S is a partition of V into subsets

$$V_i = \{a \in V \mid d(a, s_i) < d(a, s_j), j \neq i\}.$$

(We ignore the degenerate case where equality of path lengths does occur, which can be handled consistently.) Depending on the application, we might want to store some more information on these 'Voronoi regions' V_i, like the distance from each vertex $a \in V_i$ to its defining site s_i, or the corresponding set of shortest paths.

Some properties of graph Voronoi diagram are discussed in the book by Okabe *et al.* [571], who call it the *network Voronoi diagram*. For example, and unlike the situation in the Euclidean plane, regions lack a convexity property. Following the standard definition of graph convexity, a vertex set $U \subset V$ is called *convex* (with respect to G) if, for any $a, b \in U$, also all the vertices on shortest paths between a and b are in U. In fact, the sets V_i are not convex in general, but rather are *star-shaped*, in the sense that V_i contains all vertices on shortest paths from $a \in V_i$ to s_i.

Using an extension of *Dijkstra's* shortest path *algorithm*, Mehlhorn [532] computes the graph Voronoi diagram in time $O(e + v \log v)$, when G has e edges and v vertices. Erwig [329] improves the running time slightly, to $O(e + (v - n)\log(v - n))$, where n denotes the number of sites in S, and extends the algorithm to work on *directed* graphs. Note that, in the directed setting, unreachable vertices might exist, which thus do not belong to any Voronoi region; see Figure 9.10. Also, we have to distinguish between the *outward* and *inward graph Voronoi diagram*, respectively, depending on whether we want to consider path lengths from sites to vertices, or from vertices to sites. Parallel Dijkstra is used in the construction algorithms, based on sophisticated dynamic data structures like *Fibonacci heaps*. A lower bound for computing the graph Voronoi diagram is $\Omega(\max\{v, (v - n) \log(v - n)\})$.

Hurtado *et al.* [412] consider Voronoi diagrams on *geometric graphs*, mostly trees. Euclidean shortest path lengths are taken as distances, but weighted multiplicatively, modeling the attractiveness of a site (see Section 7.4). Sites may be placed anywhere on edges. They also discuss the *farthest-site* variant, where vertices are allotted to sites farthest away (rather than closest), and colored versions thereof.

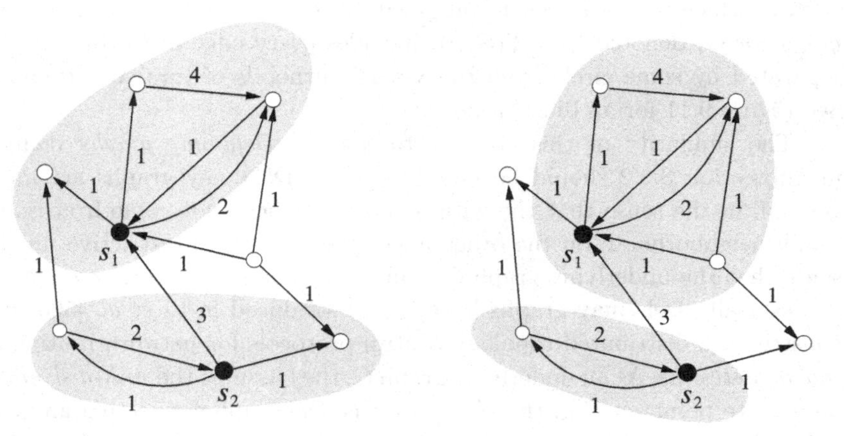

Figure 9.10. Outward and inward Voronoi diagrams on a weighted directed graph, with two graph vertices s_1 and s_2 as sites.

One motivation for studying graph Voronoi diagrams stems from handling typical queries in databases for networks; see e.g. Erwig and Güting [330]. For example, the graph G may model a *transportation network*, and the set S of sites corresponds to vertices of G that represent post offices, supermarkets, or fire stations. The graph versions of various distance problems, like the *post office problem*, *nearest neighbors* and *closest pair*, or '*largest empty circle*' (Section 8.1) can now be solved efficiently.

In particular, the last problem is relevant for *facility location* and leads to the notion of *anti-center* of a graph G with respect to S, which is the subset of vertices farthest away from any site in S. Such vertices are potential candidates for building properly located service plants. Given the graph Voronoi diagram, the anti-center can be computed in $O(v)$ time. Note that the *center* of G, that is, the subset of vertices with the smallest maximum distance to any vertex of G, corresponds to the 'smallest enclosing circle' for G.

As one more application, we mention secure *path planning* (cf. Section 8.5) in graphs, where a path from a source to a destination is to be found that keeps clear as much as possible from a set S of 'dangerous' sites.

9.4.2. *Delaunay structures for graphs*

Consider a geometric straight-line graph G with n vertices in the plane, not necessarily planar. We now measure distances as the *number* of edges along shortest paths in G (not as the weighted path length, as done for the graph Voronoi diagram in Subsection 9.4.1). The *k-neighborhood* of a vertex p is the subset of G's vertices at distance at most k from p, that is, the set of vertices which can be reached from p with k 'hops'. The *k-locally Delaunay graph* for G, denoted by $\mathrm{DT}(G, k)$, includes every edge of G that can be separated by some circle from the k-neighborhoods of its two endpoints. See Figure 9.11 for an illustration.

The similarity of this concept to *witness Delaunay graphs* defined in Subsection 8.3.2 should be noted. Witness Delaunay graphs are more general, in the sense that the witness point set need not come from some graph neighborhood; on the other hand, they are more restrictive, in the sense that the underlying graph G is just the *complete geometric graph*.

k-locally Delaunay graphs have been introduced in Li *et al.* [505], for constructing distributed topology control protocols for network routing in *mobile networks*. As an underlying graph G, they assume the *unit-disk graph* on a finite point set S in the plane, that is, the graph connecting any two points in S at distance less than 2 (see also Section 10.5). This models the situation where all wireless nodes have the same transmission range.

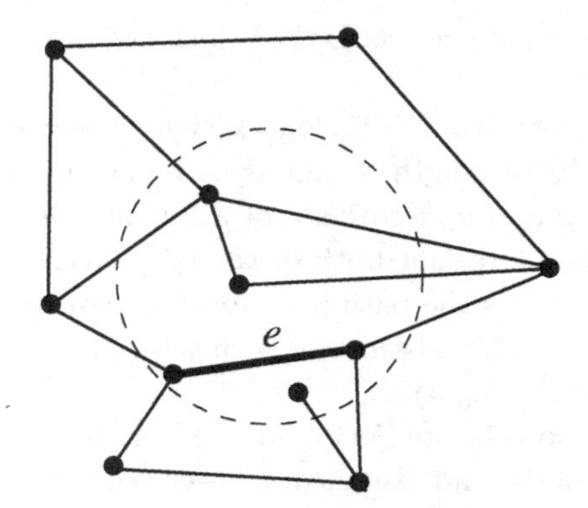

Figure 9.11. Edge e is included in DT$(G, 1)$, because e can be separated by some circle from its direct neighboring vertices. No circle exists, however, that can separate e from vertices at two hops away; therefore, e is not part of DT$(G, 2)$.

By removing this restriction, and thus allowing individual transmission radii, Kapoor and Li [444] study the structure of DT(G, k) for general geometric graphs G.

The size of DT(G, k) is an interesting quantity, because dense graphs G, though providing many edges as possible candidates for inclusion, also have rich vertex neighborhoods which cause exclusion of edges. Obviously, the graph DT$(G, k+1)$ is a subgraph of DT(G, k), that is, k-locally Delaunay graphs get sparser with increasing k. Note the difference to *higher-order Delaunay graphs* in Section 8.3.1, which show an opposite behavior.

If G is connected and k is sufficiently large, DT(G, k) becomes a subgraph of the (classical) Delaunay triangulation DT(S) for the vertex set S of G. In the extreme case where G is the complete geometric graph K_n on S, we have DT$(K_n, 1) =$ DT$(K_n, k) =$ DT(S), for all k.

As the number of edges in DT(G, k) usually happens to decrease rapidly with k, only small values of k are interesting. Basically, only the cases $k = 1$ and 2 are relevant in most applications. For unit-disk graphs, DT$(G, 2)$ is already a *planar graph*, and DT$(G, 1)$ can be decomposed into two planar graphs; see [505]. Bounds that are valid for arbitrary geometric graphs G are given in Kapoor and Li [444], based on small forbidden subgraphs, as we briefly describe below.

For example, they show that DT$(G, 1)$ has $O(n^{\frac{8}{5}})$ edges, by applying a result by Pinchasi and Radoičić [592], that any geometric graph with no self-crossing cycle of length 4 has a size of $O(n^{\frac{8}{5}})$. To see that DT$(G, 1)$ is indeed such a graph, consider some self-crossing cycle, C_4. All what C_4 contributes to DT$(G, 1)$ must be a subgraph of DT$(S) \cap G$, because for each edge e of C_4, the remaining two vertices are reachable from at least one of e's endpoints within one hop. So, as $\mathrm{DT}(S)$ does not have edge crossings, C_4 can have neither.

$\mathrm{DT}(G, 2)$ has already only $O(n \log n)$ edges, because this graph is free of self-crossing paths of length 3. Let $pqrs$ be such a path. Then, either any circle through p and q contains r or s, or any circle through r and s contains p or q. Therefore, not both of the edges $\overline{pq}$ and $\overline{rs}$ can belong to $\mathrm{DT}(G, 2)$, which excludes the path $pqrs$ from appearing in $\mathrm{DT}(G, 2)$. Pach *et al.* [576] showed that the absence of such self-crossing paths restricts the number of edges to $O(n \log n)$.

Pinchasi and Smorodinsky [593] improved the two upper bounds above, using different methods, and also gave a lower bound. We summarize their results.

Theorem 9.7. $\mathrm{DT}(G, 1)$ *has* $O(n^{\frac{3}{2}})$ *edges, and there are examples where* $\Omega(n^{\frac{4}{3}})$ *edges arise. Furthermore,* $\mathrm{DT}(G, 2)$ *has at most* $32n$ *edges.*

The graph $\mathrm{DT}(G, 1)$ has a surprising application, concerning *arrangements* of so-called *pseudo-lines* and *pseudo-parabolas*; see Tamaki and Tokuyama [674]. These are dissections of the plane, induced by open curves which pairwise cross at most once (or at most twice, for pseudo-parabolas). In fact, the number of curve segments that n pseudo-parabolas need to be cut into, in order to obtain a pseudo-line arrangement, is bounded by the number of edges in $\mathrm{DT}(G, 1)$.

Naturally, k-locally graphs for G can be based on *subgraph* concepts of the Delaunay triangulation (Section 8.2). For example, the *k-locally Gabriel graph* for G consists of all edges of G whose diametrical circle separates the respective edge from the k-neighborhoods of its two endpoints. This graph is sparser than $\mathrm{DT}(G, k)$, and has been preferably used in wireless network routing applications [505, 444]. Gabriel graphs are also easier to deal with algorithmically, because separating circles are fixed now to be diametrical circles.

Chapter 10

ALTERNATIVE SOLUTIONS IN $\mathbf{R}^d$

10.1. Exponential lower size bound

In the preceding chapters we have repeatedly stated that high-dimensional Voronoi diagrams suffer from an exponential worst-case complexity; a precise formula based on results by Seidel [625] is given in Theorem 6.1.

To give a quick argument for this fact, let us consider a set S of n points on the *moment curve* M in d-space $\mathbf{R}^d$, which is parametrized by

$$m(t) = (t, t^2, \ldots, t^d).$$

Consider a subset F of size d of S with the following *evenness property*: Between any two points of $S \backslash F$ there is an even number of points of F on M. Such subsets F of S can be chosen in $\Theta(n^{\lfloor \frac{d}{2} \rfloor})$ many different ways; see Matoušek [520] for a precise analysis. Let H_F denote a hyperplane in $\mathbf{R}^d$ containing F; see Figure 10.1.

Hyperplane H_F equals the set of all points $(x_1, \ldots, x_d)$ satisfying the equation

$$a_1 x_1 + a_2 x_2 + \cdots + a_d x_d = 0$$

with certain real coefficients $a_1, \ldots, a_d$. For each point $m(t)$ of M that lies on H_F, the parameter t is a zero of the polynomial

$$a_1 t + a_2 t^2 + \cdots + a_d t^d.$$

Since a polynomial of degree d has at most d roots, we obtain $M \cap H_F = F$. Moreover, all points in $S \backslash F$ must be situated on the same side of H_F, by the evenness property. Hence, H_F is a supporting hyperplane of the *convex hull* of S (which is a so-called *neighborly polytope* [384] in this case), and the points in F form one of its $(d-1)$-dimensional facets.

For each such subset F, there is an (unbounded) Voronoi edge of $V(S)$. Namely, this edge is the union of the centers of all d-spheres that pass

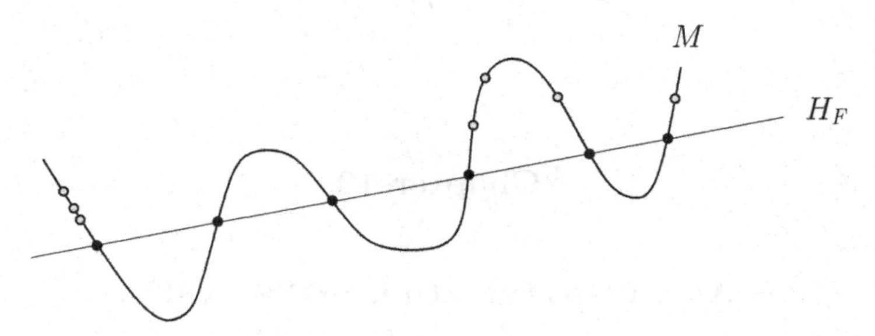

Figure 10.1. Points in F are drawn in black, points of $S\backslash F$ in gray.

through the points of F but do not contain any point of $S\backslash F$; cf. the proof of Lemma 2.2. Thus, $V(S)$ has an exponential number of unbounded edges.

A worst-case complexity of $\Theta(n^{\lfloor \frac{d}{2} \rfloor})$ renders Voronoi diagrams useless for computing, for instance, all nearest neighbors within a set of n points, which could trivially be done in $\Theta(n^2)$ time. Trusting that the given input points will not lead to the worst case, one could apply an output-sensitive method for computing the Voronoi diagram. Miller and Sheehy [543] recently presented an algorithm that runs in time $O(f \log n \log \Delta)$, where f is the total number of faces in the resulting Voronoi diagram $V(S)$, and Δ denotes the so-called *spread* of S, i.e., the ratio of the largest and the smallest pairwise distances in S. This method first constructs the Voronoi diagram of a superset of S for which a small complexity can be guaranteed. Then the extraneous points are carefully removed.

The notion of spread of a point set leads to an interesting view of various geometric problems. We refer to Erickson [327, 328] and references given there. For instance, the size of the Delaunay triangulation DT(S) in $\mathbf{R}^3$ is bounded by $O(\Delta^3)$ if S has spread Δ, a quantity independent from the size n of S; see [328].

In the subsequent sections, we will mention some alternative techniques for solving distance problems for point sets in high dimensions.

10.2. Embedding into low-dimensional space

A natural question is if high dimensions can be avoided altogether by mapping the given point set to some lower-dimensional space, and solving all distance problems there. Such *dimension reduction techniques* would work perfectly well if the pairwise distances among the points could be exactly preserved under such mapping. The example of the regular tetrahedron shows that this is too much to hope for: Its vertices have pairwise distance 1 in $\mathbf{R}^3$, but no four points in $\mathbf{R}^2$ have this property.

However, the above goal can be achieved with a small *distortion*, as a famous result by Johnson and Lindenstrauss [432] shows. It claims the existence of a global constant C such that the following holds. As usual, d denotes the Euclidean distance.

Theorem 10.1. *For each set S of n points $p_1, \ldots, p_n$ in $\mathbf{R}^n$, and for each $\varepsilon \in (0, \frac{1}{2})$, there exists a function $f : \mathbf{R}^n \longrightarrow \mathbf{R}^k$, where $k = \lfloor C \cdot \frac{\log n}{\varepsilon^2} \rfloor$, such that*

$$(1 - \varepsilon) \cdot d(p_i, p_j) \leq d(f(p_i), f(p_j)) \leq (1 + \varepsilon) \cdot d(p_i, p_j)$$

holds for all $1 \leq i, j \leq n$.

In other words, any n-point set in $\mathbf{R}^n$ can be mapped to $\mathbf{R}^{O(\log n)}$ such that the interpoint distances change by a factor of at most $1 \pm \varepsilon$.

Proof of Theorem 10.1. (Sketch) In order to find low-distortional mappings, Johnson and Lindenstrauss [432] considered the function

$$\begin{aligned} g : S^{n-1} &\longrightarrow \mathbf{R} \\ (x_1, \ldots, x_n) &\longmapsto \sqrt{{x_1}^2 + \cdots + {x_k}^2}, \end{aligned}$$

which, for each point on the sphere S^{n-1} in $\mathbf{R}^n$, computes the Euclidean length of its projection onto the first k coordinates. Let $m = \sqrt{\frac{n}{k}}$. One can show that m is the median of g, which means that the set A of all $a \in S^{n-1}$ for which $g(a) \leq m$ holds just covers half the surface area of S^{n-1}. Now a surprising property of high-dimensional spheres applies: If a 'hemisphere' set like A is extended by a belt of width only $t > 0$, the resulting set

$$A_t = \{q \in S^{n-1} \mid \exists\, a \in A : d(q, a) \leq t\}$$

covers all of S^{n-1}, except for an $O(\exp(-\frac{t^2 n}{4}))$th part. Therefore, a random point q of S^{n-1} lies, with high probability, within distance t of some point a satisfying $g(a) \leq m$. Since the function g is *1-Lipschitz*, $g(q) \leq g(a) + d(q, a) \leq m + t$ follows and, by a symmetric argument,

$$|g(q) - m| \leq t.$$

In other words, function g is sharply concentrated around m. Instead of projecting a random point onto a fixed k-dimensional subspace (given by the first k coordinates), one can fix the point and choose a random linear subspace H of dimension k in $\mathbf{R}^n$, with orthogonal projection f onto H. For two fixed points p_1, p_2 in $\mathbf{R}^n$, let $q = \frac{p_1 - p_2}{d(p_1,p_2)} \in S^{n-1}$ and $t = \frac{\varepsilon}{3} m$. By linearity of projection f, the previous inequality implies that

$$(1 - 3\varepsilon) \cdot m \cdot d(p_1, p_2) \le d(f(p_1), f(p_2)) \le (1 + 3\varepsilon) \cdot m \cdot d(p_1, p_2)$$

holds with probability at least $1 - \frac{1}{n^2}$ for dimension $k = 300 \frac{1}{\varepsilon^2} \log n$.

So, if we are given n points $p_1, \ldots, p_n$ in $\mathbf{R}^n$, the probability that one of the pairs p_i, p_j violates the previous estimates is less than $\binom{n}{2} \cdot \frac{1}{n^2} < 1$. Thus, the projection f onto a random subspace H is an embedding of low distortion, with a positive probability. A complete proof can be found in Matoušek [520]. $\square$

For algorithmic applications, one would repeatedly generate a random coordinate matrix of a mapping f, and test whether it is of low distortion. In order to do this efficiently, the matrix should have a simple structure and be sparse, that is, contain as many zero entries as possible. On the other hand, the sharp concentration property must be preserved. Recent work by Ailon and Chazelle [45] and by Matoušek [521] shows that these goals can be achieved to some extent. Instead of a projection mapping, they use a product $M \cdot H \cdot D$ of three matrices, where, after scaling, D is a diagonal matrix with independent random entries in $\{\pm 1\}$, H is an isometry given by a *Walsh matrix* with $\{\pm 1\}$ entries, and M is a sparse $k \times n$ matrix each of whose entries equals zero with probability $1 - q$, and ± 1 with probability $\frac{q}{2}$ each, independently. The result of applying the product $M \cdot H \cdot D$ to some vector can be computed in time $O(n \log n)$, by fast Fourier transform.

More information on metric space embeddings can be found, e.g., in Indyk and Matoušek [425].

10.3. Well-separated pair decomposition

In this section, we discuss a structure that solves some distance problems in high dimensions very efficiently; it has other interesting applications, too. Throughout this section, S is a set of n points in $\mathbf{R}^d$, and $s > 1$ denotes the so-called *separation constant.* Both s and the dimension d are considered fixed, while the number of points, n, is variable. Given a finite point set A, we denote by $R(A)$ the *bounding box* of A, that is, the smallest axis-parallel *hyperrectangle* in $\mathbf{R}^d$ containing A.

(A hyperrectangle is the Cartesian product of d mutually orthogonal line segments in $\mathbf{R}^d$. If the lengths of all segments are the same, a d-dimensional *hypercube* is obtained.)

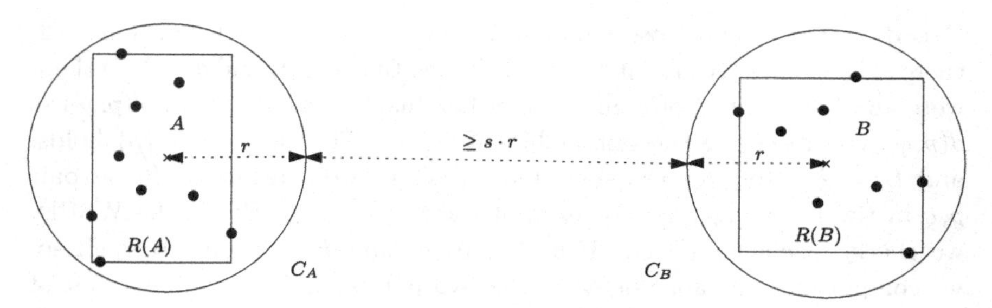

Figure 10.2. Sets A and B are well-separated with respect to $s = \frac{5}{2}$.

The results in this section are based on work by Callahan and Kosaraju [177].

Definition 10.1. Two point sets A and B in $\mathbf{R}^d$ are said to be *well-separated* (with respect to the separation constant s) if there exist two balls C_A and C_B of radius r, each containing the bounding box of its respective set, and being at least a distance of $s \cdot r$ apart.

A two-dimensional example is shown in Figure 10.2. It illustrates a simple, but important consequence of Definition 10.1. Namely, points within the same set of a well-separated pair are close, while points on opposite sides are far, and have almost the same distance between them.

Lemma 10.1. *Let A, B be well-separated with respect to s. Let $a, a' \in A$ and $b, b' \in B$. Then,*

(1) $d(a, a') < 2\,r \le 2\frac{d(C_A, C_B)}{s} \le \frac{2}{s}\,d(a, b)$ *and*
(2) $d(a', b') \le d(a', a) + d(a, b) + d(b, b') \le (\frac{2}{s} + 1 + \frac{2}{s}) \cdot d(a, b) = (1 + \frac{4}{s}) \cdot d(a, b)$.

The next definition is of central importance.

Definition 10.2. A sequence $(A_1, B_1), (A_2, B_2), \ldots, (A_m, B_m)$, where A_i, $B_j \subset S$, is called a *well-separated pair decomposition* of S with respect to s if

(1) each pair (A_i, B_i) is well-separated with respect to s, and
(2) for any two points $p \ne q$ of S there is exactly one index $i, 1 \le i \le m$, such that $p \in A_i$ and $q \in B_i$ hold, or vice versa.

The following lemma gives a first application of well-separated pair decompositions (WSPDs, for short).

Lemma 10.2. *Consider a well-separated pair decomposition of S with respect to $s \ge 2$, consisting of m pairs of sets. Then the* closest pair *of points in S can be found in time $O(m)$.*

Proof. Let a, b be a closest pair in S. By Property 2 of Definition 10.2, there exists a pair (A_i, B_i) in the WSPD such that $p \in A_i$ and $q \in B_i$. If there were another point p' of S in A_i then Lemma 10.1 would imply $d(p, p') < d(p, q)$, contradicting the minimality of $d(p, q)$. Therefore, $A_i = \{p\}$ holds, and $B_i = \{q\}$ too, for the same reason. Thus, we can find a closest pair p, q in the following way. For each of the m pairs (A_i, B_i) in the WSPD, we check, in constant time, if both sets A_i and B_i are singletons. If so, we compute the distance between the two points, and determine a pair of minimum value. □

The interesting question is about the number m of pairs in a WSPD. While the conditions in Definition 10.2 could be trivially met by forming all possible pairs of singleton sets, this would result in $m = \binom{n}{2} = \Theta(n^2)$ many pairs, making the statement of Lemma 10.2 pointless. The following result was shown in [177].

Theorem 10.2. *There exists a well-separated pair decomposition of S that consists of only $O(n)$ many pairs. It can be constructed in time $O(n \log n)$.*

Proof. (Sketch) Computing a WSPD is quite easy; the difficulty lies with the analysis. First, a *split tree* $T(S)$ is constructed that does not depend on the separation constant s. One splits the bounding box $R(S)$ of the given point set S into two parts of equal size, by a hyperplane cutting orthogonally through the midpoint of the longest edge of $R(S)$. This hyperplane splits S into subsets A and B, on which one recurs until only singleton sets are left. Each node u is associated with the subset S_u of all points stored in the leaves of $T(S)$ that have u as a predecessor.

Even though the resulting tree $T(S)$ need not be balanced, it can be constructed in time $O(n \log n)$, using the following techniques. In a preprocessing step, the points in S are sorted by all coordinates, and the resulting lists are pairwise doubly linked. When a box is split along coordinate j, the corresponding list is searched for the split value, starting from both ends simultaneously; this takes time proportional to the size of the smaller subset. The building of $T(S)$ is organized into $O(\log n)$ many phases. In each phase, all subsets stored at the leaves of the current tree are further split until their sizes are halved. One phase needs no more than linear time.

Once $T(S)$ is available, for each node with children v and w a procedure *findpairs*(v, w) is invoked. If the subsets S_v and S_w are well-separated with respect to s, the pair (S_v, S_w) gets reported, and control returns. Otherwise assume, without loss of generality, that $R(S_v)$ has a larger maximal extension than $R(S_w)$ does. If v_l, v_r denote the children of node v in $T(S)$, two procedure calls *findpairs*(v_l, w) and *findpairs*(v_r, w) are carried out.

The crucial step of the analysis is in proving that only $m = O(n)$ many well-separated pairs are reported during all the recursive calls of procedure *findpairs*. Since $T(S)$ has only a linear number of nodes, only $O(n)$ many subsets of S can occur in the output. If, for a subset A, pairs $(A, B_1), \ldots, (A, B_r)$ are reported, the partner sets $B_1, \ldots, B_r$ and their bounding boxes must be pairwise separated by hyperplanes. A packing argument shows that r is bounded by some constant. Hence, $m = O(n)$ holds. Since each recursion tree of an invocation of *findpairs* reports a well-separated pair at each of its leaves, the total number of procedure calls is linear, too. □

While the split tree $T(S)$ is clearly of linear size, the same holds for the WSPD only if all sets A_i, B_i occurring in the decomposition are stored implicitly, for example by a link to their corresponding nodes in $T(S)$. Although the number of pairs in the WSPDs just constructed is linear in the size of point set S, the sum of the sizes of all sets A_i, B_i can be quadratic. To overcome this disadvantage, a variant of the WSPD called a *semi-separated pair decomposition* has been studied. The total size of all subsets is now only $O(n \log n)$; see Abam *et al.* [3].

As an immediate consequence of Theorem 10.2 and Lemma 10.2, one can use a WSPD to determine, in time $O(n \log n)$, the *closest pair* in a set of n points. With extra effort, even more information can be extracted from a WSPD, as the next theorem shows; cf. Subsection 8.1.2 for analogous results in dimension 2.

Theorem 10.3. *The k-nearest neighbors of each point in S can be determined in time $O(n \log n + nk)$ by means of a well-separated pair decomposition of S.*

A main application of the WSPD is the construction of good *spanners*; see Subsection 8.2.4 for definitions. The following surprising result has an easy proof by induction on the ranks of the interpoint distances.

Theorem 10.4. *Let $(A_1, B_1), \ldots, (A_m, B_m)$ be a well-separated pair decomposition of point set S with respect to a constant $s > 4$. For each index i, where $1 \leq i \leq m$, let e_i be a line segment connecting a point of A_i to a point of B_i. Then the collection of all edges e_i forms an $\frac{s+4}{s-4}$-spanner of S.*

That is, not only is the collection of edges e_i a connected graph on S; we can even find, for any two points $p, q \in S$, a path in this graph whose Euclidean length does not exceed $t \cdot d(p, q)$, where $t = \frac{s+4}{s-4}$. Clearly, t tends to 1 as s grows to infinity. But increasing the precision comes at a cost, because the constants hidden in the asymptotic estimates stated in Theorem 10.2 are proportional to s^d.

An $O(n \log n)$ algorithm for t-spanners can be used to efficiently solve other problems like, e.g., approximating the *minimum spanning tree* (Subsection 8.2.1) or approximating the *dilation of graphs*; see Narasimhan and Smid [561] for these and many other results. They also discuss how to apply the WSPD concept to metrics different from the Euclidean.

10.4. Post office revisited

Even though the well-separated pair decomposition presented in Section 10.3 solves the closest pair and k-nearest neighbor problem in higher dimensions quite efficiently, it does not offer a quick solution to the *post office problem* introduced in Subsection 8.1.1.

Given a set S of n points in a space M equipped with a distance measure, one wants to construct a data structure that supports the following query: For an arbitrary query point q in M, report a point $p \in S$ closest to q. This problem is also called *proximity*, *similarity*, or *nearest neighbor search* problem in the literature.

10.4.1. *Exact solutions*

In dimension 2 the Voronoi diagram's *locus approach* provides an elegant and optimal solution, cf. Subsection 8.1.1. In higher dimensions, Voronoi diagrams can be too complex to be useful, as we discussed at the beginning of this chapter.

One would hope for an alternative data structure in d-dimensional Euclidean space $\mathbf{R}^d$, that is of $O(n)$ size and allows nearest neighbor queries to be answered in time $O(\log n)$. Despite many efforts, no structure has been found whose performance gets even close to these expectations. Clarkson [225] and Meiser [540] proposed solutions that achieve logarithmic query time but need space exponential in d. Alt and Heinrich-Litan [54] considered the L_∞-*norm* and point sets randomly drawn from the unit cube. Their algorithm needs only linear space but has an expected running time of roughly $O(\frac{dn}{\log n} + n)$, which improves on the brute force $O(dn)$ time bound if d is considered a variable.

It is widely conjectured that the 'curse of dimensionality' does not even permit achieving a query time in $O(n^{1-\varepsilon})$ without using space exponential in d. However, no lower bound that strong has been shown so far.

Some lower bounds have been established for the post office problem in *Hamming space* $\{0,1\}^d$, where the distance between a and b equals the number of coordinates $a_i \neq b_i$. One has to assume that $\log n \leq d$ holds because the whole space contains only 2^d many elements (corresponding to the vertices of the *unit hypercube* in $\mathbf{R}^d$). A recent result by Pătraşcu and Thorup states the following. If $d = \log^c n$, where $c > 1$, then any data structure of size $O(n \operatorname{polylog} n)$ must have a super-logarithmic query time of

$$\Omega\left(\frac{\log^c n}{\log\log n}\right)$$

even for randomized algorithms; see [587], also for a definition of the cell-probe model used, and for a discussion of previous results by other authors. One should observe that results for the Hamming space carry over to the L_1-*norm* in $\mathbf{R}^d$.

10.4.2. *Approximate solutions*

Since efficient exact solutions to the post office problem seem out of reach, *approximative algorithms* have received a lot of attention. Let us assume that, together with the point set S, an error bound $\varepsilon > 0$ is given. One is interested in a data structure that provides the following approximate answers. Given a query point q, let s denote a nearest neighbor of q in S. Then a point $s' \in S$ must be reported such that

$$d(q, s') \leq (1 + \varepsilon) \cdot d(q, s)$$

holds. Such a point s' is called a $(1 + \varepsilon)$-*nearest neighbor* of q in S.

With this relaxation, the post office problem becomes tractable. Arya *et al.* [74] obtained the following result.

Theorem 10.5. *In time $O(n \log n)$ one can build a data structure of size $O(n)$ that allows approximate nearest neighbor queries in $\mathbf{R}^d$ to be answered within time $O(\log n)$.*

In [74] the constant in the query time bound is in $O(d(1 + 6\frac{d}{\varepsilon})^d)$, while preprocessing is independent of ε. The bounds on size and preprocessing depend linearly on d. The construction works for all L_p-*norms* where $p \geq 1$. The data structure is based on a balanced *box decomposition tree*. Each vertex corresponds to a *rectangular annulus*, i.e., to the complement of a box within some larger, enclosing box.

Other approaches employed *dimension reduction techniques* as described in Section 10.2. Kleinberg [468] obtained logarithmic query time using space exponential in d. Indyk and Motwani [426] presented an algorithm whose query time is logarithmic in n and only linear in d. The space consumption, however, grows polynomially in n. They also introduced

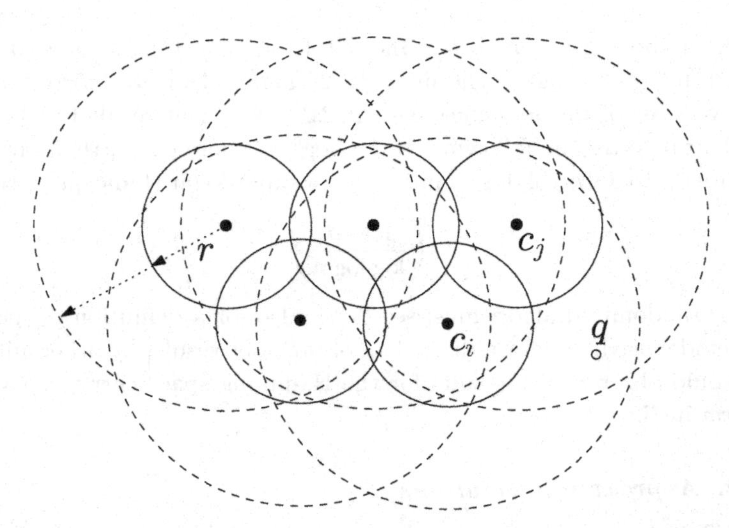

Figure 10.3. Point location in equal balls (PLEB). Query point q is contained in the balls of radius $2r$ centered at c_i and c_j, but in no ball of radius r.

an interesting reduction to a decision problem for *point-location in equal balls* (PLEB). Given n balls of radius r centered at points $c_1, \ldots, c_n$, and a query point q, is q contained in one of the balls? If so, return 'yes' and a center c_j whose r-ball contains q. Intuitively, the (approximate) distance to the nearest neighbor of q can be determined by a binary search on a sequence of exponentially growing radii; see Figure 10.3.

An approach of different flavor can be based on *dynamic spanners*; see, e.g., Gao *et al.* [351] or Gottlieb and Roddity [375]. Suppose that a $(1+\varepsilon)$-spanner, G, of the point set S is available, as mentioned at the end of Section 10.3. To obtain an approximate nearest neighbor in S for a query point q, one can simply insert q into G, check all points of S adjacent to q in $G \cup \{q\}$, and report an adjacent point s' of minimum distance to q. Let $\pi = (e_1, \ldots, e_m)$ denote the shortest path in spanner $G \cup \{q\}$ from q to a true nearest neighbor s of q in S, with edges $e_i = (p_i, p_{i+1})$. Then, indeed,

$$d(q, s') \leq \sum_{i=1}^{m-1} d(p_i, p_{i+1}) \leq (1+\varepsilon) \cdot d(q, s)$$

holds, because the first edge e_1 must be of length at least $d(q, s')$. The result in [375] provides dynamic spanners of *constant degree*; it implies Theorem 10.5 for Euclidean space $\mathbf{R}^d$ and, more generally, for metric spaces of constant *doubling dimension*. (Let λ denote the number of radius-r metric balls needed to cover a metric ball of radius $2r$. Then the doubling dimension of the space is defined to be $\log_2 \lambda$.)

Har-Peled [395] introduced *approximate Voronoi diagrams.* Using PLEB queries as described above, he computes, within time $O(n \log^3 n)$, a set of $O(n \log^2 n)$ many regions. Each region C is either a hypercube or a rectangular annulus, such that there is a point p_C in S which is a $(1+\varepsilon)$-nearest neighbor to all points in C. These regions define a partition of R^d. The hypercubes involved are stored in a compressed *quadtree. Point-location* can be implemented to run in time $O(\log n)$.

This approach was further refined by Arya *et al.* [73]. They assign up to $t \geq 1$ many points of S to each region C, requiring that, for each z in C, *at least one* of these points is a $(1+\varepsilon)$-nearest neighbor of z. Now the space consumption is only linear in n, thus providing another proof of Theorem 10.5. While dimension d is considered fixed, some trade-off is possible on the dependency of the error bound ε. For example, one can achieve a query time and storage requirement of

$$O\left(\log \frac{n}{\varepsilon}\right) \quad \text{and} \quad O\left(\frac{n}{\varepsilon^{d-1}}\right),$$

respectively, or use only $O(n)$ space with a query time of

$$O\left(\log n + \frac{1}{\varepsilon^{(d-1)/2}}\right).$$

Recently, Arya *et al.* [72] presented a unifying concept for the solution of several proximity problems that matches these bounds without introducing approximate Voronoi diagrams.

For further reading we refer to the surveys by Indyk [424] and Clarkson [227], and to the recent monograph [396] by Har-Peled.

10.5. Abstract simplicial complexes

In many practical situations, the question of (re)constructing unknown structures from *data sample points* arises. Of particular importance is *manifold reconstruction*, where the objective is to preserve the topological type, and to approximate the geometry. Applications stem from diverse areas, like visualization, pattern recognition, signal processing, neural computation, and others; see, e.g., Boissonnat *et al.* [144, 149] for references. *Subcomplexes* of the Delaunay triangulation (cf. Subsections 6.6.2 and 8.2.2) offer elegant solutions in low dimensions, especially the Delaunay simplicial complex restricted to a surface, as is described in detail in Boissonnat and Teillaud's book [149].

Sample points in parameter space can be of high dimension, though. As the size of the *Delaunay simplicial complex* $\mathrm{DT}(S)$ grows too fast, methods whose algorithmic complexity depend only on the *intrinsic dimension* of the manifold (rather than on the dimension of the ambient space) are sought. To this end, more 'light-weight' structures have been investigated, which we will describe to some extent in the present section.

Given a finite set S of elements, an *abstract simplicial complex* for S is a collection A of subsets of S that contains, for each member X of A, all possible subsets of X as well. The members of A are called abstract simplices. For example, if S is a set of n points in d-space, the subsets of points which span simplices in DT(S) form an abstract simplicial complex. However, not every abstract complex is geometrically realizable.

A prominent abstract simplicial complex is the *Vietoris–Rips complex*, often just called *Rips complex*, $R_\alpha(S)$, of a set S of points in d-space. Its k-simplices correspond to the $(k+1)$-tuples in S whose points are at pairwise distance less than α. See Figure 10.4 for an example. This structure was introduced by Vietoris [690] and Rips (see Gromov [382]) as a tool for studying hyperbolic groups. Ghrist and Muhammad [362], and others, have proposed Rips complexes as a light-weight representation of the topological structure of high-dimensional data. Chambers *et al.* [187] address basic homotopy problems in Rips complexes in the Euclidean plane. For example, they describe efficient algorithms for determining whether a given cycle is contractible in the Rips complex of a finite planar point set.

The *nerve* of a finite set B of balls in d-space is the collection of all subsets of B with non-empty common intersection. It is an abstract simplicial complex; see Edelsbrunner [302] for its various properties. In particular, the *Čech complex*, $C_\alpha(S)$, of a point set S is the nerve of the set of (congruent) balls of radius α around the points in S. (Note the similarity of these complexes to the (*weighted*) *α-shapes* defined in Subsection 6.6.2.) Čech complexes are closely related to Rips complexes, which is made explicit in the following lemma.

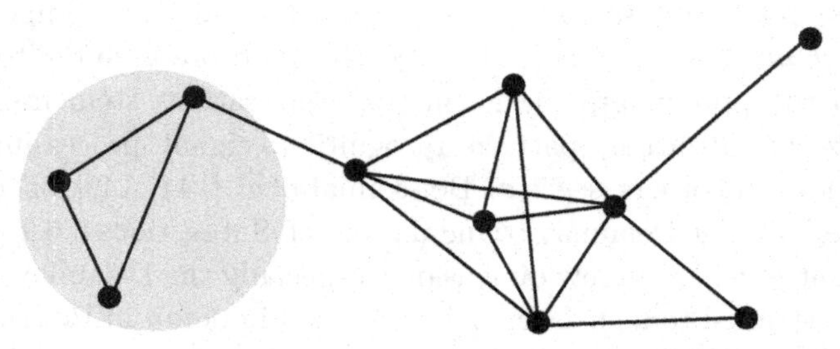

Figure 10.4. This Rips complex consists of ten 0-simplices (the defining points), seventeen 1-simplices (the shown edges), twelve 2-simplices (spanned triangles), five 3-simplices (spanned 'tetrahedra'), and one 4-simplex (the complete graph K_5). The distance α is visualized by a disk of this diameter.

Lemma 10.3. *For any finite point set S, we have $C_{\frac{\alpha}{2}}(S) \subseteq R_\alpha(S) \subseteq C_\alpha(S)$.*

Another related class of abstract simplicial complexes has been considered in the literature. These complexes are defined by two sets of points, namely, the underlying vertex set S, and some set P of so-called witness points. We say that a k-simplex spanned by S is *witnessed* by a point p if its vertices are among the $k+1$ nearest neighbors of p in S. The *witness complex* for S and P now contains all simplices spanned by S that are witnessed by some point in P. More generally, for fixed $\alpha \geq 0$, the complex $W_\alpha(S, P)$ contains all simplices spanned by S, whose vertices v lie within distance $d(v, p) + \alpha$ from some witness $p \in P$. We have the following inclusion relation:

Lemma 10.4. *If each point in S lies within distance β from some point in P, then $C_{\frac{\alpha-\beta}{2}}(S) \subseteq W_\alpha(S, P)$ holds, for all $\alpha > \beta$.*

Already for the standard case $\alpha = 0$, the witness complex $W_0(S, P)$ is a versatile tool, because the choice of the witness set P can be used to control the selection of the simplices connecting the vertex set S. Based on the *empty-sphere property* of the Delaunay simplicial complex, de Silva and Carlsson [259] prove that a k-simplex Δ spanned by S is included in $\mathrm{DT}(S)$ if Δ and all its faces have a witness in P. Consequently, $W_0(S, P)$ is an approximation of $\mathrm{DT}(S)$ and, in the limit when P covers the entire d-space, both complexes coincide.

For reconstructing manifolds, restricted versions of $\mathrm{DT}(S)$ have to be considered. Given a submanifold M in d-space, the *Delaunay complex of S restricted to M*, for short $\mathrm{DT}_M(S)$, consists of all the faces of $\mathrm{DT}(S)$ whose *dual* Voronoi faces intersect M.

Several authors consider the subcomplex $\mathrm{DT}_M(S)$ in relation to the witness complex $W_0(S, P)$, for the particular case where the points in S are taken from P. Guibas and Oudot [392] report that, if badly shaped simplices (so-called slivers) occur, then $W_0(S, P)$ is not included in $\mathrm{DT}_M(S)$, and homotopy equivalence is lost, even for dense sample sets S. Informally speaking, a *sliver* is a flat simplex all of whose edges are still long. In [392], a strategy is proposed for selecting S in order to keep homotopy equivalence, in the case where M is of exactly one dimension lower than the ambient space. Attali *et al.* [84] generalize this result to arbitrary dimensions of the submanifold M, and show that if M is a smoothly embedded surface, then a sufficiently fine sample S of M will make $W_0(S, P)$ approximate $\mathrm{DT}_M(S)$.

Boissonnat *et al.* [144] assign weights to the points in S, in order to get rid of slivers, building on results in Cheng *et al.* [201]. If $w(p) \geq 0$ denotes the weight of a point $p \in S$, then the condition

$$\max_{p \neq q \in S} \frac{w(p)}{d(p,q)} \leq \frac{1}{2}$$

ensures that no cell in the resulting *power diagram* of S (see Section 6.2) is empty, and that each point in S stays in its power cell.

(This is one more possibility of defining so-called *self-centered* diagrams, different from and opposed to the concepts of a centroidal Voronoi diagram or power diagram in Subsections 9.3.4 and 6.4.3, and of a self-centered simplicial complex in Section 6.1. Self-centered power diagrams whose cells contain their owning sites find further applications, for example in data visualization; see Andrews *et al.* [62] and Granitzer *et al.* [377].)

Self-centering weights can be found such that, in the dual *regular simplicial complex* for S (see Section 6.3) no slivers do occur, and the witness complex is included in this complex, and is homeomorphic to the underlying manifold, for suitable sample sets S. The algorithm in [144] runs in $O(c_1(\delta) \cdot n^2)$ time, where $n = |S|$ and $c_1(\delta)$ is a constant depending only on the *intrinsic dimension* δ of the manifold, though super-exponentially. More precisely, $c_1(\delta) = 2^{O(\delta^2)}$.

A thorough study of the properties and use of the various abstract simplicial complexes defined above is given in Chazals and Oudot [193]. They obtain new results on Rips, Čech, and witness *complex filtrations* (i.e, nested sequence of complexes of the same type). Based on such data structures, an algorithm is developed that, for smooth and sufficiently densely sampled manifolds, derives the homology in time $O(c_2(\delta) \cdot n^5)$, where $c_2(\delta) = O(8^{35^\delta})$. Topological estimation can be performed at a user-defined scale. Details are involved and go beyond the scope of this section. Let us just state the following nice result, concerning regular simplicial complexes and witness complexes.

Given some subset M of d-space, a finite point set $S \subset M$ is said to be an *ε-sample* of M if every point of M lies within distance ε of S. Moreover, we call S an *ε-sparse sample* if its points guarantee a minimum distance of ε from each other.

Lemma 10.5. *Let M be a compact manifold in d-space. Assume that $P \subset M$ is an ε_1-sample of M, and that $S \subset P$ is an ε_2-sample of P which is ε_2-sparse. For some weighting of S, consider the regular simplicial complex of S restricted to M, for short $\mathrm{RC}_M(S)$. Then, if the weighting fulfills the maximum condition stated above, we have $\mathrm{RC}_M(S) \subseteq W_\alpha(S,P)$ whenever $\alpha \geq \frac{8}{3}(\varepsilon_1 + \frac{1}{4}\varepsilon_2)$.*

If S is an unweighted point set (with $w(p) = 0$ for all $p \in S$) then the maximum condition on weights is automatically fulfilled. Moreover, $\mathrm{RC}_M(S)$ becomes the restricted Delaunay simplicial complex $\mathrm{DT}_M(S)$, and Lemma 10.5 can be strengthened to $\mathrm{DT}_M(S) \subseteq W_\alpha(S, P)$, for $\alpha \geq 2\,\varepsilon_1$.

We mention that the concept of witness complexes is different from the concept of *witness Delaunay graphs* [70] discussed in Subsection 8.3.2, which is based on the empty-circle property.

Very recently, Sheehy [644] provided a (computationally more tractable) approximation method for *Rips complex filtrations.* Recall that the Rips complex, $R_\alpha(S)$, of a set S of n points in d-space consists of all simplices spanned by S whose diameter is less than α. The so-called *persistence diagram* of the Rips filtration represents the changes in topology that correspond to the changes in $R_\alpha(S)$, when the parameter α is varied. A simplicial complex of size only $O(n)$ now can be constructed in time $O(n \log n)$, such that its persistence diagram is a good approximation to that of the Rips filtration. The constants depend only on the *doubling dimension* (which is $O(d)$ for Euclidean d-space) and, of course, on the accuracy of the approximation. The main tools used are metric *spanner graphs* and the efficient data structures for computing them; see Section 10.3.

The paper [644] also gives a nice overview on the topic of abstract simplicial complexes, with various further references.

Finally, we mention that a new method for storing and simplifying simplicial complexes has been proposed recently, in Attali *et al.* [85]. The encoding seems to be particularly effective for Rips complexes in higher dimensions.

Chapter 11

CONCLUSIONS

In this book, we have strived to present a reasonably complete and timely treatment of Voronoi diagrams and their relatives, from the point of view of a computational geometer. We hope that this book will be of use and help to people in various fields of science and engineering where space partitioning structures and algorithms for computing them are of importance. We also wish that we had conveyed the elegance and beauty of Voronoi diagrams — be it as a structure per se, or concerning the methods for analyzing and constructing them — and that interested readers from education, design, and art will also benefit from our expositions.

At the time of this writing, an internet search for '*Voronoi*' generates a list of about $7 \cdot 10^5$ articles, more than a single survey could cover. We have tried to focus on structural and algorithmic aspects of Voronoi diagrams, more than on their applications in other sciences. But even in this area, not every important and interesting topic could be given the attention it deserves.

11.1. Sparsely covered topics

Let us mention a few noteworthy aspects of Voronoi diagrams that have not been treated in detail in the present book.

A significant stream of research concerns the *stochastic properties* of Voronoi diagrams for randomly distributed sites. Motivation stems from their relevance in the natural and social sciences. The interested reader is referred to Chapters 5 and 8 of Okabe *et al.* [571] for a comprehensive treatment. Also, the survey paper [93] could be consulted for a brief overview. In the present book, randomization mainly affects the choices made by an *algorithm* during execution, whereas the input is usually considered fixed, but arbitrary.

Only in a minority of cases, expected quantities that result from assuming some underlying probability distribution of the *input sites* have been treated. Examples are the expected runtime of a Voronoi diagram construction algorithm for n sites uniformly distributed in the square, or the expected combinatorial size of a high-dimensional Delaunay complex whose sites are restricted to lie on a lower-dimensional manifold. We decided not to include other stochastic aspects, like metric properties of Voronoi regions (e.g., expected perimeter, or variance of their volume), or combinatorial quantities (e.g., expectations for the exact number of Voronoi edges and facets), as their relevance to computational geometry is less direct.

In view of the broad scope of applications of Voronoi diagrams, where the input data to be processed tend to rapidly increase in quantity, fast construction is an important issue. *Parallel algorithms* are a tool to reach this goal. We have mentioned a few, but did not detail and analyze parallel methods in this book. In many cases, divide & conquer methods for computing Voronoi diagrams and Delaunay triangulations lend themselves to efficient parallelization, but sometimes also the plane sweep technique does. We refer the reader to the handbook chapter [372] by Goodrich for an exposition of this material. Also, the chapter by Atallah and Chen devoted to parallel computational geometry, in the handbook [82], might partially cover this gap.

There are two radically different ways of representing a Voronoi diagram in a computer: By exact vector geometry, which we have been dealing with in this book, opposed to a representation as a binary image or pixel map, i.e., an integer square grid in $\mathbf{Z}^2$ each of whose cells represents a pixel, or a voxel for $\mathbf{Z}^3$. We have mentioned such *digital (pixel) Voronoi diagrams*, or *Voronoi diagrams on the grid*, only sporadically at important occasions. Such diagrams can be computed either directly on the pixel map, or by any exact geometric algorithm (see Chapter 3) followed by pixel extraction. They play an important role in image processing, computer vision and graphics, pattern matching, and also in various other fields where the *discrete medial axis* (often called the *medial axis transform*, MAT) of a planar shape has been applied with advantage over the years.

Pixel Voronoi diagrams can be trivially constructed in $O(m^2n)$ time on an $m \times m$ pixel map where n individual pixels are marked as sites, by assigning to each pixel the 'color' of its owning site in $O(n)$ time. In principle, this works for any distance function and, except its reliable calculation, no further geometric or structural properties are needed.

Enhanced information is comprised in the *(Euclidean) distance transform*, EDT, where each pixel holds the (squared Euclidean) distance to its closest site. The EDT can be computed in optimal time $O(m^2)$ with several methods. We refrain from details, which can be found in the comprehensive survey article by Fabbri *et al.* [333], and also in the overview of skeleton extraction methods in Siddiqi and Pizer [654].

The Voronoi diagram on the grid should not be confused with the *Voronoi diagram for a lattice*, in Subsection 7.1.2. There all the lattice points are sites, rather than just a few, and the diagram is a cell complex of congruent polytopes (prototiles) in $\mathbf{R}^d$.

Finally, let us admit that — while a host of applications of the Voronoi diagram have been presented — applications in the *natural and social sciences* have not been described in great detail in most cases. This would have had to include a thorough treatment of stochastic properties of Voronoi diagrams as well. We are aware that the appearance of Voronoi diagrams in natural phenomena is an indubitable indication of their inherent elegance and their usefulness, also in other areas of science. The book by Okabe *et al.* [571] puts strong emphasis on these important applications, which led us, being no experts in these areas, to encourage the reader to consult this book instead.

11.2. Implementation issues

Readers who want to apply Voronoi diagrams in practice might be interested in *implementation aspects.* Since implementing a geometric algorithm is an important and demanding task, we include some comments on this topic. For example, numerical problems are a big (but not the only) critical issue in this context, in computational geometry and in geometric and scientific computing in general. We have already given a few citations of relevant work at appropriate places in the book.

Are two straight lines exactly parallel, or do they intersect at a far remote point? Can we be sure that this line segment touches the circumference of that circle? Decisions based on queries of this kind might critically influence the subsequent control flow of a geometric algorithm — and with it — its accuracy or even worse, its correctness.

Theoretical work on geometric algorithms is often based on the idealistic, and sometimes unrealistic, assumptions that (i) inputs are in *general position*, and (ii) that real number arithmetic is *exact* and can be performed in *constant time*.

Assumption (i) is used to rule out 'degenerate' input configurations that deviate from the 'generic' case and tend to complicate matters. For example, if we assume that no three among n points in set S lie on a line, and no four points on a circle, then the Delaunay tessellation $\mathrm{DT}(S)$ is always a triangulation of S, nice to look at and easy to construct incrementally; see Section 3.2. That the configurations ruled out are contained in a lower-dimensional subset of the *configuration space* $\mathbf{R}^{2n}$ of the point coordinates seems to be justification for ignoring them.

However, in a VLSI application, input sets may consist of grid points that fail to be in general position. Consequently, we need to implement algorithms in such a way that they can cope with all inputs possible. There are different ways to proceed.

Starting with an algorithm $\mathcal{A}$ that works for input in general position, one could extend $\mathcal{A}$ into an algorithm that can handle degenerate input, too. Sometimes this can be done in an organic way, without implementing complicated case analysis. For example, the divide & conquer algorithm presented in Section 3.3 can be extended to gracefully handle Voronoi vertices of degree larger than 3; see [461]. It computes the Voronoi diagram $V(S)$ for arbitrary n-point sets S in the plane, hence $\mathrm{DT}(S)$ by dualization.

Another elegant approach is based on the idea that moving the points in S by an infinitesimal amount will remove degeneracies like the ones mentioned above. Thus, one first applies a *perturbation* to the input in such a way that general position is achieved. Then algorithm $\mathcal{A}$ is run on the perturbed input. One difficulty is in recovering the true output on input S, from the output which $\mathcal{A}$ has computed on the perturbed input. See Seidel [632] for a unifying presentation of the perturbation approach, a discussion of various methods, and for further references.

Now let us turn to assumption (ii) on perfect real arithmetic. Most programming languages offer standardized fixed or floating point reals, with a predefined number of digits, fixed size integers, and integers of (potentially) unbounded length. While floating point arithmetic is hardware-supported and therefore quite fast, it is prone to rounding errors. Computing with long integers, on the other hand, is exact but slow. A *floating point filter* approach tries to avoid long integer computations, using fast floating point arithmetic in combination with error bounds; see Mehlhorn and Näher [536].

If the input objects are specified by floating point or rational coordinates, they can be exactly represented by the above number types. As soon as an algorithm generates new geometric objects from given ones, their coordinates need no longer be rational numbers, however. For example, the plane sweep algorithm presented in Section 3.4 generates Voronoi vertices as intersection points of parabolas. Their coordinates involve square roots. Bisectors of more general curved objects, as discussed in Section 5.5, can be of larger algebraic degree; see Emiris *et al.* [319] for the case of ellipses. In general, one needs to find the real zeros of polynomial equations

$$f(x) = \sum_{i=0}^{n} a_i x^i$$

with rational coefficients a_i. If the degree n exceeds 4, the zeroes of f can in general not be expressed by nested radicals in the numbers a_i. Instead, the real zeroes of f must be isolated by rational intervals, and then be numerically approximated. The problem becomes even more challenging if the coefficients a_i themselves are algebraic numbers, only available by approximation.

We recommend the book by Boissonnat and Teillaud [149] that offers several solutions on how to deal with such problems, especially in the context of curves and surfaces arising in the construction of Voronoi diagrams and arrangements, and the book by Yap [713] for the algebraic background.

Both issues, degeneracy and accuracy, are interweaved. Namely, many geometric algorithms proceed by evaluating *geometric predicates*. For example, let InCircle(p, q, r, s) return a value < 0, $= 0$, or > 0 if point s lies inside, on, or outside of the circle defined by p, q, r, respectively. Value 0 characterizes a degenerate situation, and the other two cases require sufficient accuracy to return the proper result. *Controlled perturbation* is a new approach that uses perturbation to avoid degeneracies and to bypass precision problems at the same time; see Mehlhorn *et al.* [537].

Over the years, a lot of *geometric software* has been developed, including implementations of Voronoi diagrams and Delaunay triangulations.

Large algorithm libraries are LEDA (Library of Efficient Data types and Algorithms; Mehlhorn and Näher [536]) and CGAL (Computational Geometry Algorithms Library [185]). The interested reader may also consult the handbook chapter [452] by Kettner and Näher, and the books by Boissonnat and Teillaud [149] for curves and surfaces in CGAL and, very recently, Fogel *et al.* [341] for surface arrangements in CGAL. (Recall from Subsection 7.5.1 that computing envelopes and arrangements of *Voronoi surfaces* is a powerful and flexible approach to constructing Voronoi diagrams.) An introduction to geometry software outside these libraries has been given in a handbook chapter by Joswig [434].

On the 'piecewise linear side', the book on computational geometry in C by O'Rourke [572] offers various ready-to-use pieces of code in the programming language C, covering convex hulls in 2D and 3D, Delaunay triangulations, and planar polygon treatment.

To *visualize* a Voronoi diagram seems a somewhat easier task than to fully construct it. There are a multitude of interactive applets available on the internet, that allow to experiment with many variants of the fascinating structure we have presented in this book.

11.3. Some open questions

The body of literature on Voronoi diagrams is vast, and many of the arising theoretical and practical problems have found elegant and efficient solutions. Naturally, new questions have emerged from research conducted, and quite a few problems eluded satisfactory settlement until today.

Let us list — out of the various interesting open problems on Voronoi diagrams we have presented in this book — a few that seem most important in view of their basic nature and practical relevance. They concern both the combinatorial complexity of (seemingly) simple structures and their efficient algorithmic construction.

Among the few algorithmic problems for planar Voronoi diagrams which lack a worst-case optimal solution is the construction of the *order-k Voronoi diagram* for a fixed value of k. Only close-to-optimal algorithms do exist (Subsection 6.5.3). And notably, the usually very successful plane sweep approach, which has been applied to this problem recently, also leads to a runtime of only $O(k^2 n \log n)$; see [585]. An easy lower bound is $\Omega(kn + n \log n)$.

Placing an additional site such that its *region area* is optimally large when inserted in an existing Voronoi diagram is still an unsolved question (Subsection 9.2.1). Under which restrictions, other than Theorem 9.1, does the question become easier?

Analyzing the Voronoi diagram of *moving point sites* (Section 9.1) seems to be among the hardest open problems in this area. Recent achievements [611], though, kindle the hope for a near-quadratic upper bound on the number of flips in the kinetic Delaunay triangulation for certain site trajectories. In particular, the case of sites moving at constant speed on straight lines may have such a solution within reach.

Despite its importance in practice, the *minimum spanning tree* in 3-space (Subsection 8.2.1) has eluded efficient construction so far. Also, the existing subquadratic solutions [11, 710] are far from easy implementation. A partial answer is the available efficient expected-time and approximation algorithms (cf. also Chapter 10). Is there a practical and worst-case efficient algorithm for computing exact minimum spanning trees in 3-space?

A theoretical challenge is the analysis of the maximal size of the *medial axis* of a nonconvex polyhedron in 3D (Section 6.6). The known best upper and lower bounds are more than an order of magnitude apart. The same is true for the Voronoi diagram of straight lines in 3-space. Even worse is the situation for the *straight skeleton* (Section 5.3). This seemingly simpler for piecewise linear skeleton is notoriously difficult to construct, even in the plane. Also, its size in the three-dimensional case is unclear, apart from a near-quadratic lower bound [115].

Can any two tetrahedrizations on the same point set in 3D be transformed into each other by bistellar flips? Apart from theoretical interest, *flip graph connectedness* for tetrahedrizations would 'legalize' certain algorithms currently in use, which try to turn a tetrahedral mesh into the Delaunay tetrahedrization (Subsection 6.3.3).

Finally, an efficient exact solution for the *post office problem* in d dimensions is still outstanding (Section 10.4). The hope for a close-to-$O(\log n)$ solution is sparse, however, in view of the discouraging lower bound for Hamming spaces.

BIBLIOGRAPHY

[1] A. Abam, M. de Berg, and J. Gudmundsson. A simple and efficient kinetic spanner. *Computational Geometry: Theory and Applications* 43 (2010), 251–256.

[2] A. Abam, M. de Berg, S.-H. Poon, and B. Speckmann. Kinetic collision detection for convex fat objects. *Algorithmica* 53 (2009), 457–473.

[3] A. Abam, P. Carmi, M. Farshi, and M. Smid. On the power of the semi-separated pair decomposition. *Proc. 11th International Symposium on Algorithms and Data Structures*, Springer Lecture Notes in Computer Science 5664, 2009, 1–12.

[4] M. Abellanas, P. Bose, J. García, F. Hurtado, M. Nicolás, and P.A. Ramos. On structural and graph theoretic properties of higher-order Delaunay graphs. *International Journal of Computational Geometry & Applications* 19 (2009), 595–615.

[5] M. Abellanas, G. Hernandez, R. Klein, V. Neumann-Lara, and J. Urrutia. A combinatorial property of convex sets. *Discrete & Computational Geometry* 17 (1997), 307–318.

[6] M. Abellanas, F. Hurtado, and B. Palop. Transportation networks and Voronoi diagrams. *Proc. 1st International Symposium on Voronoi Diagrams in Science and Engineering*, 2004, 203–212.

[7] M. Abellanas, F. Hurtado, V. Sacristán, C. Icking, L. Ma, R. Klein, E. Langetepe, and B. Palop. Voronoi diagrams for services neighboring a highway. *Information Processing Letters* 86 (2003), 283–288.

[8] B.M. Ábrego, R. Fabila-Monroy, S. Fernández-Merchant, D. Flores-Peñaloza, F. Hurtado, V. Sacristán, and M. Saumell. On crossing numbers of geometric proximity graphs. *Computational Geometry: Theory and Applications* 44 (2011), 216–233.

[9] P.K. Agarwal, B. Aronov, T.M. Chan, and M. Sharir. On levels in arrangements of lines, segments, planes, and triangles. *Discrete & Computational Geometry* 19 (1998), 315–331.

[10] P.K. Agarwal, M. de Berg, J. Matoušek, and O. Schwarzkopf. Constructing levels in arrangements and higher order Voronoi diagrams. *SIAM Journal on Computing* 27 (1998), 654–667.

[11] P.K. Agarwal, H. Edelsbrunner, O. Schwarzkopf, and E. Welzl. Euclidean minimum spanning trees and bichromatic closest pairs. *Discrete & Computational Geometry* 6 (1991), 407–422.

[12] P.K. Agarwal, J. Gao, L.J. Guibas, H. Kaplan, V. Koltun, N. Rubin, and M. Sharir. Kinetic stable Delaunay graphs. *Proc. 26th Ann. ACM Symposium on Computational Geometry*, 2010, 27–136.

[13] P.K. Agarwal, L.J. Guibas, J. Hershberger, and E. Veach. Maintaining the extent of a moving point set. *Discrete & Computational Geometry* 26 (2001), 353–374.

[14] P.K. Agarwal, R. Klein, C. Knauer, S. Langerman, P. Morin, M. Sharir, and M. Soss. Computing the maximum detour and spanning ratio of 2- and 3-dimensional paths, trees, and cycles. *Discrete & Computational Geometry* 39 (2009), 17–37.

[15] P.K. Agarwal, J. Pach, and M. Sharir. State of the union (of geometric objects): A review. In: J.E. Goodman, J. Pach, and R. Pollack (eds.), *Surveys on Discrete and Computational Geometry: Twenty Years Later*. Contemporary Mathematics 453, American Mathematical Society, 2008, 9–48.

[16] P.K. Agarwal and M. Sharir. Efficient algorithms for geometric optimization. *ACM Computing Surveys* 30 (1998), 412–458.

[17] A. Aggarwal, L.J. Guibas, J. Saxe, and P.W. Shor. A linear-time algorithm for computing the Voronoi diagram of a convex polygon. *Discrete & Computational Geometry* 4 (1989), 591–604.

[18] A. Aggarwal, M. Hansen, and T. Leighton. Solving query-retrieval problems by compacting Voronoi diagrams. *Proc. 22nd Ann. ACM Symposium on Theory of Computing*, 1990, 331–340.

[19] A. Aggarwal, H. Imai, N. Katoh, and S. Suri. Finding k points with minimum diameter and related problems. *Journal of Algorithms* 12 (1991), 38–56.

[20] S.I. Ahmed, M. Hasan, and A. Sopan. Vindictive Voronoi games and stabbing Delaunay circles. *Proc. 7th International Symposium on Voronoi Diagrams in Science and Engineering*, 2010, 124–131.

[21] H.-K. Ahn, H. Alt, T. Asano, S.W. Bae, P. Brass, O. Cheong, C. Knauer, H.-S. Na, C.-S. Shin, and A. Wolff. Constructing optimal highways. *Proc. Computing: The Australasian Theory Symposium*, 2007, 7–14.

[22] H.-K. Ahn, O. Cheong, and R. van Oostrum. Casting a polyhedron with directional uncertainty. *Computational Geometry: Theory and Applications* 26 (2003), 129–141.

[23] H.-K. Ahn, S.-W. Cheng, O. Cheong, M. Golin, and R. van Oostrum. Competitive facility location: the Voronoi game. *Theoretical Computer Science* 310 (2004), 457–467.

[24] N. Ahuja. Dot pattern processing using Voronoi polygons as neighborhoods. *IEEE Transactions on Pattern Analysis and Machine Intelligence* 4 (1982), 336–343.

[25] O. Aichholzer, W. Aigner, F. Aurenhammer, T. Hackl, B. Jüttler, E. Pilgerstorfer, and M. Rabl. Divide-and conquer for Voronoi diagrams revisited. *Computational Geometry: Theory and Applications* 43 (2010), 688–699.

[26] O. Aichholzer, W. Aigner, F. Aurenhammer, and B. Jüttler. Exact medial axis computation for triangulated solids with respect to piecewise linear metrics. J.-D. Boissonnat *et al.* (eds.), *Proc. Curves and Surfaces 2011*, Springer Lecture Notes in Computer Science 6920, 2011, 1–27.

[27] O. Aichholzer, W. Aigner, F. Aurenhammer, T. Hackl, B. Jüttler, and M. Rabl. Medial axis computation for planar free-form shapes. *Computer-Aided Design* 41 (2009), 339–349.

[28] O. Aichholzer, D. Alberts, F. Aurenhammer, and B. Gärtner. A novel type of skeleton for polygons. *Journal of Universal Computer Science* 1 (1995), 752–761.

[29] O. Aichholzer and F. Aurenhammer. Straight skeletons for general polygonal figures. *Proc. 2nd Ann. International Computing and Combinatorics Conference*, Springer Lecture Notes in Computer Science 1090, 1996, 117–126.

[30] O. Aichholzer, F. Aurenhammer, P. Brass, and H. Krasser. Pseudo-triangulations from surfaces and a novel type of edge flip. *SIAM Journal on Computing* 32 (2003), 1621–1653.

[31] O. Aichholzer, F. Aurenhammer, D.Z. Chen, D.T. Lee, and E. Papadopoulou. Skew Voronoi diagrams. *International Journal of Computational Geometry & Applications* 9 (1999), 235–247.

[32] O. Aichholzer, F. Aurenhammer, S.-W. Cheng, N. Katoh, G. Rote, M. Taschwer, and Y.-F. Xu. Triangulations intersect nicely. *Discrete & Computational Geometry* 16 (1996), 339–359.

[33] O. Aichholzer, F. Aurenhammer, and T. Hackl. Pre-triangulations and liftable complexes. *Discrete & Computational Geometry* 38 (2007), 701–725.

[34] O. Aichholzer, F. Aurenhammer, T. Hackl, B. Jüttler, M. Oberneder, and Z. Šír. Computational and structural advantages of circular boundary representation. *International Journal of Computational Geometry & Applications* 21 (2011), 47–69.

[35] O. Aichholzer, F. Aurenhammer, T. Hackl, B. Kornberger, M. Peternell, and H. Pottmann. Approximating boundary-triangulated objects with balls. *Proc. 23rd European Workshop on Computational Geometry*, 2007, 130–133.

[36] O. Aichholzer, F. Aurenhammer, and F. Hurtado. Sequences of spanning trees and a fixed tree theorem. *Computational Geometry: Theory and Applications* 21 (2002), 3–20.

[37] O. Aichholzer, F. Aurenhammer, F. Hurtado, and H. Krasser. Towards compatible triangulations. *Theoretical Computer Science* 296 (2003), 3–13.

[38] O. Aichholzer, F. Aurenhammer, and B. Palop. Quickest paths, straight skeletons, and the city Voronoi diagram. *Discrete & Computational Geometry* 31 (2004), 17–35.

[39] O. Aichholzer, F. Aurenhammer, G. Rote, and Y.-F. Xu. Constant-level greedy triangulations approximate the MWT well. *Journal of Combinatorial Optimization* 2 (1999), 361–369.

[40] O. Aichholzer, R. Fabila-Monroy, D. Flores-Peñaloza, T. Hackl, C. Huemer, and J. Urrutia. Empty monochromatic simplices. Manuscript, Institute for Software Technology, University of Technology, Graz, Austria, 2012.

[41] O. Aichholzer, R. Fabila-Monroy, T. Hackl, M. van Kreveld, A. Pilz, P. Ramos, and B. Vogtenhuber. Blocking Delaunay triangulations. *Computational Geometry: Theory and Applications* 46 (2013), 154–159.

[42] O. Aichholzer, J. García, D. Orden, and P.A. Ramos. New results on lower bounds for the number of ($\leq k$)-facets. *European Journal of Combinatorics* 30 (2009), 1568–1574.

[43] O. Aichholzer, F. Hurtado, and M. Noy. A lower bound on the number of triangulations of planar point sets. *Computational Geometry: Theory and Applications* 29 (2004), 135–145.

[44] W. Aigner, F. Aurenhammer, and B. Jüttler. On triangulation axes of polygons. *Proc. 28th European Workshop on Computational Geometry*, 2012, 25–128.

[45] N. Ailon and B. Chazelle. Approximate nearest neighbors and the fast Johnson-Lindenstrauss transform. *SIAM Journal on Computing* 39 (2009), 302–322.

[46] S.G. Akl. A note on Euclidean matchings, triangulations, and spanning trees. *Journal of Combinatorics, Information and System Sciences* 8 (1983), 169–174.

[47] G. Albers, J.S.B. Mitchell, L.J. Guibas, T. Roos. Voronoi diagrams of moving points. *International Journal of Computational Geometry & Applications* 8 (1998), 365–380.

[48] J. Alonso, H. Martini, and M. Spirova. Minimal enclosing discs, circumcircles, and circumcenters in normed planes (Parts I and II). *Computational Geometry: Theory and Applications* 45 (2012), 258–274 and 350–369.

[49] G. Aloupis, J. Cardinal, S. Collette, F. Hurtado, S. Langerman, J. O'Rourke, and B. Palop. Highway hull revisited. *Computational Geometry: Theory and Applications* 43 (2010), 115–130.

[50] H. Alt, O. Cheong, and A. Vigneron. The Voronoi diagram of curved objects. *Discrete & Computational Geometry* 34 (2005), 439–453.

[51] H. Alt and M. Godau. Computing the Fréchet distance between two polygonal curves. *International Journal of Computational Geometry & Applications* 5 (1995), 75–91.

[52] H. Alt and C.K. Yap. Algorithmic aspect of motion planning: A tutorial, part 1. *Algorithms Reviews* 1 (1990), 43–60.

[53] H. Alt and C.K. Yap. Algorithmic aspect of motion planning: A tutorial, part 2. *Algorithms Reviews* 1 (1990), 61–77.

[54] H. Alt and L. Heinrich-Litan. Exact l_∞-nearest neighbor search in high dimensions. *Proc. 17th Ann. ACM Symposium on Computational Geometry*, 2001, 157–163.

[55] N.M. Amato and E.A. Ramos. On computing Voronoi diagrams by divide-prune-and-search. *Proc. 12th Ann. ACM Symposium on Computational Geometry*, 1996, 166–175.

[56] N.M. Amato and Y. Wu. A randomized roadmap method for path and manipulation planning. *IEEE International Conference on Robotics and Automation*, 1995, 113–120.

[57] N. Amenta, D. Attali, and O. Devillers. A tight bound for the Delaunay triangulation for points on a polyhedron. *Discrete & Computational Geometry* 48 (2012), 19–38.

[58] N. Amenta and M. Bern. Surface reconstruction by Voronoi filtering. *Discrete & Computational Geometry* 22 (1999), 481–504.

[59] N. Amenta, S. Choi, and R. Kolluri. The power crust, unions of balls, and the medial axis transform. *Computational Geometry: Theory and Applications* 19 (2001), 127–153.

[60] N. Amenta and R. Kolluri. Accurate and efficient unions of balls. *Proc. 16th Ann. ACM Symposium on Computational Geometry*, 2000, 119–128.

[61] N. Amenta and R. Kolluri. The medial axis of a union of balls. *Computational Geometry: Theory and Applications* 20 (2001), 25–37.

[62] K. Andrews, W. Kienreich, V. Sabol, J. Becker, G. Droschl, F. Kappe, M. Granitzer, P. Auer, and K. Tochtermann. The InfoSky visual explorer: exploiting hierarchical structure and document similarities. *Information Visualization* 1 (2002), 166–181.

[63] B. Angelov, J.-F. Sadoc, R. Jullien, A. Soyer, J.-P. Mornon, and J. Chomilier. Nonatomic solvent-driven Voronoi tessellation of proteins: An open tool to analyze protein folds. *Proteins: Structure, Function, and Genetics* 49 (2002), 446–456.

[64] Apollonius of Perga. Conica.

[65] E.M. Arkin, J.M. Díaz-Bañez, F. Hurtado, P. Kumar, J.S.B. Mitchell, B. Palop, P. Pérez-Lantero, M. Saumell, and R.I. Silveira. Bichromatic 2-center of pairs of points. *Proc. 10th Latin American Theoretical Informatics Symposium*, 2012, to appear.

[66] B. Aronov. On the geodesic Voronoi diagram of point sites in a simple polygon. *Algorithmica* 4 (1989), 109–140.

[67] B. Aronov. A lower bound on Voronoi diagram complexity. *Information Processing Letters* 83 (2002), 183–185.

[68] B. Aronov, M. de Berg, and S. Thite. The complexity of bisectors and Voronoi diagrams on realistic terrains. *Proc. 16th Ann. European Symposium on Algorithms*, Springer Lecture Notes in Computer Science 5193, 2008, 100–111.

[69] B. Aronov, P. Carmi, and M.J. Katz. Minimum-cost load-balancing partitions. *Algorithmica* 54 (2009), 318–336.

[70] B. Aronov, M. Dulieu, and F. Hurtado. Witness (Delaunay) graphs. *Computational Geometry: Theory and Applications* 44 (2011), 329–344.

[71] B. Aronov, M. Dulieu, and F. Hurtado. Witness Gabriel graphs. *Proc. 25th European Workshop on Computational Geometry*, 2009, 13–16.

[72] S. Arya, G. da Fonseca, and D.M. Mount. A unified approach to approximate proximity searching. *Proc. 18th Ann. European Symposium on Algorithms*, Springer Lecture Notes in Computer Science 6347, Springer, 2010, 374–385.

[73] S. Arya, T. Malamatos, and D.M. Mount. Space-efficient approximate Voronoi diagrams. *Proc. 34th Ann. ACM Symposium on Theory of Computing*, 2002, 721–730.

[74] S. Arya, D.M. Mount, N. Netanyahu, and R. Silverman. An optimal algorithm for approximate nearest neighbor searching in fixed dimensions. *Journal of the ACM* 45 (1998), 891–923.

[75] T. Asano, B.K. Bhattacharya, J.M. Keil, and F. Yao. Clustering algorithms based on minimum and maximum spanning trees. *Proc. 4th Ann. ACM Symposium on Computational Geometry*, 1988, 252–257.

[76] T. Asano, J. Matousek, and T. Tokuyama. Zone diagrams: existence, uniqueness, and algorithmic challenge. *SIAM Journal on Computing* 37 (2007), 1182–1198.

[77] T. Asano, J. Matousek, and T. Tokuyama. The distance trisector curve. *Advances in Mathematics* 212 (2007), 338–360.

[78] T. Asano, N. Katoh, H. Tamaki, and T. Tokuyama. Angular Voronoi diagram with applications. *Proc. 3rd International Symposium on Voronoi Diagrams in Science and Engineering*, IEEE Computer Society, 2006, 32–39.

[79] P.F. Ash and E.D. Bolker. Recognizing Dirichlet tessellations. *Geometriae Dedicata* 19 (1985), 175–206.

[80] P.F Ash, E.D. Bolker, H. Crapo, and W. Whiteley. Convex polyhedra, Dirichlet tessellations, and spider webs. In: M. Senechal and G. Fleck (eds.), *Shaping Space: A Polyhedral Approach*, Birkhäuser, Boston, MA, 1988, 231–250.

[81] M.J. Atallah. Some dynamic computational geometry problems. *Computers & Mathematics with Applications* 11 (1985), 1171–1181.

[82] M.J. Atallah and D.Z. Chen. Deterministic parallel computational geometry. In: J. Sack and G. Urrutia (eds.), *Handbook of Computational Geometry*, Elsevier, Amsterdam, 2000, 155–200.

[83] D. Attali, J.-D. Boissonnat, and H. Edelsbrunner. Stability and computation of medial axes—a state-of-the-art report. In: T. Müller, B. Hamann, B. Russell (eds.), *Mathematical Foundations of Scientific Visualization, Computer Graphics, and Massive Data Exploration*, Springer Series on Mathematics and Visualization, 2008, 109–125.

[84] D. Attali, H. Edelsbrunner, and Y. Mileyko. Weak witnesses for Delaunay triangulations of submanifolds. *Proc. ACM Symposium on Solid and Physical Modeling*, 2007, 143–150.

[85] D. Attali, A. Lieutier, and D. Salinas. Efficient data structure for representing and simplifying simplicial complexes in high dimensions. *International Journal of Computational Geometry & Applications* 22 (2012), 279–303.

[86] D. Attali and A. Montanvert. Computing and simplifying 2d and 3d continuous skeletons. *Computer Vision and Image Understanding* 67 (1997), 261–273.

[87] F. Aurenhammer. Power diagrams: properties, algorithms and applications. *SIAM Journal on Computing* 16 (1987), 78–96.

[88] F. Aurenhammer. A criterion for the affine equivalence of cell complexes in R^d and convex polyhedra in R^{d+1}. *Discrete & Computational Geometry* 2 (1987), 49–64.

[89] F. Aurenhammer. Recognising polytopical cell complexes and constructing projection polyhedra. *Journal of Symbolic Computation* 3 (1987), 249–255.

[90] F. Aurenhammer. Improved algorithms for discs and balls using power diagrams. *Journal of Algorithms* 9 (1988), 151–161.

[91] F. Aurenhammer. Linear combinations from power domains. *Geometriae Dedicata* 28 (1988), 45–52.

[92] F. Aurenhammer. A new duality result concerning Voronoi diagrams. *Discrete & Computational Geometry* 5 (1990), 243–254.

[93] F. Aurenhammer. Voronoi diagrams: A survey of a fundamental geometric data structure. *ACM Computing Surveys* 23 (1991), 345–405.

[94] F. Aurenhammer. Weighted skeletons and fixed-share decomposition. *Computational Geometry: Theory and Applications* 40 (2007), 93–101.

[95] F. Aurenhammer, R.L.S. Drysdale, and H. Krasser. Farthest line segment Voronoi diagrams. *Information Processing Letters* 100 (2006), 220–225.

[96] F. Aurenhammer and H. Edelsbrunner. An optimal algorithm for constructing the weighted Voronoi diagram in the plane. *Pattern Recognition* 17 (1984), 251–257.

[97] F. Aurenhammer, F. Hoffmann, and B. Aronov. Minkowski-type theorems and least-squares clustering. *Algorithmica* 20 (1998), 61–76.

[98] F. Aurenhammer and H. Imai. Geometric relations among Voronoi diagrams. *Geometriae Dedicata* 27 (1988), 65–75.

[99] F. Aurenhammer, N. Katoh, H. Kojima, M. Ohsaki, and Y.-F. Xu. Approximating uniform triangular meshes in polygons. *Theoretical Computer Science* 289 (2002), 879–895.

[100] F. Aurenhammer and R. Klein. Voronoi diagrams. In: J. Sack and G. Urrutia (eds.), *Handbook of Computational Geometry*, Elsevier, Amsterdam, 2000, 201–290.

[101] F. Aurenhammer and H. Krasser. Pseudo-simplicial complexes from maximal locally convex functions. *Discrete & Computational Geometry* 35 (2006), 201–221.

[102] F. Aurenhammer and O. Schwarzkopf. A simple on-line randomized incremental algorithm for computing higher order Voronoi diagrams. *International Journal of Computational Geometry & Applications* 2 (1992), 363–381.

[103] F. Aurenhammer and G. Stöckl. On the peeper's Voronoi diagram. *ACM SIGACT News* 22 (1991), 50–59.

[104] F. Aurenhammer, J. Wallner, M. Peternell, and H. Pottmann. Voronoi diagrams for oriented spheres. *Proc. 4th International Conference on Voronoi Diagrams in Science and Engineering*, IEEE Computer Society, 2007, 33–37.

[105] F. Aurenhammer and G. Walzl. Structure and computation of straight skeletons in 3-space. Manuscript, Institute for Theoretical Computer Science, University of Technology, Graz, Austria, 2012.

[106] F. Aurenhammer and Y.-F. Xu. Optimal triangulations. In: P.M. Pardalos and C.A. Floudas (eds.), *Encyclopedia of Optimization*, 2nd Edition, Springer, 2008, 2757–2764.

[107] D. Avis, B.K. Bhattacharya, and H. Imai. Computing the volume of the union of spheres. *The Visual Computer* 3 (1988), 323–328.

[108] F. Avnaim, J.-D. Boissonnat, O. Devillers, F.P. Preparata, and M. Yvinec. Evaluating signs of determinants using single-precision arithmetic. *Algorithmica* 17 (1997), 111–132.

[109] S.W. Bae and K.-Y. Chwa. Voronoi diagrams for a transportation network on the Euclidean plane. *International Journal of Computational Geometry & Applications* 16 (2006), 117–144.

[110] C. Bajaj. The algebraic degree of geometric optimization problems. *Discrete & Computational Geometry* 3 (1988), 177–191.

[111] M. Balzer and O. Deussen. Voronoi treemaps. *Proc. IEEE Symposium on Information Visualization*, 2005, 49–56.

[112] I. Bárány, Z. Füredi, and L. Lovász. On the number of halving planes. *Combinatorica* 10 (1990), 175–183.

[113] G. Barequet, A. Briggs, M. Dickerson, C. Dima, and M. Goodrich. Animating the offset polygon distance function. *Proc. 13th Ann. ACM Symposium on Computational Geometry*, 1997, 479–480.

[114] G. Barequet, M.T. Dickerson, and R.L.S. Drysdale. 2-Point site Voronoi diagrams. *Proc. 7th Ann. European Symposium on Algorithms*, Springer Lecture Notes in Computer Science 1663, 1999, 219–230.

[115] G. Barequet, D. Eppstein, M.T. Goodrich, and A. Vaxman. Straight skeletons of three-dimensional polyhedra. *Proc. 16th Ann. European Symposium on Algorithms*, Springer Lecture Notes in Computer Science 5193, 2008, 148–160.

[116] T.M. Barrett. Voronoi tessellation methods to delineate harvest units for spatial forest planning. *Canadian Journal of Forest Research* 27 (1997), 903–910.

[117] J. Basch, L.J. Guibas, C.D. Silverstein, and L. Zhang. A practical evaluation of kinetic data structures. *Proc. 13th Ann. ACM Symposium on Computational Geometry*, 1997, 388–393.

[118] J. Basch, L.J. Guibas, and L. Zhang. Proximity problems on moving points. *Proc. 13th Ann. ACM Symposium on Computational Geometry*, 1997, 344–351.

[119] P. Belleville, M. Keil, M. McAllister, and J. Snoeyink. On computing edges that are in all minimum-weight triangulations. *Proc. 12th Ann. ACM Symposium on Computational Geometry*, 1996, V7–V8.

[120] R. Benedetti and J.-J. Risler. Real algebraic and semi-algebraic sets. *Actualités Mathématiques*, Hermann, Paris, 1990.

[121] J.L. Bentley and T.A. Ottmann. Algorithms for reporting and counting geometric intersections. *IEEE Transactions on Computing* C-28 (1979), 643–647.

[122] J.L. Bentley and M.I. Shamos. Divide-and-conquer in multidimensional space. *Proc. 8th Ann. ACM Symposium on Theory of Computing*, 1976, 220–230.

[123] W. Benz. *Classical Geometries in Modern Contexts; Geometry of Real Inner Product Spaces.* Birkhäuser, 2005.

[124] J.L. Bentley, B.W. Weide, and A.C. Yao. Optimal expected-time algorithms for closest-point problems. *ACM Transactions on Mathematical Software* 6 (1980), 563–580.

[125] S. Bereg, K. Buchin, M. Buchin, M. Gavrilova, and B. Zhu. Voronoi diagram of polygonal chains under the discrete Fréchet distance. *International Journal of Computational Geometry & Applications* 20 (2010), 471–484.

[126] M. Bern and D. Eppstein. Mesh generation and optimal triangulation. In: D.-Z. Du and F. K. Hwang (eds.), *Computing in Euclidean Geometry*, Lecture Notes Series on Computing, World Scientific, Singapore, 1992, 23–90.

[127] M. Bern and D. Eppstein. Optimal Möbius transformations for information visualization and meshing. *Proc. 5th Workshop on Algorithms and Data Structures*, Springer Lecture Notes in Computer Science 2125, 2001, 14–25,

[128] M. Bern, D. Eppstein, and J. Gilbert. Provably good mesh generation. *Journal of Computer and Systems Sciences* 48 (1994), 384–409.

[129] D.P. Bertsekas. *Constrained Optimization and Lagrange Multiplier Methods*. Academic Press, New York, 1982.

[130] B.K. Bhattacharya and G.T. Toussaint. On geometric algorithms that use the furthest-point Voronoi diagram. In: G.T. Toussaint (ed.), *Computational Geometry*, North-Holland, Amsterdam, Netherlands, 1985, 43–61.

[131] M. Bienkowski, V. Damerow, F. Meyer auf der Heide, and C. Sohler. Average case complexity of Voronoi diagrams of n sites from the unit cube. *Proc. 21st European Workshop on Computational Geometry*, 2005, 167–170.

[132] L.J. Billera, P. Filliman, and B. Sturmfels. Constructions and complexity of secondary polytopes. *Advances in Mathematics* 83 (1990), 155–179.

[133] G. Blelloch, G.L. Miller, and D. Talmor. Developing a practical projection-based parallel Delaunay algorithm. *Proc. 12th Ann. ACM Symposium on Computational Geometry*, 1996, 186–195.

[134] H. Blum. Biological shape and visual science (Part I). *Journal of Theoretical Biology* 38 (1973), 205–287.

[135] M. Bock, A.K. Tyagi, J.U. Kreft, and W. Alt. Generalized Voronoi tessellation as a model of two-dimensional cell tissue dynamics. *Bulletin of Mathematical Biology* 72 (2010), 1696–1731.

[136] M. Bock. *Beyond whole-cell motion—reactive interpenetrating flow and elliptic Voronoi tessellation in two dimensions*. Ph.D. thesis, Department of Theoretical Biology, University of Bonn, 2013.

[137] C. Bohler, P. Cheilaris, R. Klein, C.-H. Liu, E. Papadopoulou, and M. Zavershynskyi. On the complexity of higher order abstract Voronoi diagrams. `http://archive.org/details/OnTheComplexityOfHigherOrder-AbstractVoronoiDiagrams`, 2013.

[138] M. Bogdanov, O. Devillers, and M. Teillaud. Hyperbolic Delaunay complexes and Voronoi diagrams made practical. *Proc. 29th Ann. ACM Symposium on Computational Geometry*, 2013, to appear.

[139] J.-D. Boissonnat, A. Cérézo, O. Devillers, and M. Teillaud. Output-sensitive construction of the 3-d Delaunay triangulation of points lying in two planes. *International Journal of Computational Geometry & Applications* 6 (1996), 1–14.

[140] J.-D. Boissonnat and F. Cazals. Smooth surface reconstruction via natural neighbor interpolation of distance functions. *Computational Geometry: Theory and Applications* 22 (2002), 185–203.

[141] J.-D. Boissonnat and C. Delage. Convex hull and Voronoi diagram of additively weighted points. *Proc. 13th Ann. European Symposium on Algorithms*, Springer Lecture Notes in Computer Science 3669, 2005, 367–378.

[142] J.-D. Boissonnat, O. Devillers, R. Schott, M. Teillaud, and M. Yvinec. Applications of random sampling to on-line algorithms in computational geometry. *Discrete & Computational Geometry* 8 (1992), 51–71.

[143] J.-D. Boissonnat and B. Geiger. Three dimensional reconstruction of complex shapes based on the Delaunay triangulation. In: R.S. Acharya and D.B. Goldgof (eds.), *Biomedical Image Processing and Biomedical Visualization* 1905, 1993, 964–975.

[144] J.-D. Boissonnat, L.J. Guibas, and S.Y. Oudot. Manifold reconstruction in arbitrary dimensions using witness complexes. *Discrete & Computational Geometry* 42 (2009), 37–70.

[145] J.-D. Boissonnat, F. Nielsen, and R. Nock. Bregman Voronoi diagrams. *Discrete & Computational Geometry* 44 (2010), 281–307.

[146] J.-D. Boissonnat and M.I. Karavelas. On the combinatorial complexity of Euclidean Voronoi cells and convex hulls of d-dimensional spheres. *Proc. 14th Ann. ACM-SIAM Symposium on Discrete Algorithms*, 2003, 305–312.

[147] J.-D. Boissonnat, M. Sharir, B. Tagansky, and M. Yvinec. Voronoi diagrams in higher dimensions under certain polyhedral distance functions. *Discrete & Computational Geometry* 19 (1998), 485–519.

[148] J.-D. Boissonnat and M. Teillaud. On the randomized construction of the Delaunay tree. *Theoretical Computer Science* 112 (1993), 339–354.

[149] J.-D. Boissonnat and M. Teillaud. *Effective Computational Geometry for Curves and Surfaces*. Springer, Berlin, 2007.

[150] J.-D. Boissonnat, C. Wormser, and M. Yvinec. Anisotropic diagrams: Labelle Shewchuk approach revisited. *Theoretical Computer Science* 408 (2008), 163–173.

[151] J.-D. Boissonnat and M. Yvinec. *Algorithmic Geometry.* Cambridge University Press, 1998.

[152] B. Boots. Weighting Thiessen polygons. *Economic Geography* 56 (1979), 248–259.

[153] F.L. Bookstein. The line skeleton. *Computer Graphics and Image Processing* 11 (1979), 123–137.

[154] S. Borgwardt, A. Brieden, and P. Gritzmann. Constrained minimum-k-star clustering and its application to the consolidation of farmland. *Operational Research* 11 (2011), 1–17.

[155] E. Boros and P.L. Hammer. On clustering problems with connected optima in Euclidean spaces. Technical report, RUTCOR, Ruters University, New Brunswick, New Jersey, 1988.

[156] P. Bose, P. Carmi, S. Collette, and M. Smid. On the stretch factor of convex Delaunay graphs. *Journal of Computational Geometry* 1 (2010), 41–56.

[157] P. Bose, S. Collette, F. Hurtado, M. Korman, S. Langerman, V. Sacristán, and M. Saumell. Some properties of k-Delaunay and k-Gabriel graphs. *Computational Geometry: Theory and Applications* 46 (2013), 131–139.

[158] P. Bose and L. Devroye. Intersections with random geometric objects. *Computational Geometry: Theory and Applications* 10 (1998), 139–154.

[159] P. Bose, L. Devroye, M. Löffler, J. Snoeyink, and V. Verma. Almost all Delaunay triangulations have stretch factor greater than $\pi/2$. *Computational Geometry: Theory and Applications* 44 (2011), 121–127.

[160] P. Bose, W. Lenhart, and G. Liotta. Characterizing proximity trees. In: G. Di Battista, P. Eades, H. de Fraysseix, P. Rosenstiehl, and R. Tamassia (eds.), *Proc. ALCOM International Workshop on Graph Drawing*, 1993, 9–11.

[161] P. Bose, M. Smid, and D. Xu. Delaunay and diamond triangulations contain spanners of bounded degree. *International Journal of Computational Geometry & Applications* 19 (2009), 119–140.

[162] J.E. Boyce, D.P. Dobkin, R.L.S. Drysdale, and L.J. Guibas. Finding extremal polygons. *SIAM Journal on Computing* 14 (1985), 134–147.

[163] P. Brass. On the size of higher-dimensional triangulations. *Combinatorial and Computational Geometry* 52, MSRI Publications, 2005, 147–153.

[164] C. Brenner. Towards fully automatic generation of city models. *International Archives of Photogrammetry and Remote Sensing* XXXIII, Part B3, Amsterdam 2000.

[165] M. Brévilliers, N. Chevallier, and D. Schmitt. Triangulations of line segment sets in the plane. *Proc. 27th International Conference on Foundations of Software Technology and Theoretical Computer Science*, Springer Lecture Notes in Computer Science 4855, 2007, 388–399.

[166] M. Brévilliers, N. Chevallier, and D. Schmitt. Constructing the segment Delaunay triangulation by flip. *Proc. 24th European Workshop on Computational Geometry*, 2008, 63–66.

[167] A. Brieden and P. Gritzmann. On optimal weighted balanced clusterings: gravity bodies and power diagrams. *SIAM Journal on Discrete Mathematics* 26 (2012), 415–434.

[168] E. Brisson. Representing geometric structures in d dimensions: Topology and order. *Proc. 5th Ann. ACM Symposium on Computational Geometry*, 1989, 218–227.

[169] K.Q. Brown. Voronoi diagrams from convex hulls. *Information Processing Letters* 9 (1979), 223–228.

[170] K.Q. Brown. *Geometric transforms for fast geometric algorithms*. Ph.D. thesis, Department of Computer Science, Carnegie-Mellon University, Pittsburgh, PA, 1980.

[171] H. Brugesser and P. Mani. Shellable decompositions of cells and spheres. *Mathematica Scandinavica* 29 (1971), 197–205.

[172] K. Buchin, M. Löffler, P. Morin, and W. Mulzer. Processing imprecise points for Delaunay triangulation: Simplified and extended. *Algorithmica* 61 (2011), 674–693.

[173] C. Burnikel, K. Mehlhorn, and S. Schirra. How to compute the Voronoi diagram of line segments: Theoretical and experimental results. *Proc. 2nd Ann. European Symposium on Algorithms*, Springer Lecture Notes in Computer Science 855, 1994, 227–239.

[174] H. Busemann. *The Geometry of Geodesics*. Academic Press, New York, 1955.

[175] L.A. Caffarelli, M. Feldman, and R.J. McCann. Constructing optimal maps for Monge's transportation problem as a limit of strictly convex costs. *Journal of the American Mathematical Society* 15 (2001), 1–26.

[176] L. Calabi and W.E. Hartnett. Shape recognition, prairie fires, convex deficiencies and skeletons. *American Mathematical Monthly* 75 (1968), 335–342.

[177] P.B. Callahan and S.R. Kosaraju. A decomposition of multidimensional point sets with applications to k-nearest neighbors and n-body potential fields. *Journal of the ACM* 42 (1995), 67–90.

[178] G.D. Canas and S.J. Gortler. Orphan-free anisotropic Voronoi diagrams. *Discrete & Computational Geometry* 46 (2011), 526–541.

[179] G.D. Canas and S.J. Gortler. Duals of orphan-free anisotropic Voronoi diagrams are triangulations. *Proc. 28th Ann. ACM Symposium on Computational Geometry*, 2012, 219–228.

[180] J. Canny. A Voronoi method for the piano-movers problem. *Proc. IEEE Int. Conference on Robotics and Automation*, 1985, 530–535.

[181] J. Canny and B.R. Donald. Simplified Voronoi diagrams. *Discrete & Computational Geometry* 3 (1988), 219–236.

[182] V. Capoyleas, G. Rote, and G. Wöginger. Geometric clusterings. *Journal of Algorithms* 12 (1991), 341–356.

[183] M. Caroli and M. Teillaud. Computing 3D periodic triangulations. *Proc. 17th Ann. European Symposium on Algorithms*, Springer Lecture Notes in Computer Science 5757, 2009, 59–70.

[184] J.G. Carlsson, E. Carlsson, and R. Devulapalli. Equitable partitioning with obstacles. Report, Industrial and Systems Engineering, University of Minnesota, 2012.

[185] CGAL (Computational Geometry Algorithms Library). `www.cgal.org`

[186] P. Cheilaris, E. Khramtcova, and E. Papadopoulou. Randomized incremental construction of the Hausdorff Voronoi diagram of non-crossing clusters. *Proc. 29th European Workshop on Computational Geometry*, 2013, 159–162.

[187] E.W. Chambers, J. Erickson, and P. Worah. Testing contractibility in planar Rips complexes. *Proc. 24th Ann. ACM Symposium on Computational Geometry*, 2008, 251–259.

[188] T.M. Chan. Approximating the diameter, width, smallest enclosing cylinder, and minimum-width annulus. *International Journal of Computational Geometry & Applications* 12 (2002), 67–85.

[189] T.M. Chan. Semi-online maintenance of geometric optima and measures. *SIAM Journal on Computing* 32 (2003), 700–716.

[190] T.M. Chan. A dynamic data structure for 3-d convex hull and 2-d nearest neighbor queries *Proc. 17th ACM-SIAM Symposium on Discrete Algorithms*, 2006, 1196–1202.

[191] T.M. Chan, J. Snoeyink, and C.K. Yap. Output sensitive construction of polytopes in four dimensions and clipped Voronoi diagrams in three. *Proc. 6th Ann. ACM-SIAM Symposium on Discrete Algorithms*, 1995, 282–291.

[192] B. Chandra, G. Das, G. Narasimhan, and J. Soares. New sparseness results on graph spanners. *International Journal of Computational Geometry & Applications* 5 (1995), 125–144.

[193] F. Chazals and S.Y. Oudot. Towards persistence-based reconstruction in Euclidean spaces. *Proc. 24th Ann. ACM Symposium on Computational Geometry*, 2008, 232–241.

[194] B. Chazelle. An optimal convex hull algorithm in any fixed dimension. *Discrete & Computational Geometry* 10 (1993), 377–409.

[195] B. Chazelle. *The Discrepancy Method: Randomness and Complexity.* Cambridge University Press, 2000.

[196] B. Chazelle, R.L.S. Drysdale, and D.T. Lee. Computing the largest empty rectangle. *SIAM Journal on Computing* 15 (1986), 300–315.

[197] B. Chazelle and H. Edelsbrunner. An improved algorithm for constructing kth-order Voronoi diagrams. *IEEE Transactions on Computers* C-36 (1987), 1349–1354.

[198] B. Chazelle, H. Edelsbrunner, L.J. Guibas, J.E. Hershberger, R. Seidel, and M. Sharir. Selecting heavily covered points. *SIAM Journal on Computing* 23 (1994), 1138–1151.

[199] C. Chen and H.L. Cheng. Superimposing Voronoi complexes for shape deformation. *International Journal of Computational Geometry & Applications* 16 (2006), 159–174.

[200] W.T. Chen and N.F. Huang. The strongly connecting problem on multihop packet radio networks. *IEEE Transactions on Communications* 37 (1989), 293–295.

[201] S.-W. Cheng, T.K. Dey, H. Edelsbrunner, M.A. Facello, and S.-H. Teng. Sliver Exudation. *Journal of the ACM* 47 (2000), 883–904.

[202] S.-W. Cheng, T. Dey, and J. Shewchuk. *Delaunay Mesh Generation.* CRC Computer & Information Science Series, Chapman & Hall, 2013.

[203] S.-W. Cheng, H. Edelsbrunner, P. Fu, and K.-P. Lam. Design and analysis of planar shape deformation. *Computational Geometry: Theory and Applications* 19 (2001), 205–218.

[204] S.-W. Cheng, M.J. Golin, and J.C.F. Tsang. Expected case analysis of β-skeletons with applications to the construction of minimum-weight triangulations. *Proc. 7th Canadian Conference on Computational Geometry*, 1995, 279–284.

[205] S.-W. Cheng and A. Vigneron, Motorcycle graphs and straight skeletons. *Algorithmica* 47 (2007), 159–182.

[206] S.-W. Cheng and Y.-F. Xu. Approximating the largest β-skeleton within a minimum-weight-triangulation. *Proc. 12th Ann. ACM Symposium on Computational Geometry*, 1996, 196–203.

[207] O. Cheong, A. Efrat, and S. Har-Peled. On finding a guard that sees most and a shop that sells most. *Discrete & Computational Geometry* 37 (2007), 545–563.

[208] O. Cheong, H .Everett, M. Glisse, J. Gudmundsson, S. Hornus, S. Lazard, M. Lee, and H.-S. Na. Farthest-polygon Voronoi diagrams. *Computational Geometry: Theory and Applications* 44 (2011), 234–247.

[209] O. Cheong, S. Har-Peled, N. Linial, and J. Matoušek. The one-round Voronoi game. *Proc. 18th Ann. ACM Symposium on Computational Geometry*, 2002, 97–101.

[210] L.P. Chew. Building Voronoi diagrams for convex polygons in linear expected time. Technical Report PCS-TR90-147, Department of Mathematics and Computer Science, Dartmouth College, Hanover, NH, 1986.

[211] L.P. Chew. Constrained Delaunay triangulations. *Algorithmica* 4 (1989), 97–108.

[212] L.P. Chew. There are planar graphs almost as good as the complete graph. *Journal of Computer and System Sciences* 39 (1989), 205–219.

[213] L.P. Chew. Guaranteed-quality mesh generation for curved surfaces. *Proc. 9th Ann. ACM Symposium on Computational Geometry*, 1993, 274–280.

[214] L.P. Chew. Near-quadratic bounds for the L_1 Voronoi diagram of moving points. *Computational Geometry: Theory and Applications* 7 (1997), 73–80.

[215] L.P. Chew and R.L.S. Drysdale. Voronoi diagrams based on convex distance functions. *Proc. 1st Ann. ACM Symposium on Computational Geometry*, 1985, 235–244.

[216] L.P. Chew and K. Kedem. Placing the largest similar copy of a convex polygon among polygonal obstacles. *Proc. 5th Ann. ACM Symposium on Computational Geometry*, 1989, 167–174.

[217] L.P. Chew, K. Kedem, M. Sharir, B. Tagansky, and E. Welzl. Voronoi diagrams of lines in 3-space under polyhedral convex distance functions. *Proc. 6th ACM-SIAM Symposium on Discrete Algorithms*, 1995, 197–204.

[218] F. Chin, J. Snoeyink, and C.-A. Wang. Finding the medial axis of a simple polygon in linear time. *Proc. 6th Ann. International Symposium on*

Algorithms and Computation, Springer Lecture Notes in Computer Science 1004, 1995, 382–391.

[219] H.I. Choi, S.W. Choi, and H.P. Moon. Mathematical theory of medial axis transform. *Pacific Journal of Mathematics* 181 (1997), 57–88.

[220] S. Choset and J. Burdick. Sensor based planning, part I: The generalized Voronoi graph. *IEEE International Conference on Robotics and Automation*, 1995, 1643–1648.

[221] S. Choset and J. Burdick. Sensor based planning, part II: Incremental construction of the generalized Voronoi graph. *IEEE International Conference on Robotics and Automation*, 1995, 1649–1655.

[222] N. Christofides. Worst-case analysis of a new heuristic for the traveling salesman problem. In: J.F. Traub (ed.), *Symposium on New Directions and Recent Results in Algorithms and Complexity*, Academic Press, 1976, 441.

[223] P. Cignoni, C. Montani, and R. Scopigno. A merge-first divide & conquer algorithm for E^d Delaunay triangulations. Technical report, Consiglio Nazionale delle Ricerche, Pisa, Italy, 1994.

[224] K.L. Clarkson. New applications of random sampling in computational geometry. *Discrete & Computational Geometry* 2 (1987), 195–222.

[225] K.L. Clarkson. A randomized algorithm for closest-point queries. *SIAM Journal on Computing* 17 (1988), 830–847.

[226] K.L. Clarkson. An algorithm for geometric minimum spanning trees requiring nearly linear expected time. *Algorithmica* 4 (1989), 461–469.

[227] K.L. Clarkson. Nearest-neighbor searching and metric space dimensions. In: G. Shakhnarovich, T. Darrell, and P. Indyk (eds.), *Nearest-Neighbor Methods for Learning and Vision: Theory and Practice*, 2006, 15–59.

[228] K.L. Clarkson, K. Mehlhorn, and R. Seidel. Four results on randomized incremental constructions. *Computational Geometry: Theory and Applications* 3 (1993), 185–212.

[229] K.L. Clarkson and P.W. Shor. Applications of random sampling in computational geometry, II. *Discrete & Computational Geometry* 4 (1989), 387–421.

[230] R. Cole. Reported by C. Ó'Dúnlaing, 1989.

[231] R. Connelly, E.D. Demaine, and G. Rote. Every polygon can be untangled. *Proc. 16th European Workshop on Computational Geometry*, 2000, 62–65.

[232] A.G. Corbalan, M. Mazon, T. Recio, and F. Santos. On the topological shape of planar Voronoi diagrams. *Proc. 9th Ann. ACM Symposium on Computational Geometry*, 1993, 109–115.

[233] A.G. Corbalan, M. Mazon, and T. Recio. Geometry of bisectors for strictly convex distance functions. *International Journal of Computational Geometry & Applications* 6 (1996), 45–58.

[234] T.H. Cormen, C.E. Leiserson, R.L. Rivest, and C. Stein. *Introduction to Algorithms*, 3rd Edition. MIT Press, Cambridge, MA, 2009.

[235] H. Crapo and W. Whiteley. Plane stresses and projected polyhedra I: The basic pattern. *Structural Topology* 20 (1993), 55–68.

[236] J.A. Cuesta-Albertos and A. Tuero-Diaz. A characterization for the solution of the Monge-Kantorovich mass transference problem. *Statistics and Probability Letters* 16 (1993) 147–152.

[237] T. Culver, J. Keyser, and D. Manocha. Exact computation of the medial axis of a polyhedron. *Computer Aided Geometric Design* 21 (2004), 65–98.

[238] G. Das and D. Joseph. Which triangulations approximate the complete graph. *Proc. International Symposium on Optimal Algorithms*, Springer Lecture Notes in Computer Science 401, 1989, 168–192.

[239] G. Das, A. Mukhopadhyay, S.C. Nandy, S. Patil, and S.V. Rao. Computing the straight skeleton of a monotone polygon in $O(n \log n)$ time. *Proc. 22nd Canadian Conference on Computational Geometry*, 2010, 207–210.

[240] E.F. D'Azevedo and R.B. Simpson. On optimal interpolation triangle incidences. *SIAM Journal on Scientific and Statistical Computing* 10 (1989), 1063–1075.

[241] M. de Berg, D. Gerrits, A. Khosravi, I. Rutter, C. Tsirogiannis, and A. Wolff. How Alexander the Great brought the Greeks together while inflicting minimal damage to the barbarians. *Proc. 26th European Workshop on Computational Geometry*, 2010, 73–76.

[242] M. de Berg, J. Matoušek, and O. Schwarzkopf. Piecewise linear paths among convex obstacles. *Proc. 25th Ann. ACM Symposium on Theory of Computing*, 1993, 505–514.

[243] M. de Berg, A.F. van der Stappen, J. Vleugels, and M.J. Katz. Realistic input models for geometric algorithms. *Algorithmica* 34 (2002), 81–97.

[244] M. de Berg, M. van Krefeld, M. Overmars, and O. Schwarzkopf. *Computational Geometry. Algorithms and Applications*, 2nd Edition. Springer Verlag, Berlin, 2000.

[245] L. De Floriani, B. Falcidieno, G. Nagy, and C. Pienovi. On sorting triangles in a Delaunay tessellation. *Algorithmica* 6 (1991), 522–532.

[246] D. Defays. An efficient algorithm for a complete link method. *The Computer Journal* 20 (1977), 364–366.

[247] F. Dehne and R. Klein. A sweepcircle algorithm for Voronoi diagrams. *Proc. 13th International Workshop on Graph-Theoretical Concepts in Computer Science*, Springer Lecture Notes in Computer Science 314, 1987, 59–69.

[248] F. Dehne and R. Klein. "The big sweep": On the power of the wavefront approach to Voronoi diagrams. *Algorithmica* 17 (1997), 19–32.

[249] F. Dehne, R. Klein, and R. Seidel. Maximizing a Voronoi region: the convex case. *International Journal of Computational Geometry & Applications* 15 (2005), 463–475.

[250] F. Dehne, A. Maheshwari, and R. Taylor. A coarse grained parallel algorithm for Hausdorff Voronoi diagrams. *Proc. International Conference on Parallel Processing*, 2006, 497–504.

[251] F. Dehne and H. Noltemeier. A computational geometry approach to clustering problems. *Proc. 1st Ann. ACM Symposium on Computational Geometry*, 1985, 245–250.

[252] B. Delaunay. Neue Darstellung der geometrischen Kristallographie. *Zeitschrift für Kristallographie* 84 (1932), 109–149.

[253] B. Delaunay. Sur la sphère vide. A la memoire de Georges Voronoi. *Izvestiya Akademii Nauk SSSR, Otdelenie Matematicheskih i Estest- vennyh Nauk* 7 (1934), 793–800.

[254] E.D. Demaine, M.L. Demaine, J.F. Lindy, and D.L. Souvaine. Hinged dissection of polypolyhedra. *Proc. 9th Workshop on Algorithms and Data Structures*, Springer Lecture Notes in Computer Science 3608, 2005, 205–217.

[255] E.D. Demaine, M.L. Demaine, and J.S.B. Mitchell, Folding flat silhouettes and wrapping polyhedral packages: New results in computational origami. *Computational Geometry: Theory and Applications* 16 (2000), 3–21.

[256] M. Demuth, F. Aurenhammer, and A. Pinz. Straight skeletons for binary shapes. *3rd Workshop on non-rigid shape analysis and deformable image alignment, NORDIA'10*, San Francisco, 2010.

[257] R. Descartes. *Principia Philosophiae.* Ludovicus Elzevirius, Amsterdam, 1644.

[258] S.C. De Biasi, B. Kalantari, and I. Kalantari. Mollified zone diagrams and their computation. *Transactions on Computational Science* 14 (2011), 31–59.

[259] V. De Silva and G. Carlsson. Topological estimation using witness complexes. *Proc. Symposium on Point-Based Graphics*, 2004, 157–166.

[260] S.L. Devadoss and J. O'Rourke. *Discrete and Computational Geometry.* Princeton University Press, 2011.

[261] O. Devillers. Randomization yields simple $O(n \log^* n)$ algorithms for difficult $\Omega(n)$ problems. *International Journal of Computational Geometry & Applications* 2 (1992), 97–111.

[262] O. Devillers. On deletion in Delaunay triangulations. *Proc. 15th Ann. ACM Symposium on Computational Geometry*, 1999, 181–188.

[263] O. Devillers and X. Goaoc. Random sampling of a cylinder yields a not so nasty Delaunay triangulation. Research Report RR-6323, INRIA, Sophia Antipolis, France, 2007.

[264] O. Devillers, S. Meiser, and M. Teillaud. Fully dynamic Delaunay triangulation in logarithmic expected time per operation. *Proc. 2nd Workshop on Algorithms and Data Structures*, Springer Lecture Notes in Computer Science 519, 1991, 42–53.

[265] O. Devillers, S. Meiser, and M. Teillaud. The space of spheres, a geometric tool to unify duality results on Voronoi diagrams. *Proc. 4th Canadian Conference on Computational Geometry*, 1992, 263–268.

[266] O. Devillers, S. Pion, and M. Teillaud. Walking in a triangulation. *International Journal of Foundations of Computer Science* 13 (2002), 181–199.

[267] L. Devroye, E.P. Mücke, and B. Zhu. A note on point location in Delaunay triangulations of random points. *Algorithmica* 22 (1998), 477–482.

[268] A.K. Dewdney and J.K. Vranch. A convex partition of R^3 with applications to Crum's problem and Knuth's post-office problem. *Utilitas Mathematica* 12 (1977), 193–199.

[269] T.K. Dey. Improved bounds on planar k-sets and k-levels. *Discrete & Computational Geometry* 19 (1997), 156–161.

[270] T.K. Dey and H. Edelsbrunner. Counting triangle crossings and halving planes. *Discrete & Computational Geometry* 12 (1994), 281–289.

[271] T. Dey and W. Zhao. Approximating the medial axis from the Voronoi diagram with a convergence guarantee. *Algorithmica* 38 (2004), 179–200.

[272] M.T. Dickerson, R.L.S. Drysdale, and J.R. Sack. Simple algorithms for enumerating interpoint distances and finding k nearest neighbors. *International Journal of Computational Geometry & Applications* 2 (1992), 221–239.

[273] M.T. Dickerson and D. Eppstein. Algorithms for proximity problems in higher dimensions. *Computational Geometry: Theory and Applications* 5 (1996), 277–291.

[274] M.T. Dickerson and M.H. Montague. A (usually?) connected subgraph of the minimum-weight triangulation. *Proc. 12th Ann. ACM Symposium on Computational Geometry*, 1996, 204–213.

[275] R. Diestel. *Graph Theory* 4th Edition. Graduate Texts in Mathematics 173, Springer Verlag, Heidelberg, 2010.

[276] M.B. Dillencourt. A non-Hamiltonian, nondegenerate Delaunay triangulation. *Information Processessing Letters* 25 (1987), 149–151.

[277] M.B. Dillencourt. Toughness and Delaunay triangulations. *Discrete & Computational Geometry* 5 (1990), 575–601.

[278] M.B. Dillencourt. Realizability of Delaunay triangulations. *Information Processing Letters* 33 (1990), 283–287.

[279] M.B. Dillencourt. Finding Hamiltonian cycles in Delaunay triangulations is NP-complete. *Proc. 4th Canadian Conference on Computational Geometry*, 1992, 223–228.

[280] P.G.L. Dirichlet. Über die Reduction der positiven quadratischen Formen mit drei unbestimmten ganzen Zahlen. *Journal für die reine und angewandte Mathematik* 40 (1850), 209–227.

[281] H. Djidjev and A. Lingas. On computing the Voronoi diagram for restricted planar figures. *Proc. 2nd Workshop on Algorithms and Data Structures*, Springer Lecture Notes in Computer Science 519, 1991, 54–64.

[282] Y. Dobashi, T. Haga, H. Johan, and T. Nishita. A method for creating mosaic images using Voronoi diagrams. *Proc. 23rd Ann. Eurographics Conference*, 2002, 341–348.

[283] D.P. Dobkin, S.J. Friedman, and K.J. Supowit. Delaunay graphs are almost as good as complete graphs. *Discrete & Computational Geometry* 5 (1990), 399–407.

[284] D.P. Dobkin and M.J. Laszlo. Primitives for the manipulation of three-dimensional subdivisions. *Algorithmica* 4 (1989), 3–32.

[285] D.P. Dobkin and R.J. Lipton. Multidimensional searching problems. *SIAM Journal on Computing* 5 (1976), 181–186.

[286] J.-M. Drappier. Envelopes convexes. Technical report, Rapport Centre CPAO, ENSTA Palaiseau, France, 1983.

[287] A. Driemel, S. Har-Peled, and B. Raichel. On the expected complexity of Voronoi diagrams on terrains. *Proc. 28th Ann. ACM Symposium on Computational Geometry*, 2012, 101–110.

[288] R.L.S. Drysdale. A practical algorithm for computing the Delaunay triangulation for convex distance functions. *Proc. 1st Ann. ACM–SIAM Symposium on Discrete Algorithms*, 1990, 159–168.

[289] Q. Du, V. Faber, and M. Gunzburger. Centroidal Voronoi tessellations: Applications and algorithms. *SIAM Review* 41 (1999), 637–676.

[290] Q. Du and D. Wang. Anisotropic centroidal Voronoi tessellations and their application. *SIAM Journal on Scientific Computing* 26 (2005), 737–761.

[291] A. Dumitrescu, S. Har-Peled, and C.D. Tóth. Minimum convex partitions and maximum empty polytopes. *Proc. 13th Scandinavian Symposium and Workshops on Algorithm Theory*, Springer Lecture Notes in Computer Science 7357, 2012, 213–224.

[292] A. Dumitrescu and C.D. Tóth. Minimum weight convex Steiner partitions. *Algorithmica* 60 (2011), 627–652.

[293] M. Dutour and K. Rybnikov. A new algorithm in geometry of numbers. *Proc. 4th International Symposium on Voronoi Diagrams in Science and Engineering*, 2007, IEEE Computer Society, 182–188,

[294] M. Dutour, A. Schürmann, and F. Vallentin. Complexity and algorithms for computing Voronoi cells of lattices. *Mathematics of Computation* 79 (2009), 1713–1731.

[295] R.A. Dwyer. A faster divide-and-conquer algorithm for constructing Delaunay triangulations. *Algorithmica* 2 (1987), 137–151.

[296] R.A. Dwyer. Higher-dimensional Voronoi diagrams in linear expected time. *Discrete & Computational Geometry* 6 (1991), 343–367.

[297] R.A. Dwyer. Voronoi diagrams of random lines and flats. *Discrete & Computational Geometry* 17 (1997), 123–136.

[298] B.C. Eaton and R.G. Lipsey. The principle of minimum differentiation reconsidered: some new developments in the theory of spatial competition. *Review of Economic Studies* 42 (1975), 27–49.

[299] H. Ebara, H. Nakano, Y. Nakanishi, and T. Sanada. A practical algorithm for computing the roundness. *IEICE Transactions on Information and Systems* E75-D, 1992, 253–257.

[300] H. Edelsbrunner. *Algorithms in Combinatorial Geometry*, EATCS Monographs on Theoretical Computer Science 10, Springer, Heidelberg, Germany, 1987.

[301] H. Edelsbrunner. An acyclicity theorem for cell complexes in d dimensions. *Combinatorica* 10 (1990), 251–260.

[302] H. Edelsbrunner. The union of balls and its dual shape. *Discrete & Computational Geometry* 13 (1995), 415–440.

[303] H. Edelsbrunner. Deformable smooth surface design. *Discrete & Computational Geometry* 21 (1999), 87–115.

[304] H. Edelsbrunner, L.J. Guibas, and M. Sharir. The upper envelope of piecewise linear functions: algorithms and applications. *Discrete & Computational Geometry* 4 (1989), 311–336.

[305] H. Edelsbrunner, L.J. Guibas, and J. Stolfi. Optimal point location in a monotone subdivision. *SIAM Journal on Computing* 15 (1986), 317–340.

[306] H. Edelsbrunner, D.G. Kirkpatrick, and R. Seidel. On the shape of a set of points in the plane. *IEEE Transactions on Information Theory* IT-29 (1983), 551–559.

[307] H. Edelsbrunner and E.P. Mücke. Three-dimensional alpha shapes. *ACM Transactions on Graphics* 13 (1994), 43–72.

[308] H. Edelsbrunner and H.A. Maurer. Finding extreme points in three dimensions and solving the post-office problem in the plane. *Information Processing Letters* 21 (1985), 39–47.

[309] H. Edelsbrunner, J. O'Rourke, and R. Seidel. Constructing arrangements of lines and hyperplanes with applications. *SIAM Journal on Computing* 15 (1986), 341–363.

[310] H. Edelsbrunner and R. Seidel. Voronoi diagrams and arrangements. *Discrete & Computational Geometry* 1 (1986), 25–44.

[311] H. Edelsbrunner, R. Seidel, and M. Sharir. On the zone theorem for hyperplane arrangements. *SIAM Journal on Computing* 22 (1993), 418–429.

[312] H. Edelsbrunner, F.P. Preparata, and D.B. West. Tetrahedrizing point sets in three dimensions. *Journal of Symbolic Computation* 10 (1990), 335–347.

[313] H. Edelsbrunner and N.R. Shah. Incremental topological flipping works for regular triangulations. *Algorithmica* 15 (1996), 223–241.

[314] H. Edelsbrunner and W. Shi. An $O(n \log^2 h)$ time algorithm for the three-dimensional convex hull problem. *SIAM Journal on Computing* 20 (1991), 259–277.

[315] H. Edelsbrunner and T.S. Tan. A quadratic time algorithm for the minmax length triangulation. *SIAM Journal on Computing* 22 (1993), 527–551.

[316] H. Edelsbrunner and T.S. Tan. An upper bound for conforming Delaunay triangulations. *Discrete & Computational Geometry* 10 (1993), 197–213.

[317] H. Edelsbrunner, T.S. Tan, and R. Waupotitsch. $O(N^2 \log N)$ time algorithm for the minmax angle triangulation. *SIAM Journal on Scientific and Statistical Computing* 13 (1992), 994–1008.

[318] P.E. Ehrlich and H.-C. Im Hof. Dirichlet regions in manifolds without conjugate points. *Commentarii Mathematici Helvetici* 54 (1979), 642–658.

[319] I. Emiris, G. Tzoumas, and E. Tsigaridas. The predicates for the Voronoi diagram of ellipses. *Proc. 22th Ann. ACM Symposium on Computational Geometry*, 2006, 227–236.

[320] P. Engel. Geometric crystallography. In: P.M. Gruber and J.M. Wills (eds.), *Handbook of Convex Geometry B*, Elsevier, Amsterdam, 2003, 989–1041.

[321] D. Eppstein. The farthest point Delaunay triangulation minimizes angles. *Computational Geometry: Theory and Applications* 3 (1992), 143–148.

[322] D. Eppstein. Faster construction of planar two-centers. *Proc. 8th Ann. ACM–SIAM Symposium on Discrete Algorithms*, 1997, 131–138.

[323] D. Eppstein. Beta-skeletons have unbounded dilation. *Computational Geometry: Theory and Applications* 23 (2002), 43–52.

[324] D. Eppstein. Planar Lombardi drawings for subcubic graphs. *Proc. 20th International Symposium on Graph Drawing*, Springer Lecture Notes in Computer Science 7704, 2012, 126–137.

[325] D. Eppstein. The graphs of planar soap bubbles. *Proc. 29th Ann. ACM Symposium on Computional Geometry*, 2013, to appear.

[326] D. Eppstein and J. Erickson, Raising roofs, crashing cycles, and playing pool: Applications of a data structure for finding pairwise interactions. *Discrete & Computational Geometry* 22 (1999), 569–592.

[327] J. Erickson. Nice point sets can have nasty Delaunay triangulations. *Discrete & Computational Geometry* 30 (2003), 109–132.

[328] J. Erickson. Dense point sets have sparse Delaunay triangulations or "but not too nasty". *Discrete & Computational Geometry* 33 (2005), 83–115.

[329] M. Erwig. The graph Voronoi diagram with applications. *Networks* 36 (2000), 156–163.

[330] M. Erwig and R.H. Güting. Explicit graphs in a functional model for spatial databases. *IEEE Transactions on Knowledge and Data Engineering* 5 (1994), 787–804.

[331] M. Etzion and A. Rappoport. Computing the Voronoi diagram of a 3-d polyhedron by separate computation of its symbolic and geometric parts. *Proc. 5th ACM Symposium on Solid Modeling and Applications*, 1999, 167–178.

[332] H. Everett, D. Lazard, S. Lazard, and M.S.E. Din. The Voronoi diagram of three lines in R^3. *Proc. 23rd Ann. ACM Symposium on Computational Geometry*, 2007, 255–264.

[333] R. Fabbri, L. Da Fontoura Costa, J.C. Torelli, and O.M. Bruno. 2D Euclidean distance transform algorithms: A comparative survey. *ACM Computing Surveys* 40 (2008), 1–44.

[334] C. Fan, J. He, J. Luo, and B. Zhu. Moving network Voronoi diagram. *Proc. 7th International Symposium on Voronoi Diagrams in Science and Engineering*, 2010, 142–150.

[335] C. Fan, J. Luo, W. Wang, and B. Zhu. Voronoi diagram with visual restriction. *Proc. 6th International Frontiers in Algorithmics, and Proc. 8th International Conference on Algorithmic Aspects in Information and Management.* Springer Lecture Notes in Computer Science 7285, 2012, 36–46.

[336] G. Farin. Surfaces over Dirichlet tessellations. *Computer Aided Geometric Design* 7 (1990), 281–292.

[337] T. Feder and T.H. Green. Optimal algorithms for approximate clustering. *Proc. 20th Ann. ACM Symposium on Theory of Computing*, 1988, 434–444.

[338] S. Fekete and H. Meijer. The one-round Voronoi game replayed. *Computational Geometry: Theory and Applications* 30 (2005), 81–94.

[339] G. Fejes Tóth. Packing and covering. In: J.E. Goodman and J. O'Rourke (eds.), *Handbook of Discrete and Computational Geometry*, 2nd Edition. CRC Press, 2004, 25–52.

[340] D.A. Field. Implementing Watson's algorithm in three dimensions. *Proc. 2nd Ann. ACM Symposium on Computational Geometry*, 1986, 246–259.

[341] E. Fogel, D. Halperin, and R. Wein. *CGAL Arrangements and Their Applications: A Step-by-Step Guide.* Springer, 2012.

[342] S. Fortune. Numerical stability of algorithms for 2-d Delaunay triangulations and Voronoi diagrams. *Proc. 8th Ann. ACM Symposium on Computational Geometry*, 1992, 83–92.

[343] S. Fortune. Voronoi diagrams and Delaunay triangulations. In: D.-Z. Du and F.K. Hwang (eds.), *Computing in Euclidean Geometry*, Lecture Notes Series on Computing 1, World Scientific, Singapore, 1992, 193–233.

[344] S. Fortune. A sweepline algorithm for Voronoi diagrams. *Algorithmica* 2 (1987), 153–174.

[345] M. Fredman and R. Tarjan. Fibonacci heaps and their uses in improved network optimization algorithms. *Journal of the ACM* 34 (1987), 596–615.

[346] K.R. Gabriel and R.R. Sokal. A new statistical approach to geographic variation analysis. *Systematic Zoology* 18 (1969), 259–278.

[347] A. Gajentaan and M. Overmars. On a class of $O(n^2)$ problems in computational geometry. *Computational Geometry: Theory and Applications* 5 (1995), 165–185.

[348] J. Galtier, F. Hurtado, M. Noy, S. Pérennes, and J. Urrutia. Simultaneous edge flipping in triangulations. *International Journal of Computational Geometry & Applications* 13 (2003), 113–133.

[349] R. Gambini, D.L. Huff, and G.F. Jenks. Geometric properties of market areas. *Papers of the Regional Science Association* 20 (1967), 85–92.

[350] W. Gangbo and R.J. McCann. The geometry of optimal transportation. *Acta Mathematica* 177 (1996), 113–161.

[351] J. Gao, L.J. Guibas, and A. Nguyen. Deformable spanners and applications. *Proc. 20th Ann. ACM Symposium on Computational Geometry*, 2004, 179–199.

[352] M.R. Garey and D.S. Johnson. *Computers and Intractability: A Guide to the Theory of NP-Completeness.* W.H. Freeman, New York, NY, 1979.

[353] C.F. Gauß. Recursion der 'Untersuchungen über die Eigenschaften der positiven ternären quadratischen Formen' von Ludwig August Seeber. *Journal für Reine und Angewandte Mathematik* 20 (1840), 312–320.

[354] M.L. Gavrilova. *Generalized Voronoi Diagram: A Geometry-Based Approach to Computational Intelligence.* Studies in Computational Intelligence 158, Springer, Berlin, 2008.

[355] B. Geiger. 3D Modeling using the Delaunay triangulation. *Proc. 11th ACM Symposium on Computational Geometry*, 1995, V11–V12.

[356] P.-L. George and H. Borouchaki. *Delaunay Triangulation and Meshing.* Editions HERMES, Paris, 1998.

[357] I.M. Gel'fand, M.M. Kapranov, and A.V. Zelevinsky. Newton polyhedra of principal A-determinants. *Soviet Mathematics - Doklady* 308 (1989), 20–23.

[358] D. Geiß, R. Klein, R. Penninger, and G. Rote. Optimally solving a transportation problem using Voronoi diagrams. *Computational Geometry: Theory and Applications*; to appear.

[359] R. Geraerts and M. Overmars. The corridor map method: A general framework for real-time high-quality path planning. *Computer Animation and Virtual Worlds* 18 (2007), 107–119.

[360] L. Gewali, A. Meng, J.S.B. Mitchell, and S. Ntafos. Path planning in $0/1/\infty$ weighted regions with applications. *ORSA Journal on Computing* 2 (1990), 253–272.

[361] S. Ghosh and D.M. Mount. An output-sensitive algorithm for computing visibility graphs. *SIAM Journal on Computing* 20 (1991), 888–910.

[362] R. Ghrist and A. Muhammad. Coverage and hole detection in sensor networks via homology. *Proc. 4th International Symposium on Information Processing in Sensor Networks*, IEEE Press, 2005, 254–260.

[363] A. Gibbons. *Algorithmic Graph Theory*. Cambridge University Press, Cambridge, 1985.

[364] J. Giesen, B. Miklos, and M. Pauly. The medial axis of the union of inner Voronoi balls in the plane. *Computational Geometry: Theory and Applications* 45 (2012), 515–523.

[365] A. Goede, R. Preissner, and C. Frömmel. Voronoi cell: New method for allocation of space among atoms: Elimination of avoidable errors in calculation of atomic volume and density. *Journal of Computational Chemistry* 18 (1997), 1113–1123.

[366] C.M. Gold, P.M. Remmele, and T. Roos. Voronoi diagrams of line segments made easy. *Proc. 7th Canadian Conference on Computational Geometry*, 1995, 223–228.

[367] C.M. Gold, P.M. Remmele, and T. Roos. Voronoi methods in GIS. In: M. van Kreveld, J. Nievergelt, T. Roos, and P. Widmayer (eds.), *Algorithmic Foundations of Geographic Information Systems*, Springer Lecture Notes in Computer Science 1340, 1997, 21–35.

[368] M. Golin and H.-S. Na. On the average complexity of 3d Voronoi diagrams of random points on convex polytopes. *Computational Geometry: Theory and Applications* 25 (2003), 197–231.

[369] M. Golin, R. Raman, C. Schwarz, and M. Smid. Simple randomized algorithms for closest pair problems. *Nordic Journal of Computing* 2 (1995), 3–27.

[370] T. Gonzalez. Clustering to minimize the maximum intercluster distance. *Theoretical Computer Science* 38 (1985), 293–306.

[371] J.E. Goodman and J. O'Rourke (eds.), *Handbook of Discrete and Computational Geometry*, 2nd Edition. CRC Press, 2004.

[372] M.T. Goodrich. Parallel algorithms in geometry. In: J.E. Goodman and J. O'Rourke (eds.), *Handbook of Discrete and Computational Geometry*, (2nd Edition). CRC Press, 2004, 953–967.

[373] M.T. Goodrich, C. Ó'Dúnlaing, and C.K. Yap. Constructing the Voronoi diagram of a set of line segments in parallel. *Algorithmica* 9 (1993), 128–141.

[374] R. Görke, C.-S. Shin, and A. Wolff. Constructing the city Voronoi diagram faster. *International Journal of Computational Geometry & Applications* 4 (2008), 275–294.

[375] L.-A. Gottlieb and L. Roddity. An optimal dynamic spanner for doubling metric spaces. *Proc. 16th Ann. European Symposium on Algorithms*, Springer Lecture Notes in Computer Science 5193, 2008, 478–489.

[376] I.G. Gowda, D.G. Kirkpatrick, D.T. Lee, and A. Naamad. Dynamic Voronoi diagrams. *IEEE Transactions on Information Theory* 29 (1983), 724–731.

[377] M. Granitzer, W. Kienreich, V. Sabol, K. Andrews, and W. Klieber. Evaluating a system for interactive exploration of large, hierarchically structured document repositories. *Proc. 4th IEEE Symposium on Information Visualization* 2004, 127–134.

[378] P.J. Green and R.R. Sibson. Computing Dirichlet tessellations in the plane. *The Computer Journal* 21 (1978), 168–173.

[379] C.I. Grima, A. Márquez, and L. Ortega. Motion planning and visibility problems using the polar diagram. In: M. Chover, H. Hagen, and D. Tost (eds.), *Proc. 24th Ann. Eurographics Conference*, 2003, 58–74.

[380] C.I. Grima, A. Márquez, and L. Ortega. A new 2D tessellation for angle problems: The polar diagram. *Computational Geometry: Theory and Applications* 34 (2006), 58–74.

[381] N. Grislain and J.R. Shewchuk. The strange complexity of constrained Delaunay triangulation. *Proc. 15th Canadian Conference on Computational Geometry*, 2003, 89–93.

[382] M. Gromov. Hyperbolic groups. In: M.S. Gersten (ed.), *Essays in Group Theory*, Mathematical Sciences Research Institute Publications 8, Springer, 1987, 75–265.

[383] P.M. Gruber and J.M. Wills. *Handbook of Convex Geometry A & B*. Elsevier, Amsterdam, 2003.

[384] B. Grünbaum. *Convex Polytopes*. Interscience, New York, 1967.

[385] A. Grüne, T.-C. Lin, T.-K. Yu, R. Klein, D.T. Lee, E. Langetepe, and S.-H. Poon. Maximum detour and spanning ratio of rectilinear paths in L_1-plane. *Proc. 21st International Symposium on Algorithms and Computation*, Springer Lecture Notes in Computer Science 6507, 2010, 121–131.

[386] J. Gudmundsson, M. Hammar, and M. van Kreveld. Higher order Delaunay triangulations. *Computational Geometry: Theory and Applications* 23 (2002), 85–98.

[387] J. Gudmundsson, H.J. Haverkort, and M. van Kreveld. Constrained higher order Delaunay triangulations. *Computational Geometry: Theory and Applications* 30 (2005), 271–277.

[388] L.J. Guibas. Kinetic data structures: A state of the art report. *Proc. 3rd Workshop on Algorithmic Foundations of Robotics*, 1998, 191–209.

[389] L.J. Guibas. Modeling motion. In: J.E. Goodman and J. O'Rourke (eds.), *Handbook of Discrete and Computational Geometry*, 2nd Edition. CRC Press, 2004, 1117–1134.

[390] L.J. Guibas, D.E. Knuth, and M. Sharir. Randomized incremental construction of Delaunay and Voronoi diagrams. *Algorithmica* 7 (1992), 381–413.

[391] L.J. Guibas, J.S.B. Mitchell, and T. Roos. Voronoi diagrams of moving points in the plane. *Proc. 17th International Workshop on Graph-Theoretical Concepts in Computer Science*, Springer Lecture Notes in Computer Science 570, 1991, 113–125.

[392] L.J. Guibas and S.Y. Oudot. Reconstruction using witness complexes. *Proc. 18th ACM-SIAM Symposium on Discrete Algorithms*, 2007, 1076–1085.

[393] L.J. Guibas and D. Russel. An empirical comparison of techniques for updating Delaunay triangulations. *Proc. 20th Ann. Symposium on Computational Geometry*, 2004, 170–179.

[394] L.J. Guibas and J. Stolfi. Primitives for the manipulation of general subdivisions and the computation of Voronoi diagrams. *ACM Transactions on Graphics* 4 (1985), 74–123.

[395] S. Har-Peled. A replacement for Voronoi diagrams of near linear size. *Proc. 42nd IEEE Symposium on Foundations of Computer Science*, 2001, 94–105.

[396] S. Har-Peled. *Geometric Approximation Algorithms.* American Mathematical Society, 2011.

[397] J.A. Hartigan. *Clustering Algorithms.* John Wiley, New York, 1975.

[398] D. Hartvigsen. Recognizing Voronoi diagrams with linear programming. *ORSA Journal on Computing* 4 (1992), 369–374.

[399] M. Held. VRONI: An engineering approach to the reliable and efficient computation of Voronoi diagrams of points and line segments. *Computational Geometry: Theory and Applications* 18 (2001), 95–123.

[400] M. Held, G. Lukacs, and L. Andor. Pocket machining based on contour-parallel tool paths generated by means of proximity maps. *Computer-Aided Design* 26 (1994), 189–203.

[401] M. Heller. Triangulation algorithms for adaptive terrain modeling. *Proc. 4th International Symposium on Spatial Data Handling*, 1990, 163–174.

[402] M. Hemmer, O. Setter, and D. Halperin. Constructing the exact Voronoi diagram of arbitrary lines in three-dimensional space with fast

point-location. *Proc. 18th Ann. European Symposium on Algorithms*, Springer Lecture Notes in Computer Science 6346, 2010, 398–409.

[403] M. Henze, R. Jaume, and B. Keszegh. On the complexity of the partial least-squares matching Voronoi diagram. *Proc. 29th European Workshop on Computational Geometry*, 2013, 193–196.

[404] J. Hershberger and S. Suri. An optimal algorithm for euclidean shortest paths in the plane. *SIAM Journal on Computing* 28 (1999), 2215–2256.

[405] J. Hertz, A. Krogh, and R.G. Palmer. *Introduction to the Theory of Neural Computation.* Addison-Wesley, New York, 1991.

[406] H. Heusinger and H. Noltemeier. On separable clusterings. *Journal of Algorithms* 10 (1989), 212–227.

[407] K. Hinrichs, J. Nievergelt, and P. Schorn. Plane-sweep solves the closest pair problem elegantly. *Information Processing Letters* 26 (1988), 255–261.

[408] J.-M. Ho, D.T. Lee, C.-H. Chang, and C.K. Wong. Minimum diameter spanning trees and related problems. *SIAM Journal on Computing* 20 (1991), 987–997.

[409] A. Hubard and B. Aronov. Convex equipartitions of volume and surface area. Submitted for publication, 2012.

[410] S. Huber and M. Held. Theoretical and practical results on straight skeletons of planar straight-line graphs. *Proc. 27th Ann. ACM Symposium on Computational Geometry*, 2011, 171–178.

[411] F. Hurtado, M. Noy, and J. Urrutia. Flipping edges in triangulations. *Discrete & Computational Geometry* 22 (1999), 333–346.

[412] F. Hurtado, R. Klein, E. Langetepe, and V. Sacristán. The weighted farthest color Voronoi diagram on trees and graphs. *Computational Geometry: Theory and Applications* 27 (2004), 13–26.

[413] F. Hurtado, G. Liotta, and H. Meijer. Optimal and sub-optimal robust algorithms for proximity graphs. *Computational Geometry: Theory and Applications* 25 (2003), 35–49.

[414] D.P. Huttenlocher, K. Kedem, and J.M. Kleinberg. On dynamic Voronoi diagrams and the minimum Hausdorff distance for point sets under Euclidean motion in the plane. *Proc. 8th Ann. ACM Symposium on Computational Geometry*, 1992, 110–119.

[415] D.P. Huttenlocher, K. Kedem, and M. Sharir. The upper envelope of Voronoi surfaces and its applications. *Discrete & Computational Geometry* 9 (1993), 267–291.

[416] F.K. Hwang. An $O(n \log n)$ algorithm for rectilinear minimal spanning tree. *Journal of the ACM* 26 (1979), 177–182.

[417] C. Icking and L. Ma. A tight bound for the complexity of Voronoi diagrams under polyhedral convex distance functions in 3D. *Proc. 33rd Ann. ACM Symposium on Theory of Computing*, 2001, 316–321.

[418] C. Icking, R. Klein, N.-M. Lê, and L. Ma. Convex distance functions in 3-space are different. *Fundamenta Informaticae* 22 (1995), 331–352.

[419] K. Imai, A. Kawamura, J. Matoušek, D. Reem, and T. Tokuyama. Distance k-sectors exist. *Computational Geometry: Theory and Applications* 43 (2010), 713–720.

[420] H. Imai, M. Iri, and K. Murota. Voronoi diagram in the Laguerre geometry and its applications. *SIAM Journal on Computing* 14 (1985), 93–105.

[421] M. Inaba, N. Katoh, and H. Imai. Applications of weighted Voronoi diagrams and randomization to variance-based k-clustering. *Proc. 10th Ann. ACM Symposium on Computational Geometry*, 1994, 332–339.

[422] H. Inagaki, K. Sugihara, and N. Sugie. Numerically robust incremental algorithm for constructing three-dimensional Voronoi diagrams. *Proc. 4th Canadian Conference on Computational Geometry*, 1992, 334–339.

[423] C. Indermitte, T.M. Liebling, M. Troyanova, and H. Clemencon. Voronoi diagrams on piecewise flat surfaces and an application to biological growth. *Theoretical Computer Science* 263 (2001), 263–274.

[424] P. Indyk. Nearest neighbors in high-dimensional spaces. In: J.E. Goodman and J. O'Rourke (eds.), *Handbook of Discrete and Computational Geometry*, 2nd Edition. CRC Press, 2004, 877–892.

[425] P. Indyk and J. Matoušek. Low-distortion embeddings of finite metric spaces. In: J.E. Goodman and J. O'Rourke (eds.), *Handbook of Discrete and Computational Geometry*, 2nd Edition. CRC Press, 2004, 177–196.

[426] P. Indyk and R. Motwani. Approximate nearest neighbor: towards removing the curse of dimensionality. *Proc. 30th Ann. ACM Symposium on Theory of Computing*, 1998, 604–613.

[427] J.W. Jaromczyk, M. Kowaluk, and F. Yao. An optimal algorithm for constructing β-skeletons in the L_p-metric. Manuscript, University of Warsaw, Poland, 1989.

[428] M. Jiang. An inequality on the edge lengths of triangular meshes. *Computational Geometry: Theory and Applications* 44 (2011), 100–103.

[429] B. Joe. 3-dimensional triangulations from local transformations. *SIAM Journal on Scientific and Statistical Computing* 10 (1989), 718–741.

[430] B. Joe. Construction of three-dimensional Delaunay triangulations using local transformations. *Computer Aided Geometric Design* 8 (1991), 123–142.

[431] B. Joe. Geompack. A software package for the generation of meshes using geometric algorithms. *Advances in Engineering Software and Workstations* 13 (1991), 325–331.

[432] W.B. Johnson and J. Lindenstrauss. Extensions of Lipschitz mappings into a Hilbert space. *Contemporary Mathematics* 26 (1984), 189–206.

[433] W.A. Johnson and R.F. Mehl. Reaction kinetics in processes of nucleation and growth. *Transactions of the American Institute for Mining and Metallurgical Engineering* 135 (1939), 416–458.

[434] M. Joswig. Software. In: J.E. Goodman and J. O'Rourke (eds.), *Handbook of Discrete and Computational Geometry*, 2nd Edition. CRC Press, 2004, 1415–1433.

[435] L. Ju, Q. Du, and M. Gunzburger. Probabilistic methods for centroidal Voronoi tessellations and their parallel implementations. *Parallel Computing* 28 (2002), 1477–1500.

[436] M. Jünger, G. Reinelt, and D. Zepf. Computing correct Delaunay triangulations. *Computing* 47 (1991), 43–49.

[437] B. Kalantari. *Polynomial Root-Finding and Polynomiography.* World Scientific, NJ, 2008.

[438] B. Kalantari. Polynomial root-finding methods whose basins of attraction approximate Voronoi diagram. *Discrete & Computational Geometry* 46 (2011), 187–203.

[439] T.C. Kao and D.M. Mount. An algorithm for computing compacted Voronoi diagrams defined by convex distance functions. *Proc. 3rd Canadian Conference on Computational Geometry*, 1991, 104–109.

[440] T.C. Kao and D.M. Mount. Incremental construction and dynamic maintenance of constrained Delaunay triangulations. *Proc. 4th Canadian Conference on Computational Geometry*, 1992, 170–175.

[441] M. Kapl, F. Aurenhammer, and B. Jüttler. Voronoi diagrams from distance graphs. *Proc. 29th European Workshop on Computational Geometry*, 2013, 185–188.

[442] C.S. Kaplan. Voronoi diagrams and ornamental design. *Proc. 1st Ann. Symposium of the International Society for the Arts, Mathematics, and Architecture*, 1999, 277–283.

[443] H. Kaplan and M. Sharir. Finding the maximal empty disk containing a query point. *Proc. 28th Ann. ACM Symposium on Computational Geometry*, 2012, 287–292.

[444] S. Kapoor and X.-Y. Li. Proximity structures for geometric graphs. *International Journal of Computational Geometry & Applications* 20 (2010), 415–429.

[445] M.S. Karasick, D. Lieber, L.R. Nackman, and V.T. Rajan. Visualization of three-dimensional Delaunay meshes. *Algorithmica* 19 (1997), 114–128.

[446] M.I. Karavelas and M. Yvinec. The Voronoi diagram of planar convex objects. *Proc. 11th Ann. European Symposium on Algorithms*, Springer Lecture Notes in Computer Science 2832, 2003, 337–348.

[447] J. Katajainen and M. Koppinen. Constructing Delaunay triangulations by merging buckets in quadtree order. *Annales Societatis Mathematicae Polonae IV, Fundamenta Informaticae* 11 (1988), 275–288.

[448] A. Kawamura, J. Matoušek, and T. Tokuyama. Zone diagrams in Euclidean spaces and in other normed spaces. *Proc. 26th Ann. ACM Symposium on Computational Geometry*, 2010, 216–221.

[449] J.M. Keil. Computing a subgraph of the minimum weight triangulation. *Computational Geometry: Theory and Applications* 4 (1994), 13–26.

[450] J.M. Keil and C.A. Gutwin. Classes of graphs which approximate the complete Euclidean graph. *Discrete & Computational Geometry* 7 (1992), 13–28.

[451] T. Kelly and P. Wonka. Interactive architectural modeling with procedural extrusions. *ACM Transactions on Graphics* 30 (2011), 14:1–14:15.

[452] L. Kettner and S. Näher. Two computational geometry libraries: LEDA and CGAL. In: J.E. Goodman and J. O'Rourke (eds.), *Handbook of Discrete and Computational Geometry*, 2nd Edition. CRC Press, 2004, 1435–1463.

[453] D. Kim and D.K. Kim. Region-expansion for the Voronoi diagram of 3D spheres. *Computer-Aided Design* 38 (2006), 417–430.

[454] D.G. Kirkpatrick. Efficient computation of continuous skeletons. *Proc. 20th Ann. IEEE Symposium on Foundations of Computer Science*, 1979, 18–27.

[455] D.G. Kirkpatrick. A note on Delaunay and optimal triangulations. *Information Processing Letters* 10 (1980), 127–128.

[456] D.G. Kirkpatrick. Optimal search in planar subdivisions. *SIAM Journal on Computing* 12 (1983), 28–35.

[457] D.G. Kirkpatrick and J.D. Radke. A framework for computational morphology. In: G.T. Toussaint (ed.), *Computational Geometry*, North-Holland, Amsterdam, 1985, 217–248.

[458] D.G. Kirkpatrick and J. Snoeyink. Tentative prune-and-search for computing Voronoi vertices. *Proc. 9th Ann. ACM Symposium on Computational Geometry*, 1993, 133–142.

[459] D.G. Kirkpatrick, J. Snoeyink, and B. Speckmann. Kinetic collision detection for simple polygons. *International Journal of Computational Geometry & Applications* 12 (2002), 3–27.

[460] V. Klee. On the complexity of d-dimensional Voronoi diagrams. *Archiv der Mathematik* 34 (1980), 75–80.

[461] R. Klein. *Concrete and Abstract Voronoi Diagrams.* Springer Lecture Notes in Computer Science 400, 1989.

[462] R. Klein, E. Langetepe, and Z. Nilforoushan. Abstract Voronoi diagrams revisited. *Computational Geometry: Theory and Applications* 42 (2009), 885–902.

[463] R. Klein and A. Lingas. A linear-time randomized algorithm for the bounded Voronoi diagram of a simple polygon. *Proc. 9th Ann. ACM Symposium on Computational Geometry,* 1993, 124–132.

[464] R. Klein and A. Lingas. Fast skeleton construction. *Proc. 3rd Ann. European Symposium on Algorithms,* Springer Lecture Notes in Computer Science 979, 1995, 582–596.

[465] R. Klein and A. Lingas. Manhattonian proximity in a simple polygon. *International Journal of Computational Geometry & Applications* 5 (1995), 53–74.

[466] R. Klein, K. Mehlhorn, and S. Meiser. Randomized incremental construction of abstract Voronoi diagrams. *Computational Geometry: Theory and Applications* 3 (1993), 157–184.

[467] R. Klein and D. Wood. Voronoi diagrams based on general metrics in the plane. *Proc. 5th Symposium on Theoretical Aspects of Computer Science,* Springer Lecture Notes in Computer Science 294, 1988, 281–291.

[468] J. Kleinberg. Two algorithms for nearest-neighbor search in high dimensions. *Proc. 29th Ann. ACM Symposium on Theory of Computing,* 1997, 599–608.

[469] V. Koltun and M. Sharir. Polyhedral Voronoi diagrams of polyhedra in three dimensions. *Discrete & Computational Geometry* 31 (2002), 83–124.

[470] V. Koltun and M. Sharir. Three-dimensional Euclidean Voronoi diagrams of lines with a fixed number of orientations. *SIAM Journal on Computing* 32 (2003), 616–642.

[471] K. Kobayashi and K. Sugihara. Crystal Voronoi diagram and its applications to collision-free paths. *Proc. International Conference on Computational Sciences-I,* 2001, 738–747.

[472] K. Kobayashi and K. Sugihara. Crystal Voronoi diagram and its applications. *Proc. Future Generation Computer Systems,* 2002, 681–692.

[473] E. Kopecká, D. Reem, and S. Reich. Existence of zone diagrams in compact subsets of uniformly convex spaces. *Proc. 22nd Canadian Conference on Computational Geometry*, 2010, 17–20.

[474] M. Kowaluk. Planar β-skeletons via point location in monotone subdivisions of subset of lunes. *Proc. 28th European Workshop on Computational Geometry*, 2012, 225–227.

[475] J.B. Kruskal, Jr. On the shortest spanning subtree of a graph and the traveling salesman problem. *Proc. American Mathematical Society* 7 (1956), 48–50.

[476] D. Krznaric. Minimum spanning trees in d dimensions. *Nordic Journal of Computing* 6 (1999), 446–461.

[477] D. Krznaric and C. Levcopoulos. The first subquadratic algorithm for complete linkage clustering. *Proc. 6th Ann. International Symposium on Algorithms and Computation*, Springer Lecture Notes in Computer Science 1004, 1995, 392–401.

[478] R. Kunze, F.E. Wolter, and T. Rausch. Geodesic Voronoi diagrams on parametric surfaces. *Proc. Conference on Computer Graphics International*, 1997, 230–237.

[479] F. Labelle and J.R. Shewchuk. Anisotropic Voronoi diagrams and guaranteed-quality anisotropic mesh generation. *Proc. 19th Ann. ACM Symposium on Computational Geometry*, 2003, 191–200.

[480] T. Lambert. The Delaunay triangulation maximizes the mean inradius. *Proc. 6th Canadian Conference on Computational Geometry*, 1994, 201–206.

[481] T. Lambert. *Empty-shape triangulation algorithms*. Ph.D. thesis, Department of Computer Science, University of Manitoba, Winnipeg, 1994.

[482] T. Lambert. Systematic local flip rules are generalized Delaunay rules. *Proc. 5th Ann. Canadian Conference on Computational Geometry*, 1993, 352–357.

[483] R.J. Lang. A computational algorithm for origami design. *Proc. 12th Ann. ACM Symposium on Computational Geometry*, 1996, 98–105.

[484] J.-C. Latombe. *Robot Motion Planning*. Kluwer Academic Publishers, 1991.

[485] S.M. LaValle. *Planning Algorithms*. Cambridge University Press, 2006.

[486] C.L. Lawson. Software for C^1 surface interpolation. In: J.R. Rice (ed.), *Mathematical Software III*, Academic Press, New York, 1977, 161–194.

[487] N.-M. Lê. On Voronoi diagrams in the l_p-metric in higher dimensions. *Proc. 11th Symposium on Theoretical Aspects of Computer Science*, Springer Lecture Notes in Computer Science 775, 1994, 711–722.

[488] N.-M. Lê. Randomized incremental construction of simple abstract Voronoi diagrams in 3-space. *Proc. 10th International Conference on Fundamentals of Computation Theory*, Springer Lecture Notes in Computer Science 965, 1995, 333–342.

[489] V.-B. Le and D.T. Lee. Out-of-roundness problem revisited. *IEEE Transactions on Pattern Analysis and Machine Intelligence* 13 (1991), 217–223.

[490] C.W. Lee. The associahedron and triangulations of the n-gon. *European Journal of Combinatorics* 10 (1989), 173–181.

[491] C.W. Lee. Regular triangulations of convex polytopes. In: P. Gritzmann and B. Sturmfels (eds.), *Applied Geometry and Discrete Mathematics—The Victor Klee Festschrift*, DIMACS Series in Discrete Mathematics and Theoretical Computer Science 4, American Mathematical Society, 1991, 443–456.

[492] D.T. Lee. *Proximity and reachability in the plane*. Ph.D. thesis, University of Illinois at Urbana-Champaign, Illinois, 1978.

[493] D.T. Lee. Two-dimensional Voronoi diagrams in the L_p-metric. *Journal of the ACM* 27 (1980), 604–618.

[494] D.T. Lee. Medial axis transformation of a planar shape. *IEEE Transactions on Pattern Analysis and Machine Intelligence* 4 (1982), 363–369.

[495] D.T. Lee. On k-nearest neighbor Voronoi diagrams in the plane. *IEEE Transactions on Computers* C-31 (1982), 478–487.

[496] D.T. Lee and R.L.S. Drysdale. Generalization of Voronoi diagrams in the plane. *SIAM Journal on Computing* 10 (1981), 73–87.

[497] D.T. Lee and A.K. Lin. Generalized Delaunay triangulation for planar graphs. *Discrete & Computational Geometry* 1 (1986), 201–217.

[498] D.T. Lee, C.S. Liao, and W.B. Wang. Time-based Voronoi diagrams. *Proc. 1st International Symposium on Voronoi Diagrams in Science and Engineering*, 2004, 229–243.

[499] D.T. Lee and C.K. Wong. Voronoi diagrams in L_1 (L_∞) metrics with 2-dimensional storage applications. *SIAM Journal on Computing* 9 (1980), 200–211.

[500] D.T. Lee and V.B. Wu. Multiplicative weighted farthest neighbor Voronoi diagrams in the plane. *Proc. Int. Workshop on Discrete Mathematics and Algorithms*, Hong Kong, 1993, 154–168.

[501] I. Lee and C. Torpelund-Bruin. Multiplicatively-weighted order-k Minkowski-metric Voronoi models for disaster decision support systems. *Proc. IEEE Int. Conference on Intelligence and Security Informatics*, 2008, 236–238.

[502] C. Lemaire and J.-M. Moreau. A probabilistic result on multi-dimensional Delaunay triangulations, and its application to the 2D case. *Computational Geometry: Theory and Applications* 17 (2000), 69–96.

[503] W. Lenhart and G. Liotta. Proximity drawings of outerplanar graphs. *Proc. 4th International Symposium on Graph Drawing*, Springer Lecture Notes in Computer Science 1190, 1996, 286–302.

[504] C. Levcopoulos and D. Krznaric. Quasi-greedy triangulations approximating the minimum weight triangulation. *Proc. 7th ACM–SIAM Symposium on Discrete Algorithms*, 1996, 392–401.

[505] X.-Y. Li, G. Calinescu, P.J. Wan, and Y. Wang. Localized Delaunay triangulation with applications in wireless Ad Hoc networks. *IEEE Transactions on Parallel and Distributed Systems* 14 (2003), 1035–1047.

[506] X.-Y. Li and Y. Wang. Efficient construction of low weighted bounded degree planar spanner. *International Journal of Computational Geometry & Applications* 14 (2004), 69–84.

[507] M. Lin and D. Manocha. Collision and proximity queries. In: J.E. Goodman and J. O'Rourke (eds.), *Handbook of Discrete and Computational Geometry*, 2nd Edition. CRC Press, 2004, 787–807.

[508] A. Lingas. A linear-time construction of the relative neighborhood graph from the Delaunay triangulation. *Computational Geometry: Theory and Applications* 4 (1994), 199–208.

[509] C.-H. Liu, E. Papadopoulou, and D.T. Lee. An output-sensitive approach for the L_1/L_∞ k-nearest neighbor Voronoi diagram. *Proc. 19th Ann. European Symposium on Algorithms*, 2011, 70–81.

[510] C.-H. Liu and D.T. Lee. Higher-order geodesic Voronoi diagrams in a polygonal domain with holes. *Proc. 24th ACM–SIAM Symposium on Discrete Algorithms*, 2013, 1633–1645.

[511] S. Lloyd. Least square quantization in PCM. *IEEE Transactions on Information Theory* 28 (1982), 129–137.

[512] T. Lozano-Pérez and M.A. Wesley. An algorithm for planning collision-free paths among polyhedral obstacles. *Communications of the ACM* 22 (1979), 560–570.

[513] A. Lubiw and V. Pathak. Flip distance between two triangulations of a point set is NP-complete. *Proc. 24th Canadian Conference on Computational Geometry*, 2012, 127–132.

[514] L. Ma. *Bisectors and Voronoi diagrams for convex distance functions*. Ph.D. thesis, FernUniversität Hagen, Department of Computer Science, Technical report 267, 2000.

[515] G. Majewska and M. Kowaluk. New sequential and parallel algorithms for computing β-spectrum. *Proc. 29th European Workshop on Computational Geometry*, 2013, 181–184.

[516] J. Martinez, M. Vigo, and N. Pla-Garcia. Skeleton computation of orthogonal polyhedra. *Computer Graphics Forum* 30 (2011), 1573–1582.

[517] H. Martini, K. Swanepoel, and G. Weiss. The geometry of Minkowski spaces—a survey, Part I. *Expositiones Mathematicae* 19 (2001), 97–142.

[518] H. Martini and K. Swanepoel. Geometry of Minkowski spaces. Survey, Part 2. *Expositiones Mathematicae* 22 (2004), 93–144.

[519] D.W. Matula and R.R. Sokal. Properties of Gabriel graphs relevant to geographic variation research and clustering of points in the plane. *Geographical Analysis* 12 (1980), 205–222.

[520] J. Matoušek. *Lectures on Discrete Geometry.* Graduate Texts in Mathematics 212, Springer, 2002.

[521] J. Matoušek. On variants of the Johnson–Lindenstrauss lemma. *Random Structures & Algorithms* 33 (2008), 142–156.

[522] J. Matoušek, M. Sharir, and E. Welzl. A subexponential bound for linear programming. *Proc. 8th Ann. ACM Symposium on Computational Geometry*, 1992, 1–8.

[523] H.A. Maurer. The post-office problem and related questions. *Proc. 7th International Workshop on Graph-Theoretical Concepts in Computer Science*, Springer Lecture Notes in Computer Science 100, 1981, 1–19.

[524] A. Maus. Delaunay triangulation and the convex hull of n points in expected linear time. *BIT* 24 (1984), 151–163.

[525] J.C. Maxwell. On reciprocal figures and diagrams of forces. *Philosophical Magazine Series* (4) 27 (1864), 250–261.

[526] M.L. Mazón and T. Recio. Voronoi diagrams coming from discrete groups on the plane. *Proc. 2nd Canadian Conference on Computational Geometry*, 1990, 223–226.

[527] M. McAllister, D.G. Kirkpatrick, and J. Snoeyink. A compact piecewise-linear Voronoi diagram for convex sites in the plane. *Discrete & Computational Geometry* 15 (1996), 73–105.

[528] D.H. McLain. Two dimensional interpolation from random data. *The Computer Journal* 19 (1976), 178–181.

[529] P. McMullen. The maximal number of faces of a convex polytope. *Mathematika* 17 (1970), 179–184.

[530] N. Megiddo. Applying parallel computation algorithms in the design of serial algorithms. *Journal of the ACM* 30 (1983), 852–865.

[531] N. Megiddo. Linear programming in linear time when the dimension is fixed. *Journal of the ACM* 31 (1984), 114–127.

[532] K. Mehlhorn. A faster approximation algorithm for the Steiner problem in graphs. *Information Processing Letters* 27 (1988), 125–128.

[533] K. Mehlhorn, S. Meiser, and C. Ó'Dúnlaing. On the construction of abstract Voronoi diagrams. *Discrete & Computational Geometry* 6 (1991), 211–224.

[534] K. Mehlhorn, S. Meiser, and R. Rasch. Furthest site abstract Voronoi diagrams. *International Journal of Computational Geometry & Applications* 11 (2001), 583–616.

[535] K. Mehlhorn, M. Müller, S. Näher, S. Schirra, M. Seel, C. Uhrig, and J. Ziegler. A computational basis for higher-dimensional computational geometry and applications. *Computational Geometry: Theory and Applications* 10 (1998), 289–303.

[536] K. Mehlhorn and S. Näher. *LEDA: A Platform for Combinatorial and Geometric Computing.* Cambridge University Press, 1999.

[537] K. Mehlhorn, R. Osbild, and M. Sagraloff. A general approach to the analysis of controlled perturbation algorithms. *Computational Geometry: Theory and Applications* 44 (2011), 507–528.

[538] K. Mehlhorn and P. Sanders. *Algorithms and Data Structures: The Basic Toolbox.* Springer Verlag, Berlin, 2008.

[539] S. Meiser. *Zur Konstruktion abstrakter Voronoidiagramme.* Ph.D. thesis, Department of Computer Science, Universität des Saarlandes, Saarbrücken, Germany, 1993.

[540] S. Meiser. Point location in arrangements of hyperplanes. *Information and Computation* 106 (1993), 286–303.

[541] K. Menger. Untersuchungen über allgemeine Metrik. *Mathematische Annalen* 100 (1928), 75–163.

[542] V. Milenkovic. Robust construction of the Voronoi diagram of a polyhedron. *Proc. 5th Canadian Conference on Computational Geometry*, 1993, 473–478.

[543] G. Miller and D. Sheehy. A new approach to output-sensite Voronoi diagrams and Delaunay triangulations. *Proc. 29th Ann. ACM Symposium on Computational Geometry*, 2013, to appear.

[544] H. Minkowski. *Diophantische Approximationen: Eine Einführung in die Zahlentheorie.* Teubner Verlag, Leipzig, 1927.

[545] D. Mitsche, M. Saumell, and R.I. Silveira. On the number of higher order Delaunay triangulations. *Theoretical Computer Science* 412 (2011), 3589–3597.

[546] J.S.B. Mitchell. Shortest paths among obstacles in the plane. *International Journal of Computational Geometry & Applications* 6 (1996), 309–332.

[547] J.S.B. Mitchell, D.M. Mount, and C.H. Papadimitriou. The discrete geodesic problem. *SIAM Journal on Computing* 16 (1987), 647–668.

[548] J.S.B. Mitchell and C.H. Papadimitriou. The weighted region problem: Finding shortest paths through a weighted planar subdivision. *Journal of the ACM* 38 (1991), 18–73.

[549] E. Moet, M. van Kreveld, and A.F. van der Stappen. On realistic terrains. *Proc. 22nd Ann. ACM Symposium on Computational Geometry,* 2006, 177–186.

[550] D.M. Mount and A. Saalfeld. Globally-equiangular triangulations of co-circular points in $O(n \log n)$ time. *Proc. 4th Ann. ACM Symposium on Computational Geometry,* 1988, 143–152.

[551] E.P. Mücke, I. Saias, and B. Zhu. Fast randomized point location without preprocessing in two- and three-dimensional Delaunay triangulations. *Computational Geometry: Theory and Applications* 12 (1999), 63–83.

[552] D.E. Muller and F.P. Preparata. Finding the intersection of two convex polyhedra. *Theoretical Computer Science* 7 (1978), 217–236.

[553] E. Müller. *Lehrbuch der Darstellenden Geometrie.* Teubner Verlag, Leipzig und Berlin, 1916.

[554] K. Mulmuley. Output sensitive construction of levels and Voronoi diagrams in R^d of order 1 to k. *Proc. 22nd Ann. ACM Symposium on Theory of Computing,* 1990, 322–330.

[555] K. Mulmuley. On levels in arrangements and Voronoi diagrams. *Discrete & Computational Geometry* 6 (1991), 307–338.

[556] K. Mulmuley. *Computational Geometry: An Introduction Through Randomized Algorithms.* Prentice-Hall, 1994.

[557] W. Mulzer and G. Rote. Minimum-weight triangulation is NP-hard. *Journal of the ACM* 55 (2008), 1–29.

[558] O.R. Musin. Properties of the Delaunay triangulation. *Proc. 13th Ann. ACM Symposium on Computational Geometry,* 1997, 424–426.

[559] H.-S. Na, C.-N. Lee, and O. Cheong. Voronoi diagrams on the sphere. *Computational Geometry: Theory and Applications* 23 (2002), 183–194.

[560] Y.I. Naberukhin, V.P. Voloshin, and N.N. Medvedev. Geometrical analysis of the structure of simple liquids: percolation approach. *Molecular Physics* 73 (1991), 917–936.

[561] G. Narasimhan and M. Smid. *Geometric Spanner Networks.* Cambridge University Press, 2007.

[562] R. Niggli. Die topologische Strukturanalyse. *Zeitschrift für Kristallographie* 65 (1927), 391–415.

[563] F. Nielsen and R. Nock. Hyperbolic Voronoi diagrams made easy. *Proc. 10th International Conference on Computational Science and its Applications*, 2010, 74–80.

[564] Z. Nilforoushan and A. Mohades. Hyperbolic Voronoi diagram. *Proc. 6th International Conference on Computational Science and its Applications*, 2006, 735–742.

[565] T. Nishida and K. Sugihara. Boat-sail Voronoi diagram and its applications. *International Journal of Computational Geometry & Applications* 19 (2009), 425–440.

[566] G. Nivasch. An improved, simple construction of many halving edges. In: J.E. Goodman, J. Pach, and R. Pollack (eds.), *Surveys on Discrete and Computational Geometry—Twenty Years later*, Contemporary Mathematics 453, American Mathematical Society 2008, 299–305.

[567] C. Ó'Dúnlaing, M. Sharir, and C.K. Yap. Generalized Voronoi diagrams for a ladder: I. Topological analysis. *Communications on Pure and Applied Mathematics* 39 (1986), 423–483.

[568] C. Ó'Dúnlaing, M. Sharir, and C.K. Yap. Generalized Voronoi diagrams for a ladder: II. Efficient construction of the diagram. *Algorithmica* 2 (1987), 27–59.

[569] C. Ó'Dúnlaing and C.K. Yap. A 'retraction' method for planning the motion of a disk. *Journal of Algorithms* 6 (1985), 104–111.

[570] T. Ohya, M. Iri, and K. Murota. Improvements of the incremental method for the Voronoi diagram with computational comparison of various algorithms. *Journal of the Operations Research Society of Japan* 27 (1984), 306–336.

[571] A. Okabe, B. Boots, K. Sugihara, and S.N. Chiu. *Spatial Tessellations: Concepts and Applications of Voronoi Diagrams*, 2nd Edition. John Wiley & Sons, Chichester, England, 2000.

[572] J. O'Rourke. *Computational Geometry in C*, 2nd Edition. Cambridge University Press, 1998.

[573] M. Overmars. Dynamization of order decomposable set problems. *Journal of Algorithms* 2 (1981), 245–260.

[574] M. Overmars. A random approach to motion planning. Technical report RUU-CS-92-32, Utrecht University, Holland, 1992.

[575] M. Overmars and E. Welzl. New methods for computing visibility graphs *Proc. 4th Ann. ACM Symposium on Computational Geometry*, 1988, 164–171.

[576] J. Pach, R. Pinchasi, G. Tardos, and G. Tóth. Geometric graphs with no self-intersecting path of length 3. *Proc. 10th International Symposium on Graph Drawing*, Springer Lecture Notes in Computer Science 2528, 2002, 295–311.

[577] P. Palfrader, M. Held, and S. Huber. On computing straight skeletons by means of kinetic triangulations. *Proc. 20th Ann. European Symposium on Algorithms*, 2012, 766–777.

[578] C.H. Papadimitriou. The Euclidean traveling salesman problem is NP-complete. *Theoretical Computer Science* 4 (1977), 237–244.

[579] E. Papadopoulou. The Hausdorff Voronoi diagram of point clusters in the plane. *Algorithmica* 40 (2004), 63–82.

[580] E. Papadopoulou and S.K. Dey. On the farthest line-segment Voronoi diagram. *Proc. 28th European Workshop on Computational Geometry*, 2012, 237–240.

[581] E. Papadopoulou and D.T. Lee. The min-max Voronoi diagram of polygons and applications in VLSI manufacturing. *Proc. 13th International Symposium on Algorithms and Computation*, Springer Lecture Notes in Computer Science 2518, 2002, 15–23.

[582] E. Papadopoulou and D.T. Lee. A new approach for the geodesic Voronoi diagram of points in a simple polygon and other restricted polygonal domains. *Algorithmica* 20 (1998), 319–352.

[583] E. Papadopoulou and D.T. Lee. Critical area computation via Voronoi diagrams. *IEEE Transactions on Computer-Aided Design* 18 (1999), 463–474.

[584] E. Papadopoulou and M. Zavershynskyi. On higher order Voronoi diagrams of line segments. *Proc. 23rd International Symposium on Algorithms and Computation*, Springer Lecture Notes in Computer Science 7676, 2012, 177–186.

[585] E. Papadopoulou and M. Zavershynskyi. A sweepline algorithm for higher-order Voronoi diagrams. *Proc. 29th European Workshop on Computational Geometry*, 2013, 189–192.

[586] I. Paschinger. *Konvexe Polytope und Dirichletsche Zellenkomplexe.* Ph.D. thesis, Institut für Mathematik, Universität Salzburg, Austria, 1982.

[587] M. Pătraşcu and M. Thorup. Higher lower bounds for near-neighbor and further rich problems. *SIAM Journal on Computing* 39 (2009), 730–741.

[588] G. Paulini. *Shape Delaunay triangulations.* Master thesis, Institute for Theoretical Computer Science, University of Technology Graz, Austria, 2011.

[589] G. Peschka. *Kotirte Ebenen und deren Anwendung.* Verlag Buschak & Irrgang, Brünn, 1877.

[590] J.L. Pfaltz and A. Rosenfeld. Computer representation of planar regions by their skeletons. *Communications of the ACM* 10 (1967), 119–122.

[591] A. Pilz. Flip distance between triangulations of a planar point set is APX-hard. Technical Report, Institute for Software Technology, TU Graz, Austria, 2012.

[592] R. Pinchasi and R. Radoičić. Topological graphs with no self-intersecting cycle of length 4. *Proc. 19th Ann. ACM Symposium on Computational Geometry*, 2003, 98–103.

[593] R. Pinchasi and S. Smorodinsky. On locally Delaunay geometric graphs. *Proc. 20th Ann. ACM Symposium on Computational Geometry*, 2004, 378–382.

[594] P.L. Powar. Minimal roughness property of the Delaunay triangulation: a shorter approach. *Computer Aided Geometric Design* 9 (1992), 491–494.

[595] L. Pournin and T.M. Liebling. Constrained paths in the flip-graph of regular triangulations. *Computational Geometry: Theory and Applications* 37 (2007), 134–140.

[596] F.P. Preparata and M.I. Shamos. *Computational Geometry: An Introduction.* Springer, New York, 1985.

[597] R.C. Prim. Shortest connection networks and some generalizations. *Bell System Technical Journal* 36 (1957), 1389–1401.

[598] V.T. Rajan. Optimality of the Delaunay triangulation in R^d. *Proc. 7th Ann. ACM Symposium on Computational Geometry*, 1991, 357–363.

[599] S. Rajasekaran and S. Ramaswami. Optimal parallel randomized algorithms for the Voronoi diagram of line segments in the plane and related problems. *Proc. 10th Ann. ACM Symposium on Computational Geometry*, 1994, 57–66.

[600] E.A. Ramos. On range reporting, ray shooting and k-level construction. *Proc. 15th Ann. ACM Symposium on Computational Geometry*, 1999, 390–399.

[601] S.V. Rao and A. Mukhopadhyay. Fast algorithm for computing β-skeletons and their relatives. *Proc. 8th Ann. International Symposium on Algorithms and Computation*, Springer Lecture Notes in Computer Science 1350, 1997, 374–383.

[602] R. Rasch. *Abstrakte inverse Voronoidiagramme.* Ph.D. thesis, Department of Computer Science, Universität des Saarlandes, Saarbrücken, Germany, 1994.

[603] D. Reem. The geometric stability of Voronoi diagrams with respect to small changes of the sites. *Proc. 27th Ann. ACM Symposium on Computational Geometry*, 2011, 254–263.

[604] R. Reitsma, S. Trubin, and S. Sethia. Information space regionalization using adaptive multiplicatively weighted Voronoi diagrams. *Proc. 8th International Conference on Information Visualisation*, 2004, 290–294.

[605] S. Rippa. Minimal roughness property of the Delaunay triangulation. *Computer Aided Geometric Design* 7 (1990), 489–497.

[606] D.J. Rosenkrantz, R.E. Stearns, and P.M. Lewis. An analysis of several heuristics for the traveling salesman problem. *SIAM Journal on Computing* 6 (1977), 563–581.

[607] G. Rote. Personal communication, 2012.

[608] G. Rote. Partial least-squares point matching under translations. *Proc. 26th European Workshop on Computational Geometry*, 2010, 249–251.

[609] G. Rote, F. Santos, and I. Streinu. Pseudo-triangulations–a survey. In: J.E. Goodman, J. Pach, and R. Pollack (eds.), *Surveys on Discrete and Computational Geometry — Twenty Years Later*, Contemporary Mathematics 453, American Mathematical Society, 2008, 343–410.

[610] B.L. Rothschild and E.G. Straus. On triangulations of the convex hull of n points. *Combinatorica* 5 (1985), 167–179.

[611] N. Rubin. On topological changes in the Delaunay triangulation of moving points. *Proc. 28th Ann. ACM Symposium on Computational Geometry*, 2012, 1–10.

[612] J. Ruppert. A Delaunay refinement algorithm for quality 2-dimensional mesh generation. *Journal of Algorithms* 18 (1995), 548–585.

[613] J. Ruppert and R. Seidel. On the difficulty of tetrahedralizing 3-dimensional non-convex polyhedra. *Discrete & Computational Geometry* 7 (1992), 227–253.

[614] K. Sadakane, H. Imai, K. Onishi, M. Inaba, F. Takeuchi, and K. Imai. Voronoi diagrams by divergences with additive weights. *Proc. 14th Ann. ACM Symposium on Computational Geometry*, 1998, 403–404.

[615] M. Sakamoto and M. Takagi. Patterns of weighted Voronoi tessellations. *Science and Form* 3 (1988), 103–111.

[616] F. Santos. A point configuration whose space of triangulations is disconnected. *Journal of the American Mathematical Society* 13 (2000), 611–637.

[617] B.F. Schaudt and R.L.S. Drysdale. Multiplicatively weighted crystal growth Voronoi diagram. *Proc. 7th Ann. ACM Symposium on Computational Geometry*, 1991, 214–223.

[618] D. Schmitt and J.-C. Spehner. Angular properties of Delaunay diagrams in any dimension. *Discrete & Computational Geometry* 5 (1999), 17–36.

[619] F.P. Schoenberg, T. Ferguson, and C. Li. Inverting Dirichlet tessellations. *The Computer Journal* 46 (2003), 76–83.

[620] A. Schönflies. *Kristallsysteme und Kristallstruktur.* Teubner Verlag, Leipzig, 1891.

[621] E. Schönhardt. Über die Zerlegung von Dreieckspolyedern in Tetraeder. *Mathematische Annalen* 98 (1928), 309–312.

[622] Y. Schreiber and M. Sharir. An optimal algorithm for shortest paths on a convex polytope in three dimensions. *Discrete & Computational Geometry* 39 (2008), 500–579.

[623] E. Schulte. Tilings. In: P.M. Gruber and J.M. Wills (eds.), *Handbook of Convex Geometry B*, Elsevier, Amsterdam, 2003, 899–932.

[624] O. Schwarzkopf. Dynamic maintenance of geometric structures made easy. *Proc. 32nd Ann. IEEE Symposium on Foundations of Computer Science*, 1991, 197–206.

[625] R. Seidel. The complexity of Voronoi diagrams in higher dimensions. *Proc. 20th Allerton Conference on Communication, Control, and Computing*, 1982, 94–95.

[626] R. Seidel. A method for proving lower bounds for certain geometric problems. In: G.T. Toussaint (ed.), *Computational Geometry*, North-Holland, Amsterdam, 1985, 319–334.

[627] R. Seidel. Constructing higher-dimensional convex hulls at logarithmic cost per face. *Proc. 18th Ann. ACM Symposium on Theory of Computing*, 1986, 404–413.

[628] R. Seidel. On the number of faces in higher-dimensional Voronoi diagrams. *Proc. 3rd Ann. ACM Symposium on Computational Geometry*, 1987, 181–185.

[629] R. Seidel. Constrained Delaunay triangulations and Voronoi diagrams with obstacles. Technical report 260, Institute for Information Processing, University of Technology Graz, Austria, 1988.

[630] R. Seidel. Small-dimensional linear programming and convex hulls made easy. *Discrete & Computational Geometry* 6 (1991), 423–434.

[631] R. Seidel. Backwards analysis of randomized geometric algorithms. In: J. Pach (ed.), *New Trends in Discrete and Computational Geometry*, Algorithms and Combinatorics 10, Springer, 1993, 37–68.

[632] R. Seidel. The nature and meaning of perturbations in geometric computing. *Discrete & Computational Geometry* 19 (1998), 1–17.

[633] R. Seidel. Convex hull computations. In: J.E. Goodman and J. O'Rourke (eds.), *Handbook of Discrete and Computational Geometry*, 2nd Edition. CRC Press, 2004, 495–512.

[634] D.Y. Seo, D.T. Lee, and T.-C. Lin. Geometric minimum diameter minimum cost spanning tree problem. *Proc. 20th International Symposium on Algorithms and Computation*, Springer Lecture Notes in Computer Science 5878, 2009, 283–292.

[635] J-K. Seong, G. Elber, and M-S. Kim. Trimming local and global self-intersections in offset curves/surfaces using distance maps. *Computer-Aided Design* 38 (2006), 183–193.

[636] M.I. Shamos. *Computational Geometry.* Ph.D. thesis, Department of Computer Science, Yale University, New Haven, Connecticut, 1978.

[637] M.I. Shamos and D. Hoey. Closest-point problems. *Proc. 16th Ann. IEEE Symposium on Foundations of Computer Science*, 1975, 151–162.

[638] M. Sharir. Intersection and closest-pair problems for a set of planar discs. *SIAM Journal on Computing* 14 (1985), 448–468.

[639] M. Sharir. Almost tight upper bounds for lower envelopes in higher dimensions. *Discrete & Computational Geometry* 12 (1994), 327–345.

[640] M. Sharir. A near-linear algorithm for the planar 2-center problem. *Discrete & Computational Geometry* 18 (1997), 125–134.

[641] M. Sharir. Algorithmic motion planning. In: J.E. Goodman and J. O'Rourke (eds.), *Handbook of Discrete and Computational Geometry*, 2nd Edition. CRC Press, 2004, 1037–1064.

[642] M. Sharir and P.K. Aggarwal. *Davenport–Schinzel Sequences and their Geometric Applications.* Cambridge University Press, 1995.

[643] M. Sharir, S. Smorodinsky, and G. Tardos. An improved bound for k-sets in three dimensions. *Discrete & Computational Geometry* 26 (2001), 195–204.

[644] D.J. Sheehy. Linear-size approximations to the Vietoris-Rips filtration. *Proc. 28th Ann. ACM Symposium on Computational Geometry*, 2012, 239–248.

[645] D.J. Sheehy, C.G. Armstrong, and D.J. Robinson. Shape description by medial surface construction. *IEEE Transactions on Visualization and Computer Graphics* 2 (1996), 62–72.

[646] E.C. Sherbrooke, N.M. Patrikalakis, and E. Brisson. An algorithm for the medial axis transform of 3d polyhedral solids. *IEEE Transactions on Visualization and Computer Graphics* 2 (1996), 44–61.

[647] J.R. Shewchuk. Triangle: Engineering a 2D quality mesh generator and Delaunay triangulator. In: M.C. Lin and D. Manocha (eds.), *Applied Computational Geometry: Towards Geometric Engineering*, Springer Lecture Notes in Computer Science 1148, 1996, 203–222.

[648] J.R. Shewchuk. Adaptive precision floating-point arithmetic and fast robust geometric predicates. *Discrete & Computational Geometry* 18 (1997), 305–363.

[649] J.R. Shewchuk. Delaunay refinement algorithms for triangular mesh generation. *Computational Geometry: Theory and Applications* 22 (2002), 21–74.

[650] J.R. Shewchuk. Star splaying: an algorithm for repairing Delaunay triangulations and convex hulls. *Proc. 21st Ann. ACM Symposium on Computational Geometry*, 2005, 237–246.

[651] J.R. Shewchuk. General-dimensional constrained Delaunay and constrained regular triangulations, I: Combinatorial properties. *Discrete & Computational Geometry* 39 (2008), 580–637.

[652] R. Sibson. Locally equiangular triangulations. *The Computer Journal* 21 (1978), 243–245.

[653] R. Sibson. A vector identity for the Dirichlet tessellation. *Mathematical Proceedings of the Cambridge Philosophical Society* 87 (1980), 151–155.

[654] K. Siddiqi and S.M. Pizer. *Medial Representations. Mathematics, Algorithms, and Applications.* Springer Series on Computational Imaging and Vision 37, 2008.

[655] R.I. Silveira and M. van Krefeld. Towards a definition of higher order constrained Delaunay triangulations. *Computational Geometry: Theory and Applications* 42 (2009), 322–337.

[656] S. Skyum. A sweepline algorithm for generalized Delaunay triangulations. Technical report DAIMI PB-373, Computer Science Department, Aarhus University, 1991.

[657] S.W. Sloan. A fast algorithm for constructing Delaunay triangulations in the plane. *Advances in Engineering Software* 9 (1987), 34–55.

[658] M. Smid. Maintaining the minimal distance of a point set in less than linear time. *Algorithms Reviews* 2 (1991), 33–44.

[659] M. Smid and R. Janardan. On the width and roundness of a set of points in the plane. *International Journal of Computational Geometry & Applications* (1999), 97–108.

[660] J. Snoeyink and M. van Kreveld. Linear-time reconstruction of Delaunay triangulations with applications. *Proc. 5th Ann. European Symposium on Algorithms*, Springer Lecture Notes in Computer Science 1284, 1997, 459–471.

[661] D.L. Souvaine, C.D. Tóth, and A. Winslow. Simultaneously flippable edges in triangulations. *Proc. XIV Spanish Meeting on Computational Geometry*, CMR Documents 8, Barcelona, Spain, 2011, 1–4.

[662] M.J. Spriggs, J.M. Keil, S. Bespamyatnikh, M. Segal, and J. Snoeyink. Computing a $(1+\varepsilon)$-approximate geometric minimum-diameter spanning tree. *Algorithmica* 38 (2004), 577–589.

[663] E. Steinitz. *Polyeder und Raumeinteilungen.* Enzyklopädie der Mathematischen Wissenschaften III AB 12, Leipzig, 1916.

[664] P. Su and R.L.S. Drysdale. A comparison of sequential Delaunay triangulation algorithms. *Proc. 11th Ann. ACM Symposium on Computational Geometry*, 1995, 61–70.

[665] A. Sud, M. Foskey, and D. Manocha. Homotopy-preserving medial axis simplification. *International Journal of Computational Geometry & Applications* 17 (2007), 423–451.

[666] K. Sugihara. A simple method for avoiding numerical errors and degeneracy in Voronoi diagram construction. *IEICE Transactions on Fundamentals of Electronics, Communications, and Computer Sciences*, E75-A (1992), 468–477.

[667] K. Sugihara. Voronoi diagrams in a river. *International Journal of Computational Geometry & Applications* 2 (1992), 29–48.

[668] K. Sugihara and M. Iri. Construction of the Voronoi diagram for 'one million' generators in single-precision arithmetic. *Proceedings of the IEEE* 80 (1992), 1471–1484.

[669] K. Sugihara, Y. Ooishi, and T. Imai. Topology-oriented approach to robustness and its applications to several Voronoi-diagram algorithms. *Proc. 2nd Canadian Conference on Computational Geometry*, 1990, 36–39.

[670] K.J. Supowit. The relative neighborhood graph with an application to minimum spanning trees. *Journal of the ACM* 30 (1983), 428–448.

[671] A. Suzuki and M. Iri. Approximation of a tessellation of the plane by a Voronoi diagram. *Journal of the Operations Research Society of Japan* 29 (1986), 69–96.

[672] K. Swanson, D.T. Lee, and V.L. Wu. An optimal algorithm for roundness determination on convex polygons. *Computational Geometry: Theory and Applications* 5 (1995), 225–235.

[673] B. Tagansky. *The complexitiy of substructures in arrangements of surfaces.* PhD. thesis, Tel Aviv University, Israel, 1996.

[674] H. Tamaki and T. Tokuyama. How to cut pseudo-parabolas into segments. *Discrete & Computational Geometry* 19 (1998), 265–290.

[675] T.-S. Tan. *Optimal two-dimensional triangulations.* PhD. thesis, University of Illinois at Urbana-Champaign, 1993.

[676] M. Tanase and R.C. Veltkamp. A straight skeleton approximating the medial axis. *Proc. 12th Ann. European Symposium on Algorithms*, Springer Lecture Notes in Computer Science 3221, 2004, 809–821.

[677] M. Tanase and R.C. Veltkamp. Polygon decomposition based on the straight skeleton. *Proc. 19th Ann. ACM Symposium on Computational Geometry*, 2003, 58–67.

[678] M. Tanemura, T. Ogawa, and W. Ogita. A new algorithm for three-dimensional Voronoi tessellation. *Journal of Computational Physics* 51 (1983), 191–207.

[679] T. Tanuma, H. Imai, and S. Moriyama. Revisiting hyperbolic Voronoi diagrams in two and higher dimensions from theoretical, applied and generalized viewpoints. *Transactions on Computational Science XIV*, Springer Lecture Notes in Computer Science 6970, 2011, 1–30.

[680] J.E. Taylor. The structure of singularities in soap-bubble-like and soap-film-like minimal surfaces. *Annals of Mathematics* 103 (1976), 489–539.

[681] H. Telley. Delaunay triangulation in the flat torus. Technical report, Swiss Federal Institute of Technology, Lausanne, Switzerland, 1992.

[682] S. Teramoto, E.D. Demaine, and R. Uehara. Voronoi game on graphs and its complexity. *Proc. IEEE Symposium on Computational Intelligence and Games*, 2006, 265–271.

[683] A.H. Thiessen. Precipitation average for large area. *Monthly Weather Review* 39 (1911), 1082–1084.

[684] A.C. Thompson. *Minkowski Geometry.* Encyclopedia of Mathematics and its Applications 63, Cambridge University Press, 1996.

[685] W.P. Thurston. The geometry of circles: Voronoi diagrams, Möbius transformations, convex hulls, Fortune's algorithm, the cut locus, and parametrization of shapes. Technical report, Princeton University, 1986.

[686] G.T. Toussaint. The relative neighbourhood graph of a finite planar set. *Pattern Recognition* 12 (1980), 261–268.

[687] P.M. Vaidya. Minimum spanning trees in k-dimensional space. *SIAM Journal on Computing* 17 (1988), 572–582.

[688] P.M. Vaidya. A sparse graph almost as good as the complete graph on points in K dimensions. *Discrete & Computational Geometry* 6 (1991), 369–381.

[689] P.M. Vaidya. Geometry helps in matching. *SIAM Journal on Computing* 18 (1989), 1201–1225.

[690] L. Vietoris. Über den höheren Zusammenhang kompakter Räume und eine Klasse von zusammenhangstreuen Abbildungen. *Mathematische Annalen* 97 (1927), 454–472.

[691] A. Vigneron and L. Yan. A faster algorithm for computing motorcycle graphs. *Proc. 29th Ann. ACM Symposium on Computational Geometry,* 2013, to appear.

[692] E. Viterbo and E. Biglieri. Computing the Voronoi cell of a lattice: The diamond-cutting algorithm. *IEEE Transactions on Information Theory* 42 (1996), 161–171.

[693] G.F. Voronoi. Nouvelles applications des paramètres continus à la théorie des formes quadratiques. deuxième Mémoire: Recherches sur les parallélloèdres primitifs. *Journal für die reine und angewandte Mathematik* 134 (1908), 198–287.

[694] G.F. Voronoi. Deuxième mémoire: recherches sur les paralléloedres primitifs. *Journal für die reine und angewandte Mathematik* 136 (1909), 67–181.

[695] K. Vyatkina. On the structure of straight skeletons. *Proc. Int. Conference on Computational Sciences and Its Applications,* Springer, Lecture Notes in Computer Science 5730, 2009, 362–379.

[696] C.A. Wang. Efficiently updating the constrained Delaunay triangulations. *BIT* 33 (1993), 238–252.

[697] C.A. Wang and L. Schubert. An optimal algorithm for constructing the Delaunay triangulation of a set of line segments. *Proc. 3rd Ann. ACM Symposium on Computational Geometry,* 1987, 223–232.

[698] C.A. Wang and Y.H. Tsin. Finding constrained and weighted Voronoi diagrams in the plane. *Computational Geometry: Theory and Applications* 10 (1998), 89–104.

[699] D.F. Watson. Computing the n-dimensional Delaunay tessellation with applications to Voronoi polytopes. *The Computer Journal* 24 (1981), 167–172.

[700] R. Wein, J.P. van den Berg, and D. Halperin. The visibility-Voronoi complex and its applications. *Computational Geometry: Theory and Applications* 36 (2005), 66–87.

[701] R. Wein, J.P. van den Berg, and D. Halperin. Planning high-quality paths and corridors amidst obstacles. *International Journal of Robotics Research* 27 (2008), 1213–1231.

[702] E. Welzl. Smallest enclosing disks (balls and ellipsoids). In: H. Maurer (ed.), *New Results and New Trends in Computer Science*, Springer Lecture Notes in Computer Science 555, 1991, 359–370.

[703] D.B. West. *Introduction to Graph Theory*, 3rd Edition. Prentice Hall, 2007.

[704] P. Widmayer, Y.F. Wu, and C.K. Wong. On some distance problems in fixed orientations. *SIAM Journal on Computing* 16 (1987), 728–746.

[705] E. Wigner and F. Seitz. On the constitution of metallic sodium. *Physical Review* 43 (1933), 804–810.

[706] H.M. Will. Fast and efficient computation of additively weighted Voronoi cells for applications in molecular biology. *Proc. 6th Scandinavian Workshop on Algorithm Theory*, Springer Lecture Notes in Computer Science 1432, 1998, 310–321.

[707] N.M.J. Woodhouse. *Special Relativity*. Springer Undergraduate Mathematics Series, 2003.

[708] G. Xia. Improved upper bound on the stretch factor of Delaunay triangulations. *Proc. 27th Ann. ACM Symposium on Computational Geometry*, 2011, 264–273.

[709] L. Xu, J. Xu, and E. Papadopoulou. Computing the map of geometric minimal cuts. *Algorithmica*, to appear, 2012.

[710] A.C. Yao. On constructing minimum spanning trees in k-dimensional spaces and related problems. *SIAM Journal on Computing* 11 (1982), 721–736.

[711] C.K. Yap. Algorithmic motion planning. In: J.T. Schwartz and C.-K. Yap (eds.), *Advances in Robotics, 1: Algorithmic and Geometric Aspects of Robotics*, Lawrence Erlbaum Associates, Hillsdale, New Jersey, 1987, 95–143.

[712] C.K. Yap. An $O(n \log n)$ algorithm for the Voronoi diagram of a set of simple curve segments. *Discrete & Computational Geometry* 2 (1987), 365–393.

[713] C.K. Yap. *Fundamental Problems of Algorithmic Algebra.* Oxford University Press, 2000.

[714] A. Yershova, L. Jaillet, T. Siméon, and S.M. LaValle. Dynamic-domain RRTs: efficient exploration by controlling the sampling domain. *Proc. IEEE International Conference on Robotics and Automation*, 2005, 3856–3861.

[715] T.-K. Yu and D.T. Lee. Time convex hull with a highway. *Proc. 4th International Symposium on Voronoi Diagrams in Science and Engineering*, 2007, 240–250.

[716] B. Zhu and A. Mirzaian. Sorting does not always help in computational geometry. *Proc. 3rd Canadian Conference on Computational Geometry*, 1991, 239–242.

[717] G.M. Ziegler. *Lectures on Polytopes.* Graduate Texts in Mathematics 152, Springer-Verlag, Berlin, 1998.

INDEX

abstract inverse Voronoi diagram, 174
abstract order-k Voronoi diagram, 174
abstract simplicial complex, 202, 216, 262
abstract Voronoi diagram, 57, 168, 181
Ackermann's function, 61, 112, 143
acyclicity property of complexes, 45, 80
additively weighted diagram, 86, 113, 121, 145, 154, 171
additively weighted distance, 154
admissible bisector system, 169, 181
affine Voronoi diagram, 86, 165
air-lift metric, 144, 148, 176
all nearest neighbors problem, 186, 248
alpha-shape, 45, 88, 119, 200
angular bisector, 53, 55
angular Voronoi diagram, 159
angularity, 42, 77, 105, 138
anisotropic Delaunay triangulation, 163
anisotropic Voronoi diagram, 159, 163
annulus, 194, 259
anti-center of a graph, 248
Apollonius circle, 157
Apollonius model, 86, 158
approximate nearest neighbor, 259
approximate Voronoi diagram, 261
approximation algorithm, 39, 99, 115, 198, 199, 216, 240, 245, 259
area of union, 119
arrangement of (hyper)planes, 106, 114
arrangement of hypersurfaces, 168
arrangement of pseudo-lines, 250
aspect ratio, 234
associahedron, 90
augmenting path, 98

balanced fractional clustering, 102
basin of attraction, 242
beta-skeleton, 202, 205
bi-criteria spanning tree, 198
biarcs, 70
bichromatic 2-center problem, 193
bichromatic closest pair, 188, 197
binary medial axis, 116, 268
bipartite graph, 210
bisecting curve, 168
bisector, 7, 75, 86, 115, 131, 168
bisector system, 169
bistellar flip, 78, 90
boat-sail distance, 156
bottom-up divide & conquer, 27, 67
bounded set, 8
bounded Voronoi diagram, 62, 65
bounded-degree Delaunay, 189
bounded-degree spanner graph, 205, 260

bounding box of point set, 254
box decomposition tree, 259
Bregman divergence, 86
bucketing technique, 23, 79

calibration figure (Eichfigur), 130
canonical Voronoi insertion, 191
cardinality of a set, 18
Cauchy sequence, 146
Cech complex, 262
cell complex, 75, 87
cell-tuple data structure, 82
center of a graph, 248
centrally symmetric polytope, 129
centroid, 44, 110, 212, 244
centroidal power diagram, 103
centroidal Voronoi diagram, 244
certificate of Delaunay edge, 227
chordale of circles, 81
circle packing, 191
circle-based β-skeleton, 202
circular arc shape, 70
circularly separable clustering, 214
circumcircle, 12, 19, 41, 43, 139, 208
circumradius, 129, 139
city metric, 176
city Voronoi diagram, 55, 176
clearance of a robot, 217
closed set, 8
closest covered set diagram, 112
closest pair problem, 188, 248, 255, 257
closest-polygon Voronoi diagram, 172
closure of a set, 149
cluster selection, 214
cluster Voronoi diagram, 112, 215
clustering, 93, 103, 160, 211
coarseness of a triangulation, 43, 76
collision detection, 58, 217
collision-free path, 218
colored 2-center problem, 193
colored closest pair, 188
compact order-k Voronoi diagram, 185
compact Voronoi diagram, 55, 185, 220
compatible triangulations, 64
competitive facility location, 233
complete graph, 76, 195, 205, 210, 248
complete-linkage clustering, 215
composite metric, 148
configuration space, 89, 98, 220, 270
conflicting triangle, 19, 21, 80
conforming Delaunay triangulation, 64
constrained Delaunay tetrahedrization, 93
constrained Delaunay triangulation, 63, 93, 192, 196, 205
constrained minimum spanning tree, 196
constrained order-k Delaunay triangulation, 208
constrained regular triangulation, 93
constrained simplicial complex, 93
construction history, 22, 27, 54, 79, 111, 184
continuous Dijkstra technique, 65, 127, 180
controlled perturbation, 271
convex distance function, 130, 141, 147, 171, 174, 219, 241
convex graph, 247
convex hull, 10, 16, 25, 32, 51, 84, 104, 109, 200, 251
convex hull algorithms, 85
convex hull of spheres, 120
convex hull-disjoint, 100
convex partition, 35, 206
convex polygon, 8, 10, 53, 60
convex polyhedron, 32, 83
convex polytope, 84, 89, 96
convex position of point set, 10, 13, 16, 32, 54, 90, 209, 231, 237
convex set, 8
covering, 238
cross-polytope, 143
cross-section of Voronoi diagram, 86
crossing-free geometric graph, 196
crust of sample points, 201
crystal growth Voronoi diagram, 162, 240
crystallographic group, 129

Davenport-Schinzel sequence, 30, 227
decomposition, 9

decomposition lemma (medial axis), 72
degree of a vertex, 9, 20, 199
Delaunay flip, 19, 90
Delaunay for sites on surfaces, 128
Delaunay neighbor, 187
Delaunay refinement, 191
Delaunay repair, 228
Delaunay simplicial complex, 76, 128, 210, 261
Delaunay stabbing problem, 236
Delaunay structure for graphs, 248
Delaunay subcomplex, 261
Delaunay tessellation, 12
Delaunay tessellation in 3-space, 76
Delaunay tetrahedrization, 76
Delaunay tree, 22, 111
Delaunay triangulation, 12
Delaunay triangulation on surfaces, 128
depth-first search, 37
diameter of a graph, 73
diameter of a point set, 140, 192
digital Voronoi diagram, 15, 162, 268
Dijkstra's algorithm, 127, 220, 247
dilation of a graph, 205, 209, 258
dilation of a path, 152
dimension reduction, 252, 259
directed acyclic graph, 22
Dirichlet tessellation, 3
discrepancy, 226
discrete group of motions, 125
discrete medial axis, 116, 268
discrete straight skeleton, 61
distance cone, 153, 155, 156, 161
distance graph, 164
distortion of an embedding, 253
divide & conquer, 22, 24, 50, 66, 71, 113, 133, 136, 152, 172, 270
domain of action, 2
doubling dimension, 260, 265
doubly connected edge list, 15, 31
dual structure, 3, 12, 16, 51, 136, 263
duality transform, 84, 109
dynamic data structure, 225
dynamic k-nearest neighbor searching, 226
dynamic point-location, 226
dynamic post office problem, 185, 226
dynamic programming, 203
dynamic spanner graph, 260
dynamic Voronoi diagram, 22, 27, 111

ear of a polygon, 105
ear-clipping algorithm, 105
edge event (straight skeleton), 56
edge exclusion property, 206
edge flip, 19, 42, 137
edge graph of Voronoi diagram, 8, 67
edge inclusion property, 206
ellipsoid, 174, 193
empty neighborhood graph, 202, 204
empty-circle property, 12, 35, 63, 136, 138, 227
empty-sphere property, 76, 78, 263
epsilon-closeness problem, 17, 188
epsilon-net, 226
epsilon-sample, 264
equiangularity, 42, 77
equilibrium state, 39, 235, 238
equipartition of a convex body, 96
Euclidean distance, 7, 131
Euclidean distance transform, 268
Euler's formula, 11, 49, 177
evenness property of point set, 251
exchanging flip, 90
expected storage requirement, 22
extreme point in a set, 10, 12, 104

face-to-face complex, 75, 82
facet of a cell complex, 75, 82
facet-edge data structure, 76
facility location problem, 35, 189, 229, 248
farthest-polygon Voronoi diagram, 52, 172
farthest-segment Voronoi diagram, 51, 106, 172
farthest-site abstract Voronoi diagram, 174
farthest-site Delaunay complex, 105
farthest-site Delaunay triangulation, 105, 200
farthest-site graph Voronoi diagram, 247

farthest-site Voronoi diagram, 51, 85, 103, 113, 160, 192, 200
fat object, 119
fatness of a triangulation, 44
Fermat point, 197, 213
Fibonacci heap, 247
filtration of cell complexes, 264, 265
finite element methods, 5, 43, 93
finite set of orientations, 115, 133, 143, 181
fixed point, 240, 245
fixed tree theorem, 196
fixed-share decomposition, 60
flat torus, 128
flip distance, 42, 91
flip graph, 209
flipping, 19, 42, 78, 90–92, 128, 137
floating point filter, 270
forest, 186, 196
Fréchet distance, 66
fractal set, 243
fractional clustering, 102
framework statics, 39
free space of a robot, 220, 222
fundamental domain, 126, 129

Gabriel graph, 141, 202, 205, 211, 250
general position of point set, 12, 15, 48, 49, 76, 117, 119, 141, 186, 269
generalized Dirichlet tessellation, 81
geodesic distance, 64, 206
geodesic path, 123, 147
geodesic Voronoi diagram, 64, 123, 162, 175, 176
geometric predicate, 271
geometric software, 271
geometric transformation, 28, 31, 83, 109
gerrymander problem, 36
gradient method, 98, 99
graph, 8
graph drawing, 160, 209, 211
graph Voronoi diagram, 246
grass fire interpretation, 154
greedy algorithm, 196
greedy triangulation, 45, 203, 206
growth model, 35, 57, 144, 161, 177, 240

Halley's method, 243
Hamiltonian cycle, 198, 209
Hamming space, 258
Hausdorff distance, 66, 112, 215
Hausdorff Voronoi diagram, 112, 171
Helly's theorem, 175
higher-order Voronoi diagram, 103
highway hull, 181
histogram, 63
homothet, 137, 143
Hotelling game, 235
hull-disjoint, 100, 112, 214, 215
hyperbolic Delaunay complex, 124
hyperbolic geometry, 124
hyperbolic Voronoi diagram, 86, 124
hypercube, 95, 143, 259
hyperplane arrangement, 106
hyperrectangle, 254

imprecise point sets, 23
in-front/behind relation, 45, 80
incircle test, 23, 27
incremental insertion, 18, 50, 77, 97, 113, 128, 136, 152, 172, 181, 184
incremental search, 23
independent edge set, 91, 208
independent vertex set, 237
inradius, 44, 129
inter-cluster measure, 212, 214
interdistance enumeration problem, 189
interference pattern, 55, 154
interior of a set, 8
intersection of balls, 82
intra-cluster measure, 212
intrinsic dimension of manifold, 261, 264
inversion transform, 32
inward graph Voronoi diagram, 247
isothetic, 80, 113, 176
iterated logarithm, 51

Johnson–Mehl model, 2, 86, 155
Jordan curve, 145, 150, 170, 171
Jordan matrix, 165

k closest pairs problem, 188
k-center problem, 193, 213

k-centroid problem, 212, 246
k-clustering, 212
k-facet of a point set, 109
k-flat, 114
k-level greedy triangulation, 204
k-level in an arrangement, 108
k-locally Delaunay graph, 248
k-locally Gabriel graph, 250
k-means clustering, 212, 246
k-median, 189
k-nearest neighbor problem, 103, 185, 226, 257
k-neighborhood in graphs, 248
k-sector curve, 241
k-set of a point set, 108
k-simplex, 87
k-triangle of a point set, 109
Karlsruhe metric, 148
kinetic Delaunay triangulation, 227
kinetic spanner graph, 206
kinetic Voronoi diagram, 27, 50, 227
Klein model of hyperbolic geometry, 124
Kruskal's algorithm, 196, 214
Kullback–Leibler divergence, 86

L_1-norm, 131, 176, 197, 206, 241, 259
L_1-Voronoi diagram, 177
L_∞-medial axis, 59, 61, 143
L_∞-norm, 59, 131, 197, 214, 258
L_p-higher order Voronoi diagram, 109
L_p-norm, 131, 142, 259
L_p-Voronoi diagram, 133
Laguerre-Voronoi diagram, 81
largest empty circle problem, 103, 190, 248
largest empty circum-shape, 139
lattice Delaunay polytope, 129
lattice prototile, 129
least-squares clustering, 94, 213
least-squares fitting, 101
least-squares matching, 101
level in an arrangement, 106
lifting map, 27, 31, 84, 92, 109, 110
light triangulation edge, 203
line skeleton, 59
line Voronoi diagram, 48
linear axis of polygon, 59
linear programming, 37, 192, 194
linearization of a Voronoi diagram, 54, 86
linearly separable clustering, 214
Lipschitz function, 253
Lloyd's method, 245
LMT-skeleton, 204
locally Delaunay, 41, 51, 79, 138
locally equiangular, 42
locally minimal triangulation, 44, 204
locally regular facet, 90
locational optimization, 4
locus approach, 183, 185, 258
Lombardi drawing, 161
Lorentz transformation, 166
low-dimensional embedding, 252
lower convex hull, 32, 84
lower envelope, 33, 99, 112, 114, 153, 168, 173
lune-based β-skeleton, 202

m-star-shaped region, 150
m-straight path, 146, 171
Möbius transformation, 161
Manhattan distance, 131
manifold reconstruction, 261
matching, 101, 198, 237
maximal inscribed ball, 117
maximal inscribed disk, 68, 71
maximal outerplanar graph, 210
maximal territory diagram, 242
maximized Voronoi region, 229
maximum norm, 131
Maxwell's theorem, 39
Maxwell–Cremona theorem, 40
medial axis, 2, 48, 50, 53, 58, 67, 71, 114, 143, 202, 223
medial axis transform, 71, 117, 268
medial ball representation, 118
merge chain, 24, 134, 152, 172
mesh generation, 43, 63, 93, 191
mesh improvement, 159, 191
metric, 144
min–max Voronoi diagram, 112
minidisk algorithm, 193
minimal Steiner tree, 197, 246
minimization diagram, 33, 168
minimum convex partition, 207

minimum convex Steiner partition, 207
minimum diameter spanning tree, 197
minimum length matching, 198
minimum spanning tree, 139, 187, 189, 195, 202, 205, 214, 258
minimum-cost flow problem, 98
minimum-cost load balancing, 155
minimum-weight triangulation, 44, 202, 206
minimum-width annulus, 194
Minkowski difference, 220
Minkowski norm, 131
Minkowski sum, 74, 140, 221
Minkowski theorem, 96
mitered offset, 57, 74
mixed weighted distance, 160, 213
molecular modeling, 117, 120
moment curve, 251
Monge–Kantorovich problem, 95, 155
monotone curve, 24, 147
monotone polygon, 57, 58
morphing of shapes, 120, 196
Moscow metric, 148
motion planning problem, 74, 216
motorcycle graph, 58
multiplicatively weighted diagram, 86, 156, 172
multiplicatively weighted distance, 156

n^2-hard problem, 119
natural neighbor interpolant, 44
nearest neighbor problem, 183, 258
neighborly polytope, 251
nerve of a set of balls, 262
network Voronoi diagram, 228, 246
Newton's method, 243
nice curve, 145, 168
nice metric, 146
norm, 130
NP-hard problem, 45, 91, 93, 103, 139, 197, 198, 212, 234, 238

offset of a shape, 57, 74, 221
on-line algorithm, 109
on-line clustering, 211
one-round Voronoi game, 234
online algorithm, 22
open set, 8
optimal clustering, 191, 211, 212
orbit of a point, 126
order-*k* abstract Voronoi diagram, 174
order-*k* Delaunay graph, 207, 249
order-*k* Delaunay triangulation, 208
order-*k* Gabriel graph, 209
order-*k* power diagram, 82, 106
order-*k* segment Voronoi diagram, 52
order-*k* Voronoi diagram, 44, 85, 103, 185, 209, 214, 226
oriented sphere, 165
orphan-free Voronoi diagram, 150, 159, 164
orthogonal dual, 39
orthogonal polygon, 59
orthogonal polyhedron, 61
orthogonal sphere, 88
orthogonal walk, 80
outerplanar graph, 210, 211
output-sensitive algorithm, 85, 108, 201
outward graph Voronoi diagram, 247
overlay of convex partitions, 102

packing, 238
parallel algorithm, 27, 51, 91, 113, 245, 268
parametric search, 66
partial least-squares matching, 101
partial order, 22, 45
partition of a set, 9, 112
partition theorems (power diagram), 96
path planning, 216, 248
path-connected, 150, 169, 177
pattern matching, 101
pattern recognition, 120, 200, 204
peeper's Voronoi diagram, 64
perfect matching, 198, 237
perturbation scheme, 270
pixel Voronoi diagram, 15, 33, 45, 162, 268
planar graph, 8, 15, 51, 134, 170, 179, 246, 249
planar straight line graph, 8, 47

plane sweep, 28, 50, 63, 66, 70, 112, 136, 152, 155, 173, 186, 188
Plateau's laws, 161
plesiohedron, 129
Poincaré model of hyperbolic geometry, 124
point-location, 20, 80, 111, 184, 218, 226, 261
point-location in equal balls, 260
point-symmetric region, 229, 234
point-symmetric set, 130, 139
Poisson Voronoi diagram, 4
polar Voronoi diagram, 219
polarity, 84, 110
pole vertex, 201
polygonal surface, 58, 92, 127
polyhedral cell complex, 75, 82, 87
polyhedral distance function, 115, 142
polyhedron, 83
polylog function, 128
polynomial root finding, 242
polynomial-time algorithm, 45, 93, 203, 209, 212
polynomiography, 244
polytope, 84, 89, 93, 251
post office problem, 183, 220, 226, 228, 248, 258
potential-field method, 223
power cell, 81
power diagram, 40, 81, 94, 117, 152, 201, 213, 264
power function, 81, 94
pre-triangulation, 92
preorder of a tree, 70, 198
Prim's algorithm, 196
priority queue, 30, 105, 189, 227
prototile of a lattice, 129
proximity problem, 186, 258
pseudo-circles, 133
pseudo-lines, 250
pseudo-parabolas, 250
pseudo-simplicial complex, 92
pseudo-triangulation, 58, 92, 217

quad-edge data structure, 15, 26, 31
quadrangle inequality, 189
quadratic-form distance, 129, 165
quadtree, 27, 261
quasi-Euclidean distance, 165
quickest path, 176
quotient space of a surface, 126

r-sample, 201
radical axis of circles, 81
radii graph, 199
random sampling, 226
randomized algorithm, 20, 22, 50, 51, 53, 66, 71, 109, 152, 172, 181, 193, 225
reciprocal figure, 39
recognition of shape Delaunay, 139
recognition of Voronoi diagrams, 35
rectilinear polygon, 59
recursive algorithm, 24
regular simplicial complex, 45, 88, 117, 264
regular tessellation, 88
regular triangulation, 44, 84, 201
regularizing flip, 91
relative neighborhood graph, 204
removing a Delaunay site, 106
restricted Delaunay complex, 263
retraction approach, 218
Riemannian manifold, 127
Rips complex, 262
roadmap, 222
robot motion planning, 216
roof construction, 59
root finding, 242
rotational motion planning, 220
roughness of triangular surface, 44
roundness of point set, 194

sampling technique, 116, 201, 222, 261
scene analysis, 40
Schönhard polytope, 93
Schauder's fixed-point theorem, 241
Schlegel diagram, 83
scope of a triangle, 207
search tree, 22, 27, 30, 225, 256
second-closest pair, 189
secondary polytope, 90, 93
segment Delaunay triangulation, 51
self-centered power diagram, 264
self-centered tetrahedrization, 76

self-centered Voronoi diagram, 244
self-edge of a site, 67
semi-separated pair decomposition, 257
sensor network, 199, 248
separating regions problem, 236
separator, 7, 168
shape Delaunay tessellation, 136, 205, 228
shape reconstruction, 117, 200, 201
sharp P-hard, 129
shelling order, 45, 106
shortest connection network, 195, 197
shortest path, 64, 123, 144, 147, 162, 175, 205, 220, 246
similarity problem, 258
simple cell complex, 82, 86, 87
simple curve, 10, 145
simple polygon, 48, 50, 52, 57, 63, 191
simplex, 87
simplicial cell complex, 87
simply connected, 49, 57, 115, 125, 150, 154, 171, 177
single-linkage clustering, 215
singular face, 117
site event (plane sweep), 28
skeleton, 48, 55
skew distance, 156
skew Voronoi diagram, 156
skin surface for balls, 119
slab method for point-location, 184
sliver simplex, 263
smallest annulus problem, 193
smallest enclosing circle, 43, 192, 213
smallest enclosing shape, 139
smallest enclosing sphere, 192, 213
smooth object, 133, 136, 221, 222
Snell's law of refraction, 175, 222
soap bubbles, 161
space-filler, 129
spanner of a point set, 205, 257, 260, 265
spanning cycle, 198, 199
spanning ratio of a graph, 205
spanning tree, 195
sparse graph, 205
spatial sampling, 116
sphere packing, 191
spike event (plane sweep), 29
spiral search, 23
spline curve, 70
split event (straight skeleton), 56
split tree, 256
spread of a finite point set, 252
stable Delaunay subgraph, 228
star-shaped region, 106, 120, 133, 154, 159, 231
star-shapedness in a graph, 247
starting tetrahedron, 79
starting triangle, 21, 23, 111
Steinitz's theorem, 92
stochastic properties of diagrams, 267
straight skeleton, 53, 55, 162, 177
straight skeleton for pixel shapes, 61
straight walk, 80
stretch factor of a graph, 205
strict triangle inequality, 133
strictly convex set, 131, 136
subgraph of the Delaunay, 194
supergraph of the Delaunay, 207
support hull, 137
surface reconstruction, 128, 201, 261
surface sampling, 5, 116, 264
sweep-circle algorithm, 125
sweep-line algorithm, 28
symmetric axis, 48, 71
symmetric distance function, 130, 139, 144

tentative prune-and-search, 186
terrain modeling, 43, 58, 208
terrain recognition, 40
territory diagram, 241
tessellation, 11
Thales' theorem, 157, 227
Thiessen polygons, 1
tilted plane, 155
top-down divide & conquer, 27, 67
topological clustering, 216
torus, 127, 231
tracing method, 116
translational motion planning, 220
transportation network, 176, 248
transportation problem, 100, 155
travelling salesman tour, 198

tree, 10, 47, 50, 53, 57, 67, 174, 210, 211, 246
triangle inequality, 130, 144
triangular network, 11, 41
triangular surface, 44, 110, 208
triangulation, 11, 41, 47
triangulation axis of polygon, 55
trisector curve, 239
two-center problem, 193
two-site Voronoi diagram, 113

unavoidable triangulation edge, 203
unbounded set, 8
uniform triangular mesh, 191
union of balls, 82, 88, 117, 201
union-find data structure, 196
unit-disk graph, 248
untangling a polygon, 41
upper bound theorem, 83
upper envelope of (hyper)planes, 83, 106, 110
upper envelope of surfaces, 112, 160

variance of a cluster, 100, 212
vertex-removing flip, 90
vertical projection, 32, 83, 84
very nice metric, 149, 171
Vietoris–Rips complex, 262
visibility graph, 220
visibility walk, 80
visibility-constrained Voronoi diagram, 64, 162, 222
visibility-Voronoi complex, 221
visible, 49
Voronoi art, 244
Voronoi diagram, 8
Voronoi diagram for a lattice, 129
Voronoi diagram for circles, 66, 145, 154
Voronoi diagram for curved objects, 66
Voronoi diagram for graphs, 228, 246
Voronoi diagram for line segments, 47, 114, 143, 185, 218
Voronoi diagram for lines, 114, 143
Voronoi diagram for moving sites, 27, 50, 136, 227
Voronoi diagram for obstacles, 64, 176
Voronoi diagram for oriented spheres, 165
Voronoi diagram for periodical sites, 129
Voronoi diagram for spheres, 86, 120
Voronoi diagram in 3-space, 75
Voronoi diagram in a river, 156
Voronoi diagram inversion, 35
Voronoi diagram of a convex n-gon, 54
Voronoi diagram on a cone, 124
Voronoi diagram on a torus, 128, 231
Voronoi diagram on polyhedra, 127
Voronoi diagram on the sphere, 123
Voronoi diagrams on surfaces, 123
Voronoi diagrams on the grid, 268
Voronoi edge, 8, 75
Voronoi facet, 75
Voronoi game, 233, 238
Voronoi region, 7
Voronoi surface, 33, 112, 114, 168, 271
Voronoi tree, 27
Voronoi tree map, 60
Voronoi vertex, 8, 75

Walsh matrix, 254
wavefront, 28, 55, 65, 177
wavefront model, 9, 144, 161
weighted α-shape, 88, 117, 201
weighted centroid, 44, 244
weighted region problem, 175
weighted skeleton, 60, 96
weighted Voronoi diagram, 86, 152, 228
well-separated pair decomposition, 255
well-separated point sets, 255
width of a point set, 119
Wigner–Seitz zones, 2
Wirkungsbereiche, 2
witness complex, 210, 263
witness Delaunay graph, 210, 236, 248, 265
witness Gabriel graph, 211

zone diagram, 238
zone in an arrangement, 114
Zorn's lemma, 242